SEQUENCES

H. Halberstam
K. F. Roth, F.R.S.

SEQUENCES

Springer-Verlag
New York Heidelberg Berlin

H. Halberstam
Department of Mathematics
University of Illinois
Urbana, Illinois 61801
U.S.A.

K. F. Roth, F.R.S.
Department of Mathematics
Imperial College of Science
and Technology
Queen's Gate, London SW7 5HH
U.K.

AMS Subject Classifications: 10LXX, 10H20, 10H30

Library of Congress Cataloging in Publication Data
Halberstam, H. (Heini)
Sequences.
Bibliography: p.
Includes index.
1. Sequences (Mathematics) 2. Sieves (Mathematics)
I. Roth, Klaus Friedrich. II. Title.
QA246.5.H3 1982 512'.72 82–19222

Softcover reprint of the hardcover 1st edition 1983

9 8 7 6 5 4 3 2 1

ISBN-13: 978-1-4613-8229-4 e-ISBN-13: 978-1-4613-8227-0
DOI: 10.1007/978-1-4613-8227-0

PREFACE

THIS volume is concerned with a substantial branch of number theory of which no connected account appears to exist; we describe the general nature of the constituent topics in the introduction. Although some excellent surveys dealing with limited aspects of the subject under consideration have been published, the literature as a whole is far from easy to study. This is due in part to the extent of the literature; it is necessary to thread one's way through a maze of results, a complicated structure of inter-relationships, and many conflicting notations. In addition, however, not all the original papers are free from obscurities, and consequently some of these papers are difficult (a few even exceedingly difficult) to master.

We try to give a readable and coherent account of the subject, containing a cross-section of the more interesting results. We felt that it would have been neither practicable nor desirable to attempt a comprehensive account; we treat each aspect of the subject from some special point of view, and select results accordingly. Needless to say, this approach entails the omission of many interesting and important results (quite apart from defects in the selection due to errors of judgement on our part). Those results selected for inclusion are, however, proved in complete detail and without the assumption of any prior knowledge on the part of the reader. Our strict attention to detail imposes a certain constraint on the style, but where this might tend to obscure the motivation of a proof, we include (in addition to the formal proof) an informal discussion of the underlying ideas.

One of the attractions of the subject-matter is that, whilst it is rich in ideas, it requires little initiation. The methods of number theory (and probability theory) certainly feature in the various proofs, but they do so in a particularly simple and basic form. For this reason the subject may appeal to the non-specialist, and we have spared no effort to make the book completely self-contained. (In particular, Chapter III contains its own account of the relevant parts of probability theory, and the Appendix contains the few number theoretic results assumed in the text.)

A third-year undergraduate would have all the knowledge required to read the book, but from some points of view it is perhaps more suitable for graduate students. We should mention that each chapter is essentially self-contained and may be read by itself.

We hope that the book may prove useful for teaching purposes; any chapter could be used as a basis for a postgraduate course.

H. H.
K. F. R.

Nottingham and London
September 1964

ACKNOWLEDGEMENTS

ANYONE who turns the pages of this book will immediately notice the predominance of results due to Paul Erdös. In so far as the substance of this book may be said to define a distinct branch of number theory—and its relevance to a wide range of topics in classical number theory appears to justify this claim—Erdös is certainly its founder. He was the first to recognize its true potential and has been the central figure in many of its developments.

The authors were indeed fortunate to have the benefit of many discussions with Dr. Erdös. His unique insight and encyclopaedic knowledge were, of course, invaluable; but the authors are no less indebted to him for his constant interest and encouragement.

The authors are grateful to Prof. A. Rényi, both for the many useful reprints and references he provided and for his prompt and helpful replies to all inquiries.

The authors read the proofs of the entire book at each stage, and are solely responsible for the errors that still remain. They are, however, much indebted to Dr. A. Baker, Dr. D. A. Burgess, and Mr. H. Kestelman for generous assistance with the proof-reading of three of the chapters.

H. H.
K. F. R.

PREFACE TO THE SPRINGER EDITION

CHANGES from the first edition have been kept to a minimum. Several misprints and some errors that have come to light in the years since the publication of the first edition have been corrected, and we acknowledge with pleasure our indebtedness to all those who have drawn our attention to them, and in particular to Professors B. Berndt, K.-L. Chung, P. A. Diananda, H.-E. Richert, S. Segal, H. Stark and to Dr. D. Singmaster. At several places in the text, and in a short postscript, we have added references to developments that have occurred since the first appearance of 'Sequences'.

We thank Springer-Verlag for making this edition possible, and for their help in its preparation.

Urbana and London H. H.
September 1982 K. F. R.

TO OUR WIVES

CONTENTS

INTRODUCTION

As is already indicated in the Preface, we have tried to make each chapter a self-contained unit, so that any single chapter can conveniently be read by itself. Accordingly, each chapter is furnished with its own introduction† and there would be little point in describing the entire content of individual chapters here. Instead, we try to give some idea of the general nature of the topics to be considered, and in particular, to indicate the unity of purpose which leads us to combine the subject-matter of the various chapters in a single book.

One important aspect of the Theory of Numbers is the study of various specific integer sequences such as the sequence of primes, the sequence of squares (or, more generally, the sequence of kth powers for some integer $k \geqslant 2$). The object of such study may be to determine the distribution of the elements of some important sequence (as in the case of the prime number theorem and its refinements), to investigate the representation of integers as sums of elements of a given sequence (as is the case in Waring's problem), or to investigate any other arithmetic property which is not an immediate consequence of the definition of the sequence in question. As might be expected, such work usually depends heavily on special properties of the particular sequence under investigation. It happens occasionally, however, and sometimes quite unexpectedly, that the results obtained for a special sequence can be shown, by other means, to hold for a remarkably wide class of sequences, or even for all integer sequences.‡ Theorem 5 of Chapter I (§ 3) and Theorem 11 of Chapter II (§ 4) provide striking examples of this phenomenon. The first of these results relates to all integer bases (cf. Definition 2, Chapter I, § 1) and the second holds for all integer sequences; but each of these theorems represents an entirely unexpected generalization of a corresponding result established, in the first instance, for the sequence of squares by means of methods based on very special characteristics of this sequence. Furthermore, the proofs of these more general

† Chapter I has only a very concise introduction, but this is compensated by the opening discussions of the individual sections of that chapter.

‡ In this discussion, which is quite informal, we sometimes sacrifice precision for readability. For example, by 'integer sequence', we really mean a subsequence of the sequence of non-negative integers.

results are very much simpler than the original proofs which related only to the sequence of squares.

The main theme of this book is the study of integer sequences with a view to finding properties common to (or laws relating to) extensive classes of such sequences. We shall only rarely consider specific sequences, except as a means of establishing the existence of a sequence having some property which is under investigation.

We have given two examples of theorems, each relating to an extensive class of sequences, which arose as generalizations of much more special results. Theorems of the desired generality may, however, be sought in entirely different ways. A possible approach is to investigate some specific aspect of sequence behaviour; for example, the possible density relationships in the addition of sequences. Alternatively, one may begin by investigating some specific class of sequences: for example, the class of all those integer sequences such that no element divides any other, or the class of all those integer sequences containing no three consecutive terms of any arithmetic progression. There are, of course, many other possible lines of investigation, and indeed, each investigation is likely to suggest others.

The subject appears to be at a fairly early stage of its development, and yet many of the really interesting open questions are difficult. Nevertheless, various powerful and beautiful results are known, and no doubt many more such results are still to be discovered. A better understanding of the general principles underlying the arithmetic properties of sequences would be valuable not only for its own sake, but also for applications to more classical problems; for in the consideration of specific sequences, these general principles could be used in conjunction with the special characteristics of the sequence under consideration. This remark is probably particularly applicable to questions concerning the distribution of the elements of a general integer sequence among (and within) congruence classes; such questions seem particularly intractable, but are highly relevant to additive number theory. In this connexion, we remark that one of the many fascinating facets of Kneser's theorem (Chapter I, §§ 7–10) is the manner in which congruence classes feature in the proof.

In Chapters I, II, and III we establish general laws relating to the addition of sequences, and, in particular, relating to the number of representations of integers as sums of a *fixed* number of elements of a given sequence. A closely related problem is that of investigating the number of 'partitions' of integers into parts from a given sequence, that is, representations as sums of an *arbitrary* (and, in particular, unbounded) number of elements of the sequence. Some surprisingly general results are known in this connexion, and indeed, in many respects such investigations have been more successful than corresponding investigations relating to sums of a fixed number of elements of the sequence (cf. the discussion concerning the question (ii) in § 1 of Chapter II). We refer the reader to 'The Theory of Partitions' (Addison Wesley, 1976) by G. E. Andrews.

Chapter IV stands in a special relationship to the main theme of the book. Of the topics considered there, only the theorems in § 10, obtained by applications of the 'large sieve', fit ideally into the framework of this book; one could hardly hope for better examples of completely general and yet very powerful results concerning integer sequences. A substantial part of the chapter is devoted to the Selberg sieve, which is applicable only to sequences of rather a special kind. Nevertheless, Selberg's method is not entirely devoid of generality, and is of such power and beauty that it would have been a pity to exclude his sieve altogether. We compromise by describing the general mechanism of the Brun–Selberg sieve method, and indicating the manner in which it can be applied without actually carrying out any specific application. We might add that it would, in any case, have been a formidable task to include a readable and self-contained account of the more interesting applications; the subject merits a book to itself. The topic of § 3 is also not in keeping with the general character of the book, but the theorem we prove there is required for applications in the subsequent chapter. Throughout Chapter IV, we endeavour to display the inter-relations between the various sieve methods.

In Chapter V we describe some interesting investigations which hinge on the multiplicative structure of the integers. Although comparatively short, the chapter includes a number of remarkable results; Theorems 13, 14, and 15 are particularly profound, and will amply repay study.

Throughout the text we shall frequently mention unsolved problems in connexion with the work we describe. We do not, however, include comprehensive collections of such problems; many further questions are raised in the original papers and surveys to which we refer. We would also like to draw attention to a very useful collection of problems which has appeared recently: namely, P. Erdös, 'Quelques problèmes de la Théorie des Nombres', *Monographies de l'Enseignement Mathématique* No. 6, Genève.

NOTATION

1. We are aware of the drawbacks of elaborate notations and have tried, in various ways, to limit the burden placed on the reader; for example, we try to avoid the introduction of formal notations where some purely temporary device (such as can be explained at the beginning of an argument and completely forgotten once the argument is concluded) would do equally well. Nevertheless, in the course of the book we have occasion to introduce quite a number of notations, although many of these are already in common use and will be familiar to most readers.

The work of the various chapters is sufficiently distinctive to justify a separate system of notations for each chapter. *In relation to notation, as in other ways, each chapter is a self-contained unit (in which all the relevant notations are fully explained*†). Although, whenever appropriate, individual notations are used in more than one chapter, the corresponding definitions will be repeated in each of the chapters concerned.

It may happen (but this is the exception rather than the rule) that the same expression has different meanings in different chapters: for example, in the early chapters we use $h\mathscr{A}$, where h is a natural number and $\mathscr{A}$ is a sequence, to denote the sequence consisting of those numbers representable as the sum of h (possibly distinct) elements of $\mathscr{A}$, whereas in Chapter V we find it convenient to use $h\mathscr{A}$ to denote the sequence obtained from $\mathscr{A}$ by multiplying each element by h. (The early chapters deal with additive questions, whilst Chapter V is concerned with the multiplicative structure of the integers.) Although this may appear confusing here, in each chapter the appropriate meaning will be made perfectly clear and will seem quite natural in its context.

There are a few basic notations and conventions which will be in constant use throughout the book, and we list these below.

2. Conventions from set theory

We indicate that an object x does or does not belong to a set X by writing $x \in X$ or $x \notin X$. If X and Y are sets of objects, we write

$$X \subset Y \quad \text{or, alternatively,} \quad Y \supset X$$

to indicate that all the elements of X are also elements of Y. We say that $X = Y$ if both $X \subset Y$ and $Y \subset X$.

† We here exclude the basic conventions listed below, although even these will mostly be repeated in the text.

The *union* of a (finite or infinite) collection of sets is the set of objects each belonging to at least one set in the collection. The union of two sets X_1, X_2 is denoted by

$$X_1 \cup X_2;$$

the union of sets X_i, with i ranging over some index set $\mathscr{I}$, is denoted by

$$\bigcup_{i \in \mathscr{I}} X_i,$$

or simply $\bigcup_i X_i$ when the index set is obvious from the context.

The *intersection* of a (finite or infinite) collection of sets is the set of objects common to all the sets of the collection. The intersection of two sets X_1, X_2 is denoted by

$$X_1 \cap X_2;$$

the intersection of sets X_i, with i ranging over some index set $\mathscr{I}$, is denoted by

$$\bigcap_{i \in \mathscr{I}} X_i,$$

or simply by $\bigcap_i X_i$ when the index set is obvious from the context. Two sets are said to be *disjoint* if their intersection is empty. If every pair among a collection of sets consists of two disjoint sets, the collection is said to consist of disjoint sets.

The *difference* $X-Y$ of two sets X, Y is the set of all objects in X that are not in Y. If all the sets X occurring in some given context are subsets of a single set (or 'space') Ω, we refer to $\Omega - X$ as the *complement* of X (with respect to Ω) and normally denote it by X^c.

It will be convenient at times to refer to the number of elements in a finite set $\mathscr{S}$ as the *order* of $\mathscr{S}$; we use $|\mathscr{S}|$ to denote the order of $\mathscr{S}$.

Suppose α, β are real numbers satisfying $\alpha < \beta$. Then, as usual, (α, β) denotes the *open* interval consisting of all numbers θ satisfying $\alpha < \theta < \beta$, and $[\alpha, \beta]$ denotes the *closed* interval consisting of all numbers θ satisfying $\alpha \leqslant \theta \leqslant \beta$. There will be occasions, noted in the text, when the open interval (α, β) will be taken to mean just the set of all *integers* n satisfying $\alpha < n < \beta$, with a similar convention in the use of $[\alpha, \beta]$; and yet another possible usage (in the case when α, β are integers) is noted below in the paragraph on arithmetical conventions.

3. Integers and sets of integers

All small Latin letters, with the exception of c, e, and x, will denote integers when printed in italic type, unless the contrary is either explicitly stated or is obvious from the context (as is the case, for example, when such letters are used to denote functions). The letter

p is normally reserved for prime numbers (although occasionally it denotes a probability).

Capital script letters are reserved for sequences of integers, or sets of integers. Any script capital may be used to denote either an integer sequence or a set of integers; but $\mathscr{A}$, $\mathscr{B}$, $\mathscr{C}$ will usually be associated with integer sequences. For the meaning of $\mathscr{Z}$, $\mathscr{Z}_0$, $\mathscr{Z}_1$, see § 1 of Chapter I.

4. Integer sequences, their counting numbers and densities

The 'integer sequences' under investigation will usually be subsequences of the sequence of non-negative integers. But from time to time it will be convenient to consider more general integer sequences; for example, in §§ 7–10 of Chapter I we will admit (monotone strictly increasing integer) sequences containing negative integers, whilst not all the sequences considered in Chapter IV are monotone. In each case, the appropriate interpretation of 'sequence' will be clearly indicated in the text.

Suppose that $\mathscr{A}$ is a monotone (strictly) increasing sequence of integers. For any integer n, we define the *counting number* (or *counting function*) of $\mathscr{A}$ up to n to be the number of *positive*† elements of $\mathscr{A}$ not exceeding n; we denote this counting number by $A(n)$ (and we use $B(n)$, $C(n)$,... for the counting numbers corresponding to sequences $\mathscr{B}$, $\mathscr{C}$,...).

We shall discuss the various concepts of density, their elementary properties and respective uses, in the text. For the present, we merely list the definitions and notations which will be relevant in this connexion.

The *Schnirelmann density* $\sigma\mathscr{A}$ of $\mathscr{A}$ is defined by

$$\sigma\mathscr{A} = \inf_n \frac{A(n)}{n} \quad (n = 1, 2, \ldots).$$

The *lower asymptotic density* $\underline{\mathbf{d}}\mathscr{A}$ and the *upper asymptotic density* $\bar{\mathbf{d}}\mathscr{A}$ are defined by

$$\underline{\mathbf{d}}\mathscr{A} = \liminf_{n\to\infty} \frac{A(n)}{n}, \qquad \bar{\mathbf{d}}\mathscr{A} = \limsup_{n\to\infty} \frac{A(n)}{n}.$$

If $\underline{\mathbf{d}}\mathscr{A} = \bar{\mathbf{d}}\mathscr{A}$, we say that $\mathscr{A}$ possesses *asymptotic density* $\mathbf{d}\mathscr{A}$, given by the common value.

We shall use 'logarithmic density' only in the context of (monotonic strictly increasing) sequences of natural numbers and, accordingly, we

† Our reasons for not 'counting' the non-positive elements, even though there may be any finite number of these, will be explained in the text.

now suppose that $\mathscr{A}$ is such a sequence. We define the *lower logarithmic density* $\underline{\delta}\mathscr{A}$ and the *upper logarithmic density* $\overline{\delta}\mathscr{A}$ by

$$\underline{\delta}\mathscr{A} = \liminf_{n\to\infty} \frac{1}{\log n} \sum_{\substack{a\in\mathscr{A}\\ a\leqslant n}} \frac{1}{a}, \qquad \overline{\delta}\mathscr{A} = \limsup_{n\to\infty} \frac{1}{\log n} \sum_{\substack{a\in\mathscr{A}\\ a\leqslant n}} \frac{1}{a}.$$

If $\underline{\delta}\mathscr{A} = \overline{\delta}\mathscr{A}$, we say that $\mathscr{A}$ possesses *logarithmic density* $\delta\mathscr{A}$, given by the common value.

5. Constants

We use π and e in the usual way, and denote Euler's constant by γ.

The letter c (also c_1, c_2,...) denotes a positive constant, unless the contrary is either explicitly stated or is obvious from the context (as is the case, for example, when c is used to denote the typical element of an integer sequence $\mathscr{C}$).

The (symbolic) statement

$$F = O(G) \quad \text{or, alternatively,} \quad F \ll G,$$

about a function F and a non-negative function G, will mean that there exists a constant c such that

$$|F| \leqslant cG;$$

unless otherwise stated the constant c will be absolute and the inequality will be valid over the entire range of definition of the functions. When the constant c is not absolute, the implicit parameters will be indicated in the text; and if the inequality is not always valid, a range of validity will be given (in fact usually displayed with the inequality).

6. Arithmetical conventions

Here we give only a few of the most common standard conventions of elementary number theory. Further standard arithmetical notations are explained in § 4 of the Appendix.

We indicate whether or not (the integer) r divides (the integer) s by writing $r \mid s$ or $r \nmid s$. If p^l is the exact power to which the prime p divides n (in the sense that $p^l \mid n$, $p^{l+1} \nmid n$), we write $p^l \| n$.

The symbols $(r, s, t,...)$ and $[r, s, t,...]$ denote respectively the highest common factor (H.C.F.) and lowest common multiple (L.C.M.) of the integers r, s, t,.... (When only two integers r, s are concerned, there is an ambiguity involving the open interval (r, s) or the closed interval $[r, s]$, but the alternative meanings are sufficiently different to exclude any risk of confusion in practice.)

If α is a real number, $[\alpha]$ denotes the integral part of α (i.e. the greatest integer not exceeding α).

I

ADDITION OF SEQUENCES: STUDY OF DENSITY RELATIONSHIPS

1. Introduction and notation

LET $\mathscr{Z}$ denote the set of all integers. Although $\mathscr{Z}$ is not itself a sequence, we shall sometimes use the phrase 'subsequence of $\mathscr{Z}$' to describe a monotone strictly increasing sequence of integers. Let $\mathscr{Z}_0$ denote the sequence of all non-negative integers and $\mathscr{Z}_1$ the sequence of all natural numbers. Throughout the first five sections of this chapter the letters $\mathscr{A}$, $\mathscr{B}$, $\mathscr{C}$,... stand for subsequences of $\mathscr{Z}_0$; the letters a, b, c,... stand for elements of $\mathscr{A}$, $\mathscr{B}$, $\mathscr{C}$,... respectively.

Definition 1. The sequence of all distinct integers of the form

$$a+b. \qquad a \in \mathscr{A}, \quad b \in \mathscr{B},$$

will be denoted by $\mathscr{A}+\mathscr{B}$, and referred to as the '*sum*' of the sequences $\mathscr{A}$ and $\mathscr{B}$. The sum of more than two sequences is defined similarly; if $\mathscr{A}_0$, $\mathscr{A}_1$,..., $\mathscr{A}_k$ are $k+1$ sequences, their sum will be denoted by

$$\sum_{r=0}^{k} \mathscr{A}_r.$$

It is convenient to write $\mathscr{A}+\mathscr{A}$ as $2\mathscr{A}$ and, more generally, the h-fold sum $\mathscr{A}+\mathscr{A}+...+\mathscr{A}$ as $h\mathscr{A}$.

If 0 belongs to $\mathscr{B}$ but not to $\mathscr{A}$, $\mathscr{A}+\mathscr{B}$ consists of all (distinct) elements of the form

$$a \quad \text{or} \quad a+b \quad (b > 0); \tag{1.1}$$

if 0 belongs to both $\mathscr{A}$ and $\mathscr{B}$, the positive part of $\mathscr{A}+\mathscr{B}$ consists of all (distinct) elements of the form

$$a,\, b \quad \text{or} \quad a+b \quad (a > 0,\, b > 0). \tag{1.2}$$

Thus the insertion of 0 in one or both of $\mathscr{A}$, $\mathscr{B}$ represents a notational device with an important effect on the precise composition of $\mathscr{A}+\mathscr{B}$; it will be important to bear this in mind in the sequel.

For any given integer t, $t+\mathscr{A}$ denotes the sequence consisting of all integers of the form

$$t+a, \qquad a \in \mathscr{A}.$$

The extent to which $\mathscr{A}+\mathscr{B}$ yields 'new' elements—that is, elements which are not in $\mathscr{A}$ when dealing with case (1.1), and elements which are not in $\mathscr{A} \cup \mathscr{B}$ in case (1.2)—may seem at first sight to depend strongly on the arithmetical character of the sequences $\mathscr{A}$ and $\mathscr{B}$. Schnirelmann† was the first to point out that there exist very general circumstances in which, on the contrary, substantial information can be gained about $\mathscr{A}+\mathscr{B}$ from the mere knowledge of the *densities of* $\mathscr{A}$ *and* $\mathscr{B}$.

The iterated sums of a single sequence $\mathscr{A}$, containing 0, present a particularly interesting topic. Under what circumstances, for example, is $2\mathscr{A}$ substantially more numerous than $\mathscr{A}$? Clearly not always, for when $\mathscr{A}$ consists of the non-negative even integers, we have $2\mathscr{A} = \mathscr{A}$. At the other extreme, we may have $2\mathscr{A} = \mathscr{Z}_0$ without $\mathscr{A}$ itself being particularly numerous. The sequence consisting of zero and the odd natural numbers provides a trivial example of this, but we shall see later that $2\mathscr{A} = \mathscr{Z}_0$ may be satisfied by quite 'thin' sequences $\mathscr{A}$ (certainly of zero density).

Similar questions may be posed about higher multiples of $\mathscr{A}$; and a classical theorem of Lagrange, which states that every natural number is the sum of four integral squares, provides a non-trivial example of a sequence $\mathscr{A}$ satisfying $4\mathscr{A} = \mathscr{Z}_0$. We formulate the following definition.

Definition 2. If $\mathscr{A} \subset \mathscr{Z}_0$ and there exists a natural number h such that $h\mathscr{A} = \mathscr{Z}_0$, $\mathscr{A}$ is called a *basis*, of *order* h. If h_0 is the least value of h possessing this property, h_0 is called the *exact order* of $\mathscr{A}$.

Thus the sequence

$$0, 1, 3, 5, \ldots, (2r+1), \ldots$$

is a basis of exact order 2; also, since no integer congruent to 7 (mod 8) can be the sum of three integer squares, the sequence

$$0, 1, 4, 9, \ldots, r^2, \ldots$$

is, by Lagrange's theorem, a basis of exact order 4. We observe that both these bases contain 0 and 1—obviously this must always be the case, in view of our manner of defining a basis.

The present chapter is an account, in terms of density considerations, of some fundamental laws governing the addition of general integer sequences. Several important results concerning bases will form a natural part of this account.

Among the many kinds of density (of an integer sequence) which have been investigated, we shall restrict ourselves to the two most natural types—to Schnirelmann density (defined in the next section) and to (lower) asymptotic density (defined in § 6).

† See reference (1) for Theorem 1 in § 2.

2. Schnirelmann density and Schnirelmann's theorems

Let $A(n)$ denote the counting number of the positive part of $\mathscr{A}$, so that

$$A(n) = \sum_{1\leqslant a\leqslant n} 1$$

is the number of *positive* elements of $\mathscr{A}$ not exceeding n.

Furthermore, if $m \leqslant n$, we denote by $A(m,n)$, the number of elements a of $\mathscr{A}$ satisfying $m \leqslant a \leqslant n$. Then $A(1,n) = A(n)$ if $1 \leqslant n$, and

$$A(l,n) = A(l,m)+A(m+1,n) \quad \text{if } l \leqslant m < n.$$

In the early part of the chapter, where we are concerned with subsequences of $\mathscr{Z}_0$, the presence (or absence) of 0 in $\mathscr{A}$ will be stipulated according to notational convenience; it is really the positive part of $\mathscr{A}$ which is under investigation. We have therefore defined $A(n)$ so that the zero element will not be counted, even if present. In the later part of the chapter it will sometimes be convenient to permit $\mathscr{A}$ to contain negative elements (so that $\mathscr{A}$ can be any subsequence of $\mathscr{Z}$), but there again it is the positive part of $\mathscr{A}$ that will be important. Finally we note that, in particular, $A(n) = 0$ if $n \leqslant 0$.

We define the *Schnirelmann density* $\sigma\mathscr{A}$ of $\mathscr{A}$ by

$$\sigma\mathscr{A} = \inf_n \frac{A(n)}{n} \quad (n = 1,2,3,\ldots);$$

this will usually be referred to as the density of $\mathscr{A}$ where there is no danger of confusion. The following obvious consequences of this definition should be noted at the outset:

(i) $A(n) \geqslant n\,\sigma\mathscr{A}$ $(n = 1,2,3,\ldots)$;

(ii) $\sigma\mathscr{A} = 1$ if and only if $\mathscr{A} \supset \mathscr{Z}_1$;

(iii) if 1 does not belong to $\mathscr{A}$, $A(1) = 0$ and hence, by (i), $\sigma\mathscr{A} = 0$.

Thus, by contrast with asymptotic density, the early elements of $\mathscr{A}$ have a disproportionately important effect on the value of $\sigma\mathscr{A}$.

It may appear from these remarks that Schnirelmann density is a comparatively artificial concept. Nevertheless, the simple and striking form the results below assume will soon convince the reader that this is a density appropriate to the types of question considered. We shall, however, formulate analogous results in terms of (lower) asymptotic density in the later sections of the chapter.

THEOREM 1 (Schnirelmann).(1) *If* $1 \in \mathscr{A}$ *and* $0 \in \mathscr{B}$,

(2.1) $$\sigma(\mathscr{A}+\mathscr{B}) \geqslant \sigma\mathscr{A}+\sigma\mathscr{B}-\sigma\mathscr{A}\,\sigma\mathscr{B}.$$

(1) L. G. Schnirelmann, 'Über additive Eigenschaften von Zahlen', *Annals Inst. polyt. Novocherkassk.* **14** (1930) 3–28; *Math. Annalen* **107** (1933) 649–90.

Proof. Let $$1 = a_1 < a_2 < ... < a_r$$
be all the elements of $\mathscr{A}$ counted in $A(n)$. These elements are separated by gaps of 'length' $g_i = a_{i+1}-a_i-1$ $(i = 1,2,...,r-1)$ and, if n does not belong to $\mathscr{A}$, there is a gap of length $g_r = n-a_r$ between a_r and n (including n). Consider the integers in the ith gap, that is, the integers $a_i+1, a_i+2,..., a_i+g_i$. If $1 \leqslant t \leqslant g_i$ and $t \in \mathscr{B}$, then $a_i+t \in \mathscr{A}+\mathscr{B}$, and therefore at least $B(g_i)$ numbers of the ith gap lie in $\mathscr{A}+\mathscr{B}$. Hence the counting number of $\mathscr{A}+\mathscr{B}$ up to n is at least

$$\begin{aligned} A(n)+\sum_{i=1}^{r} B(g_i) &\geqslant A(n)+\Big(\sum_{i=1}^{r} g_i\Big)\sigma\mathscr{B} \\ &= A(n)+\{n-A(n)\}\sigma\mathscr{B} \\ &= (1-\sigma\mathscr{B})A(n)+n\sigma\mathscr{B} \\ &\geqslant n\{(1-\sigma\mathscr{B})\sigma\mathscr{A}+\sigma\mathscr{B}\}, \end{aligned}$$

and the result follows since this inequality holds for every natural n.

Clearly, Theorem 1 is non-trivial only when neither one of $\sigma\mathscr{A}$, $\sigma\mathscr{B}$ is equal to 0 or 1, i.e. if

$$0 < \sigma\mathscr{A} < 1, \qquad 0 < \sigma\mathscr{B} < 1,$$

and in this event (2.1) implies

$$\sigma(\mathscr{A}+\mathscr{B}) > \sigma\mathscr{A}. \tag{2.2}$$

The example described after the statement of Theorem 2 below shows Theorem 1 to be best possible. As one might expect, (2.1) can be substantially improved if 0 is adjoined to $\mathscr{A}$, although even this is surprisingly hard to prove. The following theorem is an indication of the possible extent of such an improvement, although the case considered is a special one.

THEOREM 1′ (Schnirelmann).† *If* $0 \in \mathscr{A} \cap \mathscr{B}$ *and* $\sigma\mathscr{A}+\sigma\mathscr{B} \geqslant 1$ *then* $\mathscr{A}+\mathscr{B} = \mathscr{Z}_0$.

Proof. We may assume that $\sigma\mathscr{A} > 0$, so that $1 \in \mathscr{A}$.

Suppose the result is false, so that there exists a natural number n which does not lie in $\mathscr{A}+\mathscr{B}$. Since $\mathscr{A}+\mathscr{B} \supset \mathscr{A}$, n cannot lie in $\mathscr{A}$ and, in particular, $n > 1$. Hence

$$A(n-1) = A(n) \geqslant n\sigma\mathscr{A} > (n-1)\sigma\mathscr{A};$$

also $$B(n-1) \geqslant (n-1)\sigma\mathscr{B}$$

and therefore, by hypothesis,

$$A(n-1)+B(n-1) > n-1.$$

† Schnirelmann; see ref. (1).

Let $a_1, a_2,..., a_r$ be all the elements counted in $A(n-1)$ and $b_1, b_2,..., b_s$ all the elements counted in $B(n-1)$, so that $r = A(n-1)$ and $s = B(n-1)$. Then each of the $r+s$ numbers

$$a_1, a_2,..., a_r, n-b_1, n-b_2,..., n-b_s$$

is positive and none exceeds $n-1$. Since $r+s > n-1$ it follows from Dirichlet's Box Argument that there exists an i $(1 \leqslant i \leqslant r)$ and a j $(1 \leqslant j \leqslant s)$ such that $a_i = n-b_j$. Hence $n = a_i+b_j \in \mathscr{A}+\mathscr{B}$, contrary to our original supposition. Thus $\mathscr{A}+\mathscr{B}$, which certainly contains 0, also contains every natural number.

Theorem 1′ and Theorem 2 (to be stated later) were pointers towards the following result.

Theorem 3 (Mann). *If* $0 \in \mathscr{A} \cap \mathscr{B}$,

$$\sigma(\mathscr{A}+\mathscr{B}) \geqslant \min(1, \sigma\mathscr{A}+\sigma\mathscr{B}). \tag{2.3}$$

A fairly simple proof of this result will be given† in § 4. Theorem 3 was first proved by H. B. Mann(2) in 1942, thirteen years after Schnirelmann's discovery of Theorems 1 and 1′. In 1932 Khintchin(3) had proved Theorem 3 for the important special case $\sigma\mathscr{A} = \sigma\mathscr{B}$, by means of a difficult and extremely ingenious argument; but in the ten years that followed, the general case attained the stature of a famous unsolved problem (known then as the $(\alpha+\beta)$-conjecture). Although Mann's theorem represents a major mathematical achievement, and has led to many interesting developments, for most applications to other branches of number theory the weaker Theorems 1 and 1′ are equally effective. It is remarkable, however, as in the case of several other theorems of this chapter, that non-trivial results of the kind should exist at all.

Between the time of Khintchin's special result and Mann's solution of the general case, Besicovitch(4) proved a theorem (Theorem 2 below) which was the nearest approach to the $(\alpha+\beta)$-conjecture then available. Even in retrospect, however, this theorem is still of interest, quite apart from the great elegance of its proof. For $0 \in \mathscr{A}$ is not stipulated, so that Besicovitch's theorem is not superseded by that of Mann; it thus remains the best result known concerning the number of elements of type a or $a+b$ only. The condition $1 \in \mathscr{A}$ is, of course, automatically satisfied when $\sigma\mathscr{A} > 0$ and is therefore unimportant.

† See Theorem 3′.

(2) H. B. Mann, 'A proof of the fundamental theorem on the density of sums of sets of positive integers', *Ann. Math.* (2) **43** (1942) 523–7.

(3) A. Khintchin, 'Zur additiven Zahlentheorie', *Mat. Sb.* N.S. **39** (1932) 27–34.

(4) A. S. Besicovitch, 'On the density of the sum of two sequences of integers', *J. Lond. Math. Soc.* **10** (1935) 246–8.

THEOREM 2 (Besicovitch). *Let* $1 \in \mathscr{A}$ *and* $0 \in \mathscr{B}$. *If the number* $\beta \geqslant 0$ *is such that*

(2.4) $$B(m) \geqslant \beta(m+1) \quad (m = 1, 2, 3, \ldots)$$

and

(2.5) $$\sigma\mathscr{A}+\beta \leqslant 1,$$

then

(2.6) $$\sigma(\mathscr{A}+\mathscr{B}) \geqslant \sigma\mathscr{A}+\beta.$$

Before proving Theorem 2 we remark that relative to the premisses (2.4) and (2.5) the conclusion (2.6) is best possible. This can be seen at once from the following example, due to R. Rado, quoted by Besicovitch.[4]

Take $\mathscr{A}$ to be the sequence

$$1, 10, 11, 12, 13, \ldots$$

and $\mathscr{B}$ to be the sequence

$$0, 1, 9, 10, 11, 12, \ldots$$

so that $\sigma\mathscr{A} = \beta = \frac{1}{9}$. Then $\mathscr{A}+\mathscr{B}$ is the sequence

$$1, 2, 10, 11, 12, \ldots,$$

of density $\sigma(\mathscr{A}+\mathscr{B}) = \frac{2}{9}$, and (2.6) is seen to hold with equality. This example can be varied in obvious ways.

Proof. We write $\mathscr{C} = \mathscr{A}+\mathscr{B}$ and $\sigma\mathscr{A} = \alpha$. Let P_n denote the proposition

(2.7) $$C(n) \geqslant (\alpha+\beta)n,$$

which is to be proved for every natural number n. Since P_0 is true and, in view of (2.5), P_{c-1} implies P_c for every element c of $\mathscr{C}$, we may assume

(2.8) $$n \notin \mathscr{C}$$

for the purpose of proving (2.7).

In order to describe Besicovitch's idea, we recall (although this will not be required for the present proof) that the proof of Theorem 1 was based on the following fact.

(I) If $a' < a''$ are consecutive elements of $\mathscr{A}$, and $g = a''-a'-1$, then

(2.9) $$C(a', a'+g) \geqslant A(a', a'+g)+B(g).$$

The inequality (2.9) is inadequate for the present proof, in view of the fact that (2.4) is inapplicable to $B(g)$ when $g = 0$.

Besicovitch's ingenious innovation was to modify the proof of Theorem 1 in such a way that the subtle result (II) below (which we prove later) plays the role corresponding to that of (I) in the proof of Theorem 1.

(II) Let a' be an element of $\mathscr{A}$ and let a'' be the least element of $\mathscr{A}$ for which $a' < a''$ and the interval $[a', a'']$ contains an integer not in $\mathscr{C}$. Then, writing $g = a''-a'-1$, we have

$$C(a', a'+g) \geqslant A(a', a'+g)+B(g). \tag{2.10}$$

We note that, since $\mathscr{A} \subset \mathscr{C}$, the possibility $g = 0$ cannot arise in (II). Before proving (II), we shall deduce (2.7) from it; this will require only a modification of the proof of Theorem 1.

We select a subsequence of those elements of $\mathscr{A}$ counted in $A(n)$, as follows. After writing $a^{(1)} = 1$, we define $a^{(i)}$ inductively (for $i = 2, 3, \ldots$) by choosing $a^{(i)}$ to be the least element of $\mathscr{A}$ which exceeds $a^{(i-1)}$ and is such that the interval $[a^{(i-1)}, a^{(i)}]$ contains an integer not in $\mathscr{C}$. In view of (2.8) the terminating element $a^{(s)}$ will have the property (III) below.

(III) The least integer d, which lies in the interval $[a^{(s)}, n]$ but not in $\mathscr{C}$, exists and is such that the interval $[d, n]$ contains no element of $\mathscr{A}$.

As in the proof of Theorem 1, we write

$$g^{(i)} = a^{(i+1)}-a^{(i)}-1 \quad (i = 1, 2, \ldots, s-1),$$

and

$$g^{(s)} = n-a^{(s)}.$$

We note that $g^{(i)} \geqslant 1$ for $1 \leqslant i \leqslant s$.

Now, by (II), we have

$$C(a^{(i)}, a^{(i)}+g^{(i)}) \geqslant A(a^{(i)}, a^{(i)}+g^{(i)})+B(g^{(i)}) \tag{2.11}$$

for $1 \leqslant i \leqslant s-1$. In fact, (2.11) holds for $i = s$ also. For the inequality (2.11) concerns only those elements of $\mathscr{A}$ which do not exceed n; but on inserting (if necessary) the integer $n+1$ into $\mathscr{A}$, we can (in view of (III)) apply (II) with $a' = a^{(s)}$ and $a'' = n+1$.

Finally, on summing (2.11) over the range $1 \leqslant i \leqslant s$, we obtain

$$C(n) \geqslant A(n)+\sum_{i=1}^{s} B(g^{(i)}) \geqslant A(n)+\beta\sum_{i=1}^{s}(g^{(i)}+1) = A(n)+\beta n,$$

which implies (2.7).

It remains only to prove (II). Let d be the least integer in the interval $[a', a'']$ which does not lie in $\mathscr{C}$. Then, since $d = a+b$ is impossible (with $a \in \mathscr{A}$, $b \in \mathscr{B}$), the elements of $\mathscr{A}$ in the interval $[a', d-1]$ do not include any of the numbers $d-b$ (with $b \in \mathscr{B}$) in the interval (namely those for which b satisfies $1 \leqslant b \leqslant d-a'$ and hence $1 \leqslant b \leqslant d-a'-1$). Hence, since every integer in the interval $[a', d-1]$ lies in $\mathscr{C}$, we have

$$C(a', d-1) \geqslant A(a', d-1)+B(d-a'-1). \tag{2.12}$$

Furthermore, $a'+b \in \mathscr{C}$ for every element b of $\mathscr{B}$, so that we have $C(d, a''-1) \geqslant B(d-a', a''-a'-1)$. Hence, since none of the integers in the interval $[d, a''-1]$ lies in $\mathscr{A}$, we have

$$C(d, a''-1) \geqslant A(d, a''-1)+B(d-a', a''-a'-1). \tag{2.13}$$

On adding (2.12) and (2.13) we obtain (2.10), and this completes the proof of (II) and of the theorem.

Let us return for the moment to Theorems 1 and 1′. Inequality (2.1) can be expressed symmetrically in the form

$$1-\sigma(\mathscr{A}+\mathscr{B}) \leqslant (1-\sigma\mathscr{A})(1-\sigma\mathscr{B}),$$

and by induction:

COROLLARY 1. *If each of* $\mathscr{A}_0, \mathscr{A}_1,..., \mathscr{A}_{k-1}$ *contains* 0 *and* 1, *then*

$$1-\sigma(\mathscr{A}_0+\mathscr{A}_1+...+\mathscr{A}_{k-1}) \leqslant \prod_{r=0}^{k-1}(1-\sigma\mathscr{A}_r).$$

In particular, if the k sequences are identical, we find that

COROLLARY 2. *If* $0 \in \mathscr{A}$,

$$\sigma(k\mathscr{A}) \geqslant 1-(1-\sigma\mathscr{A})^k.$$

If $\sigma\mathscr{A} > 0$ and k is chosen sufficiently large, the right-hand side of this inequality can be made at least as large as $\frac{1}{2}$, and it follows from Theorem 1′ that with such a choice of k, $2k\mathscr{A} = \mathscr{Z}_0$. In other words, we have proved (cf. Definition 2, § 1):

THEOREM 4. *Every sequence which contains* 0 *and has positive Schnirelmann density is a basis.*

Theorem 4 is a result of a kind which has long been of central interest in number theory. For example, Schnirelmann himself used it to prove that the sequence $\mathscr{P}$ consisting of 0, 1 and the primes, is a basis; he first showed that $\sigma(2\mathscr{P}) > 0$ by means of a sieve method† and then applied Theorem 4. At the time, this result was an important contribution to the still unsolved Goldbach problem. Linnik(5) has also made effective use of Theorem 4 in giving a relatively simple proof of Hilbert's famous theorem that, for each natural k, the sequence

$$0,\ 1,\ 2^k,\ 3^k,...,\ r^k,...$$

is a basis.

Now $\mathscr{P}$ and the sequence of kth powers have density 0, and it is reasonable to say that, in general, the most interesting bases are those having zero density. It would therefore be extremely valuable to find conditions sufficient to ensure that a sequence $\mathscr{B}$ is a basis, which are less stringent than the condition

$$B(n) > \beta n \quad (\beta > 0;\ n = 1, 2, 3,...)$$

required for an appeal to Theorem 4. Unfortunately, known results in this direction are, up to the present, of rather a negative nature; for

† See Chapter IV.

(5) Ju. V. Linnik, 'An elementary solution of the problem of Waring by Schnirelmann's method', *Mat. Sb.* N.S. 12 (54) (1943) 225–30.

example, it has been proved[6] that for sequences $\mathscr{B}$ of zero asymptotic density, a lower bound for $B(n)$ (in terms of n) can never by itself ensure that $\mathscr{B}$ is a basis.

On the other hand, a condition of type

$$B(n) > \beta n^{\gamma} \quad (\beta > 0,\ 0 < \gamma < 1;\ n = 1, 2, 3,\ldots) \tag{2.14}$$

is certainly necessary if $\mathscr{B}$ is to be a basis. Indeed, if $\mathscr{B}$ is a basis of order h it is clear from combinatorial considerations that

$$\{B(n)+1\}^h \geqslant n+1$$

for every natural n, and therefore (2.14) holds with $\gamma = 1/h$ and some positive $\beta < 1$.

What we have just proved is that a basis cannot be too 'thin'. We shall describe later (see (5.47) and (5.48)) a method for constructing, for a given natural number h, a basis $\mathscr{B}$ of order h with

$$B(n) \leqslant \beta' n^{1/h}$$

for some $\beta' > 0$ and $n = 1, 2, 3,\ldots$. In other words, we shall construct a basis of order h which, to within a constant factor, is as 'thin' as is possible.

We shall return to the subject of bases in section 5. For the present we turn to another development stemming from Theorem 1.

3. Essential components and complementary sequences

If $\sigma\mathscr{B} = 0$ neither Theorem 1 nor Theorem 3 yields any information. Khintchin was the first to pose the problem of finding all sequences $\mathscr{B}$ of zero density, each satisfying (2.2) for every sequence $\mathscr{A}$ of positive density $\sigma\mathscr{A} < 1$.

Definition 3. If $\mathscr{W}$ is a sequence such that

$$\sigma(\mathscr{A}+\mathscr{W}) > \sigma\mathscr{A}$$

for every sequence $\mathscr{A}$ with $0 < \sigma\mathscr{A} < 1$, $\mathscr{W}$ is said to be an *essential component*.

By Theorem 1 every sequence of positive density which contains 0 is an essential component. Khintchin[7] himself proved in 1933 that the sequence $\mathscr{Q}$ consisting of 0 and the squares of the natural numbers (so that $\mathscr{Q}$ has density 0) is an essential component. Using a complicated argument based on special properties of $\mathscr{Q}$, he even proved that

$$\sigma(\mathscr{A}+\mathscr{Q}) \geqslant \sigma\mathscr{A} + \frac{1}{2.10^8}(1-\sigma\mathscr{A})^2\sigma\mathscr{A} \tag{3.1}$$

(6) A. Stöhr, 'Gelöste und ungelöste Fragen über Basen der natürlichen Zahlenreihe I', *J. reine angew. Math.* **194** (1955) 40–65. See § 7.

(7) A. Khintchin, 'Über ein metrisches Problem der additiven Zahlentheorie', *Mat. Sb.* N.S. **40** (1933) 180–9.

for every sequence $\mathscr{A}$. Three years later Erdös[8] discovered a theorem of startling generality; namely, that every basis is an essential component. In fact Erdös proved appreciably more.

THEOREM 5 (Erdös). *If $\mathscr{B}$ is a basis of exact order h, then for every*† *sequence $\mathscr{A}$*

$$\sigma(\mathscr{A}+\mathscr{B}) \geqslant \sigma\mathscr{A}+\frac{1}{2h}(1-\sigma\mathscr{A})\,\sigma\mathscr{A}. \tag{3.2}$$

The significance of Theorem 5 is primarily qualitative, and at the time Erdös was probably not especially concerned with the degree of precision that can be attained in inequalities of this type. (Even so, when $h = 4$ and $\mathscr{B} = \mathscr{Q}$, (3.2) yields a result much better than (3.1).) Landau soon pointed out, however, that only a small modification of Erdös's argument was required to obtain an improvement on (3.2). He observed that in the averaging argument used by Erdös, it was wasteful to calculate in terms of the exact order of the basis. For example, whilst the exact order of $\mathscr{Q}$ is 4, approximately five-sixths of the positive integers are already representable as sums of three squares,[9] and it is possible to take advantage of this fact. Landau's idea was, accordingly, to calculate in terms of the 'average order' h^* of $\mathscr{B}$, which he defined by

$$h^* = \sup_n \frac{1}{n}\sum_{m=1}^{n} h(m) \qquad (n = 1, 2, 3,\ldots)$$

where, for each m, $h(m)$ denotes the least number of elements of $\mathscr{B}$ with sum m. Plainly $h^* \leqslant h$ always. (It is not hard to see that the average order of $\mathscr{Q}$ is 19/6.) Landau's modification[10] resulted in the following sharper form of Theorem 5.

THEOREM 5′ (Erdös–Landau). *If $\mathscr{B}$ is a basis of average order h^*, then for every sequence $\mathscr{A}$*

$$\sigma(\mathscr{A}+\mathscr{B}) \geqslant \sigma\mathscr{A}+\frac{1}{2h^*}(1-\sigma\mathscr{A})\,\sigma\mathscr{A}. \tag{3.3}$$

Proof. Let $\mathscr{C} = \mathscr{A}+\mathscr{B}$. We have to find a suitable lower bound for $C(n)-A(n)$; in other words, to show that the elements of $\mathscr{C}$ which do not lie in $\mathscr{A}$ are sufficiently numerous.

† (3.2) is, of course, significant only if $0 < \sigma\mathscr{A} < 1$.

(8) P. Erdös, 'On the arithmetical density of the sum of two sequences one of which forms a basis for the integers', *Acta arith.* 1 (1936) 197–200.

(9) See E. Landau, *Vorlesungen über Zahlentheorie*, 1 (Hirzel, Leipzig, 1927) 114–25.

(10) E. Landau, 'Über einige neuere Fortschritte der additiven Zahlentheorie'. *Cambridge tracts in Mathematics*, No. 35 (Cambridge University Press, 1937), 60–2.

For each fixed integer m, $1 \leqslant m \leqslant n$, let $\bar{D}_n(m)$ denote the number of elements a of $\mathscr{A}$ satisfying

$$a+m \leqslant n, \qquad a+m \notin \mathscr{A}.$$

Clearly

$$\bar{D}_n(b) \leqslant C(n)-A(n) \tag{3.4}$$

for every element b ($\leqslant n$) of $\mathscr{B}$.

At first sight this does not appear to be a very promising starting-point for an averaging argument; for the average of $\bar{D}_n(b)$ over elements b of $\mathscr{B}$ depends heavily on the structure of $\mathscr{B}$. The central idea behind Erdös's method is to use (3.4) to relate $C(n)-A(n)$ not to an average over elements of $\mathscr{B}$, but to the average

$$\bar{D}_n^* = \frac{1}{n}\sum_{m=1}^{n} \bar{D}_n(m)$$

of $\bar{D}_n(m)$ over all natural $m \leqslant n$; it is a matter of routine calculation to obtain a lower bound for $\bar{D}_n^*$.

The inequality relating $C(n)-A(n)$ to $\bar{D}_n^*$, which takes the form†

$$\bar{D}_n^* \leqslant h^*\{C(n)-A(n)\}, \tag{3.5}$$

is obtained in the following strikingly simple manner.

We have for any natural m and m' that

$$\bar{D}_n(m+m') \leqslant \bar{D}_n(m)+\bar{D}_n(m'); \tag{3.6}$$

for the numbers $a+(m+m')$ counted in $\bar{D}_n(m+m')$ fall into two mutually exclusive sets according as $a+m$ does or does not belong to $\mathscr{A}$, and these sets contain at most $\bar{D}_n(m')$ and $\bar{D}_n(m)$ elements respectively. Since each natural number is representable as the sum of $h(m)$ elements of $\mathscr{B}$, it follows from (3.4) and (3.6) that

$$\bar{D}_n(m) \leqslant h(m)\{C(n)-A(n)\} \quad (1 \leqslant m \leqslant n).$$

On summing over m and recalling the definition of h^*, we obtain (3.5).

It remains to obtain a lower bound for $\bar{D}_n^*$. It is convenient to define a counting function for those elements of $\mathscr{A}$ not counted in $\bar{D}_n(m)$, and accordingly we denote by $D_n(m)$ the number of elements a of $\mathscr{A}$ satisfying

$$a+m \leqslant n, \qquad a+m \in \mathscr{A}.$$

We have
$$\bar{D}_n(m)+D_n(m) = A(n-m) \geqslant (n-m)\sigma\mathscr{A},$$
so that, summing over m,

$$n\bar{D}_n^* = \sum_{m=1}^{n} \bar{D}_n(m) \geqslant \frac{(n-1)n}{2}\sigma\mathscr{A} - \sum_{m=1}^{n} D_n(m).$$

† As modified by Landau; see ref. (10).

Now the sum on the right-hand side of this inequality represents the number of solutions in i, j, and m of

$$a_i+m = a_j \leqslant n. \tag{3.7}$$

For fixed j, the number of possibilities for i, m is $j-1$, so that the total number of solutions of (3.7) is $\frac{1}{2}(A(n)-1)A(n)$.

Hence

$$\bar{D}_n^* \geqslant \tfrac{1}{2}(n-1)\sigma\mathscr{A} - \frac{1}{2n}(A(n)-1)A(n)$$

$$\geqslant \tfrac{1}{2}n\left\{\sigma\mathscr{A} - \left(\frac{A(n)}{n}\right)^2\right\},$$

since $\sigma\mathscr{A} \leqslant n^{-1}A(n)$. By (3.5), it follows that

$$\frac{C(n)}{n} \geqslant \frac{A(n)}{n} - \frac{1}{2h^*}\left(\frac{A(n)}{n}\right)^2 + \frac{1}{2h^*}\sigma\mathscr{A}.$$

To complete the proof of the theorem we need only show that in the first two terms on the right, $A(n)/n$ may be replaced by $\sigma\mathscr{A}$. This is certainly permissible, for $\sigma\mathscr{A} \leqslant n^{-1}A(n) \leqslant 1$ and, since $h^* \geqslant 1$, $x-(2h^*)^{-1}x^2$ is an increasing function of x for $0 \leqslant x \leqslant 1$.

It may appear at first sight that the above argument can be applied to a wider class of sequences $\mathscr{B}$ than the class of all bases; for the existence of the average order h^* was the only property of $\mathscr{B}$ used. However, it is easily seen that every sequence $\mathscr{B}$ containing† 0 and 1, of average order h^*, is in fact a basis of order less than $2h^*$; for it is clear that

$$h(n) \leqslant h(m)+h(n-m) \quad (m = 0, 1, 2,..., n),$$

where $h(0)$ is interpreted to be 0, and hence, on summing over m, we obtain

$$(n+1)h(n) \leqslant 2\sum_{m=1}^{n} h(m) \leqslant 2nh^*$$

for all natural n. This yields the result asserted.

The proof of Theorem 5′ is remarkably simple. As is not surprising, more detailed analysis has led to inequalities of the type

$$\sigma(\mathscr{A}+\mathscr{B}) \geqslant \alpha+\frac{1}{h^*}f(\alpha), \qquad \sigma\mathscr{A} = \alpha, \tag{3.8}$$

more precise than (3.3). The exact extent to which (3.3) can be improved is, however, not known. Let

$$f^*(\alpha) = \sup_f f(\alpha),$$

where the upper bound is taken over all those functions f such that (3.8) holds for all sequences satisfying the premisses of Theorem 5′. Erdös

† The presence of 1 in $\mathscr{B}$ is implicit in the existence of $h(1)$; and the zero element plays a trivial role.

conjectured at one time that $f^*(\alpha) \geqslant \alpha(1-\alpha)$, but later disproved(11) his own conjecture by showing that $f^*(\alpha) < \alpha(1-\alpha)-c(\min(\alpha, 1-\alpha))^{\frac{3}{2}}$ where c is a positive absolute constant. For $0 < \alpha < \frac{1}{2}$ this leads to the inequality $f^*(\alpha) < \alpha - c'\alpha^{\frac{3}{2}}$; and Brauer(12) proved that

$$f^*(\alpha) > \alpha - c''\alpha^{\frac{3}{2}} \quad \text{for } 0 < \alpha < 1$$

(Erdös points out(11) that $c'' \leqslant 2$). Thus for $0 < \alpha < \frac{1}{2}$ the correct order of magnitude of $f^*(\alpha)$ is known. For other results the reader is referred to the literature: e.g. Kasch(13) and S. Selberg(14) (for results of the same general nature as that of Brauer), Plünnecke(15) (for a very precise estimate in the special case when the definition of f^* is modified by fixing $\mathscr{B}$ to be the sequence of squares).

The characteristic property of an essential component is that it increases (by addition) the density of *every* sequence $\mathscr{A}$, with $0 < \sigma\mathscr{A} < 1$. It is to be expected, however, that we can increase the density much more effectively if we add to any *given* $\mathscr{A}$ a suitable sequence whose structure may depend on $\mathscr{A}$. The following theorem represents a very powerful result of this type. The theorem is, in fact, best possible(16) with respect to those sequences which have positive lower asymptotic density (definition in § 6); but such sequences constitute only a subclass of the sequences for which the theorem is highly effective.

THEOREM 6 (Lorentz(17)). *To every sequence $\mathscr{A}$ (containing more than one element) there corresponds a 'complementary' sequence $\mathscr{B}$ such that $\mathscr{A}+\mathscr{B}$ contains all large integers and*

$$B(n) \ll \sum_{\substack{\nu=1 \\ A(\nu)>0}}^{n} \frac{\log A(\nu)}{A(\nu)}, \tag{3.9}$$

where the implied constant is absolute.

We note that, since the terms in the sum on the right-hand side of

(11) P. Erdös, 'Über einige Probleme der additiven Zahlentheorie', *J. reine angew. Math.* **206** (1961) 61–6.

(12) A. Brauer, 'Über die Dichte der Summe zweier Mengen, deren eine von positiver Dichte ist', *Math. Z.* **44** (1939) 212–32.

(13) F. Kasch, 'Abschätzung der Dichte von Summenmengen', I and III, ibid. **62** (1955) 368–87; **66** (1956/57) 164–72.

(14) S. Selberg, 'Note on a metrical problem in the additive theory of numbers', *Arch. Math. Naturv.* **47** (1944) 111–8.

(15) H. Plünnecke, 'Über ein metrisches Problem der additiven Zahlentheorie', *J. reine angew. Math.* **197** (1957) 97–103.

(16) P. Erdös, 'Some results on additive number theory', *Proc. Am. math. Soc.* **5** (1954) 847–53.

(17) G. Lorentz, 'On a problem of additive number theory', ibid. 838–41.

(3.9) decrease and are small for large $A(\nu)$, $\mathscr{B}$ has zero asymptotic density provided only that the sequence $\mathscr{A}$ is infinite.

As might be expected in the case of any sequence of zero asymptotic density, better results can be obtained for many such sequences which are of special interest in the theory of numbers. For example, we shall see later† that if $\mathscr{A}$ is the sequence of primes, there exists a complementary sequence $\mathscr{B}$ with $B(n) \ll (\log n)^2$. For comparison, note that (3.9) yields only $B(n) \ll (\log n)^3$ when $\mathscr{A}$ is the sequence of primes.

Theorem 6′ below is an analogue of Theorem 6 when additions are performed modulo n.

THEOREM 6′ (Lorentz[17]). *If $a_1, a_2,..., a_l$ is a set of incongruent residues modulo n, there exists a 'complementary' set of residues $b_1, b_2,..., b_k$ such that every residue modulo n is of the form a_i+b_j, and*

$$k \ll n\frac{\max(1, \log l)}{l}.$$

We shall see that Theorems 6 and 6′ are both consequences of the following lemma. We use $|\mathscr{T}|$ to denote the number of elements of a finite set $\mathscr{T}$.

LEMMA. *Let $\mathscr{A}$ be a sequence of natural numbers and let a_0 be its least element. Let n, m be natural numbers satisfying*

$$n+1 \geqslant m+a_0. \tag{3.10}$$

Let $\mathscr{I}$, $\mathscr{J}$ denote the sets of integers

$$\mathscr{I} = [n+1, 2n], \qquad \mathscr{J} = [m, 2n-1] \tag{3.11}$$

lying in the intervals $n+1 \leqslant i \leqslant 2n$, $m \leqslant j \leqslant 2n-1$ respectively.

Then there exists a subset $\mathscr{S}$ of $\mathscr{J}$ such that

(i) $$\mathscr{I} \subset \mathscr{A}+\mathscr{S},$$

(ii) $$|\mathscr{S}| \ll n\frac{\max(1, \log A(n-m+1))}{A(n-m+1)}.$$

Proof of the lemma. For any non-empty subset $\mathscr{R}$ of $\mathscr{I}$, we denote by $j^*(\mathscr{R})$ the least‡ integer j of $\mathscr{J}$ for which $|(j+\mathscr{A}) \cap \mathscr{R}|$ is maximal, and denote this maximal number by $\kappa(\mathscr{R})$.

The following simple construction provides a set $\mathscr{S}$ which will be shown to have the desired properties. Let $\mathscr{R}_0$ be the set $\mathscr{I}$. We take $s_1 = j^*(\mathscr{R}_0)$ and denote by $\mathscr{R}_1$ the set of those integers of $\mathscr{R}_0$ which do not lie in $s_1+\mathscr{A}$. We now take $s_2 = j^*(\mathscr{R}_1)$ and denote by $\mathscr{R}_2$ the

† See Theorem 3 of Chapter III. See also Erdös (16).

‡ We require an explicit rule for selecting an element j with the stated maximizing property; the fact that we choose the least such j is inconsequential.

set of those integers of $\mathscr{R}_1$ which do not lie in $s_2+\mathscr{A}$; and continue this process. Thus $\mathscr{R}_{k-1}$ consists of those elements of $\mathscr{I}$ which are not covered by $\bigcup_{\nu=1}^{k-1}(s_\nu+\mathscr{A})$, and the number of 'new' elements contributed by $s_k+\mathscr{A}$ is given by

(3.12) $$\kappa(\mathscr{R}_{k-1}) = |\mathscr{R}_{k-1}|-|\mathscr{R}_k|.$$

In view of (3.10) and (3.12), $\mathscr{R}_k$ is a proper subset of $\mathscr{R}_{k-1}$ for each k, so that the construction must terminate with an empty set $\mathscr{R}_t$.

The subset $\mathscr{S} = \{s_1, s_2,..., s_t\}$ of $\mathscr{J}$ thus has the property (i), and it remains to establish (ii). For this purpose, we require a lower bound for the number (3.12) of 'new elements' contributed by $s_k+\mathscr{A}$.

We show that for any subset $\mathscr{R}$ of $\mathscr{I}$,

(3.13) $$\kappa(\mathscr{R}) \geqslant \frac{A(n-m+1)}{2n}|\mathscr{R}|.$$

By the definition of $\kappa(\mathscr{R})$,

$$|(j+\mathscr{A})\cap\mathscr{R}| \leqslant \kappa(\mathscr{R})$$

for each element j of $\mathscr{J}$, so that

(3.14) $$E = \sum_{j\in\mathscr{J}}|(j+\mathscr{A})\cap\mathscr{R}| \leqslant |\mathscr{J}|\kappa(\mathscr{R}) \leqslant 2n\kappa(\mathscr{R}).$$

On the other hand, each element r of $\mathscr{R}$ lies in exactly $A(r-m)$ of the translations $(j+\mathscr{A})$ of $\mathscr{A}$. Accordingly,

(3.15) $$E = \sum_{r\in\mathscr{R}}A(r-m) \geqslant \sum_{r\in\mathscr{R}}A(n-m+1) = |\mathscr{R}|A(n-m+1).$$

Comparing (3.14) and (3.15), we obtain (3.13).

Applying (3.13) (with $\mathscr{R} = \mathscr{R}_{k-1}$) to (3.12), we obtain

$$|\mathscr{R}_k|/|\mathscr{R}_{k-1}| \leqslant 1-\frac{1}{\tau},$$

where

(3.16) $$\tau = 2n/A(n-m+1).$$

Since $|\mathscr{R}_0| = n$, this implies

(3.17) $$|\mathscr{R}_k| \leqslant n\left(1-\frac{1}{\tau}\right)^k \leqslant ne^{-k/\tau}$$

for $1 \leqslant k < t$.

We recall that, for each k, $\mathscr{R}_k$ is a proper subset of $\mathscr{R}_{k-1}$; so that, for any h satisfying $1 \leqslant h < t$,

(3.18) $$t \leqslant |\mathscr{R}_h|+h.$$

We now choose the least integer h satisfying

(3.19) $$h/\tau > \log A(n-m+1)$$

so that h is defined by

$$h = \left[2n\frac{\log A(n-m+1)}{A(n-m+1)}\right]+1. \tag{3.20}$$

For the purpose of proving (ii) we may assume that $A(n-m+1) \geqslant 2$ and (in view of (3.20)) $h < t$. Hence, on applying (3.17) (with $k = h$) to the term $|\mathscr{R}_h|$ in (3.18), and using (3.19), we obtain the desired bound

$$|\mathscr{S}| = t \ll n\frac{\log A(n-m+1)}{A(n-m+1)}.$$

This proves (ii), and thus completes the proof of the lemma.

Proof of Theorem 6. Let l_0 be the least l for which $A(2^{l-2}) \geqslant 2$. For each $l = l_0, l_0+1,\ldots$, we apply the lemma with $n = 2^l$ and $m = 2^{l-1}+1$, thus obtaining a subset $\mathscr{S} = \mathscr{B}_l$ of $[2^{l-1}+1, 2^{l+1}-1]$ such that

(i) $$[2^l+1, 2^{l+1}] \subset \mathscr{A}+\mathscr{B}_l,$$

(ii) $$|\mathscr{B}_l| \ll \frac{2^l \log A(2^{l-1})}{A(2^{l-1})}.$$

We show that the sequence

$$\mathscr{B} = \bigcup_{l=l_0}^{\infty} \mathscr{B}_l,$$

which is obviously 'complementary' to $\mathscr{A}$, has the desired property (3.9).

For a given natural number N, we choose L to satisfy $2^{L-1} \leqslant N < 2^L$. We note that if $l \geqslant L+1$, then all the integers of $\mathscr{B}_l$ exceed N. Hence, if $L \geqslant l_0$ (and $B(N) = 0$ if this is not the case),

$$B(N) \leqslant \sum_{l=l_0}^{L} |\mathscr{B}_l| \ll \sum_{l=l_0}^{L} 2^l \frac{\log A(2^{l-1})}{A(2^{l-1})}$$

$$\ll \sum_{l=l_0}^{L} \sum_{2^{l-2}<\nu\leqslant 2^{l-1}} \frac{\log A(\nu)}{A(\nu)} \ll \sum_{\nu=2^{l_0-2}}^{N} \frac{\log A(\nu)}{A(\nu)}.$$

This establishes Theorem 6.

Proof of Theorem 6$'$. We may suppose that the a_ν are represented by natural numbers not exceeding n. Taking $\mathscr{A}$ to be this set of representatives and applying the lemma with $m = 1$, we obtain a set $\mathscr{S}$ with

$$|\mathscr{S}| \ll n\frac{\max(1, \log l)}{l}$$

such that $\mathscr{A}+\mathscr{S}$ contains the complete set of residues $[n+1, 2n]$. The distinct residues modulo n represented by the elements of $\mathscr{S}$ provide the desired set $b_1,\ldots, b_k$.

4. The theorems of Mann, Dyson, and van der Corput

Mann's original proof (1942; see ref. (2)) of Theorem 3 was complicated. In 1943 Artin and Scherk[18] found a modified proof which, while very elegant, was still far from easy. In 1945 Dyson[19] gave a (comparatively) very simple proof, at the same time obtaining a non-trivial generalization of Mann's theorem. Dyson's method, while based on that of Mann, introduced a new idea which subsequently led to further interesting developments culminating in the profound theorem of Kneser (see § 7).

As is immediately seen by induction, it is a trivial consequence of Theorem 3, that if $\mathscr{A}_0, \mathscr{A}_1,..., \mathscr{A}_k$ are $k+1$ subsequences of $\mathscr{Z}_0$, each containing 0, then

$$\sigma\Big(\sum_{r=0}^{k}\mathscr{A}_r\Big) \geqslant \min\Big(1, \sum_{r=0}^{k}\sigma\mathscr{A}_r\Big). \tag{4.1}$$

Dyson's theorem, stated below, is an improvement of this result.

THEOREM 7 (Dyson). *Let $\mathscr{A}_0, \mathscr{A}_1,..., \mathscr{A}_k$ be $k+1$ subsequences of $\mathscr{Z}_0$, each containing* 0, *and let $\mathscr{C}$ denote their sum. If, for a positive integer n, the number η is chosen so that*

$$\sum_{r=0}^{k} A_r(m) \geqslant \eta m \quad \textit{for } m = 1,2,..., n, \tag{4.2}$$

then

$$C(m) \geqslant \min(1,\eta)m \quad \textit{for } m = 1,2,..., n. \tag{4.3}$$

To see that this result implies (4.1) we need only note that, since $A_r(m) \geqslant m\sigma\mathscr{A}_r$, (4.2) is satisfied, for any n, with $\eta = \sum_{r=0}^{k}\sigma\mathscr{A}_r$.

It should be made clear that Mann's paper[2] included the case $k = 1$ of Theorem 7. Indeed, Mann's theorem took the sharper form (with $\mathscr{C} = \mathscr{A}+\mathscr{B}$)

$$\frac{C(n)}{n} \geqslant \min_{\substack{1\leqslant m\leqslant n \\ m\notin\mathscr{C}}} \frac{A(m)+B(m)}{m} \quad (n = 1,2,3,...;\ n \notin \mathscr{C}),$$

and the method of Artin and Scherk led to a slightly more precise version of the same system of inequalities. In fact, Mann's method has since been adapted to yield further refinements of this kind. We do

(18) E. Artin and P. Scherk, 'On the sum of two sets of integers', *Ann. Math.* (2) **44** (1943) 138–42.

(19) F. Dyson, 'A theorem on the densities of sets of integers', *J. Lond. math. Soc.* **20** (1945) 8–14.

not, however, treat this aspect of the subject here, and, instead, refer the reader to the literature.[20]

The following variant on Dyson's theorem, first proved by van der Corput[21] in 1947, will be useful in connexion with Kneser's theorem later in the chapter.

THEOREM 8 (van der Corput). *Let* $\mathscr{A}_0, \mathscr{A}_1, \ldots, \mathscr{A}_k$ *be* $k+1$ *subsequences of* $\mathscr{Z}_0$, *each containing* 0, *and let* $\mathscr{C}$ *denote their sum. If, for a positive integer* n, *the number* η *is chosen so that*

$$1+\sum_{r=0}^{k} A_r(m) \geqslant \eta(m+1) \quad \textit{for } m = 0,1,2,\ldots,n, \tag{4.4}$$

then

$$1+C(m) \geqslant \eta(m+1) \quad \textit{for } m = 0,1,2,\ldots,n. \tag{4.5}$$

Comparing Theorems 7 and 8, we see that Theorem 8 has a slightly weaker hypothesis and a correspondingly weaker conclusion (since $\eta \leqslant 1$).

We shall give an independent proof of Theorem 3, despite the fact that Theorem 7 (which embodies Theorem 3 as the special case $k = 1$) will be proved later on. We do this to provide a simple proof of Mann's theorem for those readers who do not wish to study its generalizations; in one respect it is possible to make a simplification which does not carry over to the more general case. Theorems 7 and 8 will be proved simultaneously after being restated in the form of a single theorem (Theorem 9). Before proceeding with these proofs, however, it is convenient to introduce some further notation.

We shall be concerned with sets of $k+1$ sequences $\mathscr{A}_0, \mathscr{A}_1, \ldots, \mathscr{A}_k$; and the sum of the numbers of elements of these sequences lying in some given interval will often be relevant. We therefore find it convenient to introduce the concept of a '*strong union*' of $\mathscr{A}_0, \mathscr{A}_1, \ldots, \mathscr{A}_k$ in which each element (apart from the exceptional element 0 discussed below) is counted as often as it occurs among $\mathscr{A}_0, \mathscr{A}_1, \ldots, \mathscr{A}_k$. To enable us to deal with Theorems 7 and 8 simultaneously, we treat the zero element as exceptional; even though it lies in each $\mathscr{A}_\nu$, the zero element is to be counted exactly once in the strong union. The distinction between 'union' (which retains its usual meaning) and 'strong union' is required for counting purposes only; we may regard the two as identical except

(20) See Chapter III (and Bibliography) of H. B. Mann, *Addition Theorems* (Interscience Publishers, New York, 1965).

(21) J. van der Corput, 'On sets of integers', I, II, III, *Proc. Sect. Sci. K. ned. Akad. Wet.* **50** (1947) 252–61, 340–50, 429–35.

in the one respect that the elements of the latter are 'weighted' for counting purposes. More formally:

Definition 4. The strong union†

$$\bigvee_{r=0}^{k} \mathscr{A}_r$$

of $k+1$ subsequences $\mathscr{A}_0, \mathscr{A}_1,..., \mathscr{A}_k$ of $\mathscr{Z}_0$, each containing 0, is defined to be the aggregate of the elements of $\mathscr{A}_0, \mathscr{A}_1,..., \mathscr{A}_k$; each element being counted according to its multiplicity, except that 0 is counted only once.

The 'weight' attached to the zero element is used only in the proof of Theorem 9 (which is a combination of Theorems 7 and 8). Accordingly, the special weighting of the element 0 is relevant only when this theorem is proved and, later in the chapter, applied.

In view of our motivation for introducing the concept of a 'strong union', the following definition suggests itself.

Definition 5. If $m \leqslant n$ are integers and $\underline{\mathscr{V}}$ is a strong union of sequences (or, in particular, a sequence), we denote by $\underline{V}(m,n)$ the number of elements of $\underline{\mathscr{V}}$ lying in the interval $[m,n]$. Further, we write $\underline{V}(1,n) = \underline{V}(n)$.

Thus, for example, the left-hand sides of (4.2) and (4.4) may be written as $\underline{V}(m)$ and $\underline{V}(0,m)$ respectively, where $\underline{\mathscr{V}} = \bigvee_{r=0}^{k} \mathscr{A}_r$.

Finally, the following definition will be very useful in connexion with Theorems 3′, 7, and 8.

Definition 6. By saying '*the interval* $[m,n]$ *is* $\underline{\mathscr{V}}$*-good with respect to* η', we mean that the inequality

$$\underline{V}(m,n) \geqslant \eta(n-m+1) \tag{4.6}$$

is satisfied.

This terminology will be used only with $\eta > 0$, $n \geqslant m \geqslant 0$, and (4.6) then expresses the fact that the ratio of the number of elements of $\underline{\mathscr{V}}$ in $[m,n]$ to the number of elements of $\mathscr{Z}_0$ in $[m,n]$ is at least η.

We note that being '$\underline{\mathscr{V}}$-good with respect to η' is an additive property of intervals‡ in the sense that if $[l,m]$ and $[m+1,n]$ have this property (for fixed $\underline{\mathscr{V}}$ and η), then so does $[l,n]$; for if $l \leqslant m < n$,

$$\underline{V}(l,n) = \underline{V}(l,m)+\underline{V}(m+1,n). \tag{4.7}$$

We shall now prove Mann's theorem (Theorem 3) in the form of the sharper Theorem 3′ below, which corresponds to the special case $k = 1$

† We shall use $\mathscr{A} \vee \mathscr{B}$ for the strong union of the two sequences $\mathscr{A}, \mathscr{B}$.

‡ In this context the interval $[m, n]$ is taken to mean the set of integers $m, m+1,..., n$.

of Theorem 7. The reader will notice that, in the statement of Theorem 3′, the restriction $\mathscr{A}, \mathscr{B} \subset [0, n]$ is entirely superfluous; this convention is introduced for the sole purpose of simplifying the notation in the proof. The theorem may therefore be applied to infinite sequences $\mathscr{A}, \mathscr{B}$. (For the meaning of $\mathscr{A} \vee \mathscr{B}$ see footnote to Definition 4.)

THEOREM 3′. *Let* $0 < \eta \leqslant 1$ *and let* n *be a positive integer. If* $\mathscr{A}$ *and* $\mathscr{B}$ *are two sequences, each containing* 0 *and lying entirely within the interval* $[0, n]$, *then the statement*

(4.8) $[1, m]$ *is* $\mathscr{A} \vee \mathscr{B}$*-good with respect to* η *for* $m = 1, 2, \ldots, n$

implies

(4.9) $[1, m]$ *is* $\mathscr{A} + \mathscr{B}$*-good with respect to* η *for* $m = 1, 2, \ldots, n$.

Proof. η will remain fixed throughout the proof, and we shall therefore use '$\underline{\mathscr{V}}$-good' as an abbreviation for '$\underline{\mathscr{V}}$-good with respect to η'. The proof will be of the following inductive nature.

We suppose on the contrary that the result is false, and that n is the least positive integer for which there exist a pair of sequences contained in $[0, n]$ for which (4.8) is true and (4.9) false. Among all such pairs of sequences we choose $\mathscr{A}, \mathscr{B}$ with the additional property that $B(0, n)$, the number of elements of $\mathscr{B}$, is minimal. We may clearly suppose that $\mathscr{B}$ contains at least one positive element; for if 0 is the sole element of $\mathscr{B}$, the statements (4.8) and (4.9) are identical.

We then obtain a contradiction in the following manner. We shall construct (by removing certain *positive* elements from $\mathscr{B}$ and placing certain additional elements in $\mathscr{A}$) a pair of new sequences $\mathscr{A}', \mathscr{B}'$ with the following properties:

(i) $[1, m]$ is $\mathscr{A}' \vee \mathscr{B}'$-good for $m = 1, 2, \ldots, n$.

(ii) $\mathscr{A}' + \mathscr{B}' \subset \mathscr{A} + \mathscr{B}$.

(iii) $B'(0, n) < B(0, n)$.

But properties (i) and (ii) imply that the pair $\mathscr{A}', \mathscr{B}'$ satisfy (4.8) but not (4.9), and hence (iii) contradicts the minimal property of $\mathscr{B}$.

Let a^* be the least among those elements a of $\mathscr{A}$ for which†

(4.10) $$a + \mathscr{B} \text{ is not entirely contained in } \mathscr{A}.$$

The element a^* exists, for since $\mathscr{B}$ contains a positive member, the largest element of $\mathscr{A} + \mathscr{B}$ is certainly not in $\mathscr{A}$. We note that, if $a^* > 0$ and r is any integer satisfying

(4.11) $$0 \leqslant r < a^*,$$

† Statement (4.10) is true, in particular, for any a for which $a + \mathscr{B}$ is not entirely contained in $[0, n]$.

(I) $b+\mathscr{A} \cap [0,r] \subset \mathscr{A}$ for every $b \in \mathscr{B}$;

and,† in particular,

(II) $(\mathscr{A}+\mathscr{B}) \cap [0,r] \subset \mathscr{A}$.

We define $\mathscr{B}''$ to consist of all those elements b'' of $\mathscr{B}$ for which

$$a^*+b'' \notin \mathscr{A}. \tag{4.12}$$

$\mathscr{B}''$ is non-empty, since a^* satisfies (4.10), and consists entirely of positive elements. We are now ready to construct the sequences $\mathscr{A}'$, $\mathscr{B}'$.

We form $\mathscr{B}'$ by removing from $\mathscr{B}$ all the elements of $\mathscr{B}''$; and $\mathscr{A}'$ by supplementing $\mathscr{A}$ by all those elements of $a^*+\mathscr{B}''$ which do not exceed n. It is essential to note that $\mathscr{B}'$ contains 0. We note also that

$$a^*+\mathscr{B}' \subset \mathscr{A}. \tag{4.13}$$

The sequence $\mathscr{B}'$ certainly possesses property (iii), since $\mathscr{B}''$ is non-empty. To see that $\mathscr{A}'$, $\mathscr{B}'$ satisfy (ii) we need only note that any element of $(a^*+\mathscr{B}'')+\mathscr{B}'$ may be written, by (4.13), in the form

$$(a^*+b'')+b' = (a^*+b')+b'' = a+b''.$$

It remains to verify (i).

If $a^* = 0$, $\mathscr{A}' \vee \mathscr{B}'$ and $\mathscr{A} \vee \mathscr{B}$ are identical (since in this case $\mathscr{A}'$, $\mathscr{B}'$ differ from $\mathscr{A}$, $\mathscr{B}$ only in that some of the elements of $\mathscr{B}$ have been transferred to $\mathscr{A}$), so that (4.8) and (i) are identical statements. We may thus assume that $a^* > 0$. Here $\mathscr{A}' \vee \mathscr{B}'$ and $\mathscr{A} \vee \mathscr{B}$ are no longer identical; the elements of $\mathscr{B}''$ have been translated by a distance a^* (and are taken to have disappeared if they fall outside $[0,n]$). In this translation of $\mathscr{B}''$, a^*+b'' falls outside the interval $[1,m]$ only if b'' lies in $[m-a^*+1,m]$. Thus, to confirm (i) it will suffice to prove that, for each m $(1 \leqslant m \leqslant n)$, the interval $[1,m]$ remains $\mathscr{A} \vee \mathscr{B}$-good if all the positive‡ elements of $\mathscr{B}$ in the interval $[m-a^*+1,m]$ are removed from $\mathscr{B}$.

We are no longer concerned with any translations—we have merely to show that (4.8) remains true, for each m, when certain elements of $\mathscr{B}$ are removed from $\mathscr{B}$. Let b_1 be the least positive element of $\mathscr{B}$ in $[m-a^*+1,m]$; if there is no such element there is nothing to prove. We write $m = b_1+r$, so that r satisfies (4.11). The interval $[1,b_1-1]$ is $\mathscr{A} \vee \mathscr{B}$-good (if $b_1 > 1$), so that it will suffice to prove that the interval $[b_1,m] = [b_1,b_1+r]$ is $\mathscr{A}$-good (i.e. remains $\mathscr{A} \vee \mathscr{B}$-good even when all the positive elements are removed from $\mathscr{B}$).

† According to our system of notation, $b+\mathscr{A} \cap [0,r]$ is the translate, by a distance b, of the part of $\mathscr{A}$ lying in the interval $[0,r]$.

‡ It is important to remove only positive elements since $m-a^*+1$ may be negative or 0.

By (I) the interval $[b_1, b_1+r]$ contains at least as many elements of $\mathscr{A}$ as the interval $[0,r]$. Hence, if $[0,r]$ is $\mathscr{A}$-good then so is $[b_1, b_1+r]$. Since $r < a^* \leqslant n$, it follows from the minimal choice of n (see beginning of proof) that $[1,r]$ is $\mathscr{A}+\mathscr{B}$-good. Since $0 \in \mathscr{A}+\mathscr{B}$ and $\eta \leqslant 1$, $[0,r]$ is $\mathscr{A}+\mathscr{B}$-good also.† Hence, by (II), $[0,r]$ is $\mathscr{A}$-good, and (i) has been established. This completes the proof of the theorem.

We shall now prove Theorem 9 below, which combines Theorems 7 and 8. Theorems 7 and 8 are simply the cases $\theta = 1$ and $\theta = 0$, respectively, of Theorem 9 (apart from changes of notation). The reader may be well advised to take $\theta = 1$ on a first reading of the proof; he will find later that the case $\theta = 0$ is almost the same, varying only in the slightly different role played by the zero element of the sequences. In this connexion it is worth mentioning that the exceptional treatment of the zero element in the definition of a strong union is designed to cater for the case $\theta = 0$ (Theorem 8). As far as the case $\theta = 1$ (Theorem 7) is concerned, the multiplicity of the zero element is irrelevant.

As in Theorem 3′, the superfluous condition $\mathscr{A}_r \subset [0,n]$ is introduced for notational convenience.

THEOREM 9. *Let $\theta = 0$ or 1. Let $k \geqslant 1$ be an integer, $0 < \eta \leqslant 1$, and let n be a positive integer. Let $\mathscr{A}_0$, $\mathscr{A}_1, \ldots, \mathscr{A}_k$ be $k+1$ sequences, each containing 0 and lying entirely within the interval $[0,n]$. Denote by $\underline{\mathscr{V}}$ and $\mathscr{C}$ the strong union and sum, respectively, of $\mathscr{A}_0$, $\mathscr{A}_1, \ldots, \mathscr{A}_k$. Then the statement*

(4.14) *$[\theta, m]$ is $\underline{\mathscr{V}}$-good with respect to η, for $m = \theta, \theta+1, \ldots, n$,*

implies

(4.15) *$[\theta, m]$ is $\mathscr{C}$-good with respect to η, for $m = \theta, \theta+1, \ldots, n$.*

Proof. We first choose the least k for which there exist η and n such that the theorem is false; this enables us to assume, if (4.14) is true and (4.15) false, that $\mathscr{A}_k$ contains a positive element. We next choose‡ η and n such that the theorem is false. Finally, we choose $\mathscr{A}_0$, $\mathscr{A}_1, \ldots, \mathscr{A}_k$ so that (4.14) is true, (4.15) is false, and the number $A_k(0,n)$ of elements of $\mathscr{A}_k$ is minimal.

Here again, η will remain fixed throughout, so that the words 'with respect to η' can always be omitted.

We obtain a contradiction in the following way. We shall construct (by removing certain *positive* elements from $\mathscr{A}_k$ and placing additional

† This is the sole use made, in the entire proof, of the fact $\eta \leqslant 1$.

‡ Choosing a minimal n led to a simplification in the proof of Theorem 3′, but would be pointless here.

elements in one of the other $\mathscr{A}_s$) a set of new sequences $\mathscr{A}'_0, \mathscr{A}'_1, \ldots, \mathscr{A}'_k$ with the following properties:

(i) $[\theta, m]$ is $\mathscr{V}'$-good for $m = \theta, \theta+1, \ldots, n$.

(ii) $\mathscr{C}' \subset \mathscr{C}$.

(iii) $A'_k(0, n) < A_k(0, n)$.

Properties (i) and (ii) imply that $\mathscr{A}'_0, \mathscr{A}'_1, \ldots, \mathscr{A}'_k$ satisfy (4.14) but not (4.15), and hence (iii) contradicts the minimal property of $\mathscr{A}_k$.

To exhibit clearly the special role played by $\mathscr{A}_k$ in the proof, we write $\mathscr{A}_k = \mathscr{B}$. We denote by $\mathscr{V}_1 = \mathscr{V}_1(\mathscr{A}_0, \mathscr{A}_1, \ldots, \mathscr{A}_{k-1})$ the strong union of $\mathscr{A}_0, \mathscr{A}_1, \ldots, \mathscr{A}_{k-1}$, so that

$$\mathscr{V} = \mathscr{V}_1 \vee \mathscr{B}. \tag{4.16}$$

For each $s = 0, 1, \ldots, k-1$, let a_s^* be the least among those elements a_s of $\mathscr{A}_s$ for which

$$a_s + \mathscr{B} \text{ is not entirely contained in } \mathscr{A}_s; \tag{4.17}$$

a_s^* exists, for, since $\mathscr{B}$ contains a positive element, the greatest element of $\mathscr{A}_s + \mathscr{B}$ is certainly not in $\mathscr{A}_s$. We write

$$a^* = \min(a_0^*, a_1^*, \ldots, a_{k-1}^*) \tag{4.18}$$

and we observe that if $a^* > 0$ and r is any integer satisfying

$$0 \leqslant r < a^*, \tag{4.19}$$

we have

(I) $b + \mathscr{A}_s \cap [0, r] \subset \mathscr{A}_s$ for $s = 0, 1, \ldots, k-1$ and every $b \in \mathscr{B}$;

and, in particular,

(II) if $[0, r]$ is $\mathscr{V}_1$-good, then so is $[b, b+r]$ for every $b \in \mathscr{B}$.

We shall require the following lemma.

Lemma 1. *If $a^* > 0$ then for every r satisfying* (4.19), *the interval* $[0, r]$ *is $\mathscr{V}_1$-good.*

Proof. Suppose there exist r satisfying (4.19) for which $[0, r]$ is not $\mathscr{V}_1$-good, and let r^* be the least such r. Since $0 \in \mathscr{V}_1$ and $\eta \leqslant 1$, we see that $r^* \geqslant 1$ and that $[1, r^*]$ is not $\mathscr{V}_1$-good either. Hence $[\theta, r^*]$ is not $\mathscr{V}_1$-good. However, by (4.14), $[\theta, r^*]$ is $\mathscr{V} = \mathscr{V}_1 \vee \mathscr{B}$-good, and therefore $[\theta, r^*]$ must contain a positive element b_0 of $\mathscr{B}$. Now, by the definition of r^*, the interval $[0, b_0-1]$ is $\mathscr{V}_1$-good whilst the interval $[0, r^*]$ is not. Thus the interval $[b_0, r^*]$ is not $\mathscr{V}_1$-good and, by (II) (with $r = r^*-b_0$), neither is the interval $[0, r^*-b_0]$. Since b_0 is positive, this contradicts the definition of r^*, and the lemma is proved.

We recall that a^*, defined by (4.18), lies in at least one of $\mathscr{A}_0, \mathscr{A}_1,..., \mathscr{A}_{k-1}$. Let t be the least s for which $a^* \in \mathscr{A}_s$, so that t satisfies

(4.20) $$0 \leqslant t \leqslant k-1, \qquad a^* \in \mathscr{A}_t.$$

We define $\mathscr{B}''$ to consist of all those elements b'' of $\mathscr{B}$ for which

(4.21) $$a^*+b'' \notin \mathscr{A}_t.$$

$\mathscr{B}''$ is non-empty, since a^* satisfies (4.17) with $s = t$; and consists entirely of positive elements, since $a^* \in \mathscr{A}_t$.

We are now ready to derive the sequences $\mathscr{A}'_0, \mathscr{A}'_1,..., \mathscr{A}'_k$ from $\mathscr{A}_0$, $\mathscr{A}_1,..., \mathscr{A}_k$. As the letter $\mathscr{B}$ is being used to denote $\mathscr{A}_k$, let $\mathscr{B}'$ denote $\mathscr{A}'_k$. We form $\mathscr{B}'$ by removing from $\mathscr{B}$ all the elements of $\mathscr{B}''$; and $\mathscr{A}'_t$ by supplementing $\mathscr{A}_t$ by all those elements of $a^*+\mathscr{B}''$ which do not exceed n. We leave the remaining sequences $\mathscr{A}_s$ (if $k > 1$) unchanged and write $\mathscr{A}'_s = \mathscr{A}_s$. [It will be helpful later in the chapter to point out here that taking $\mathscr{A}_0, \mathscr{A}_1,..., \mathscr{A}_k$ to be, for the moment, infinite sequences and disregarding the bounding effect of the interval $[0, n]$, we may represent the derivation of $\mathscr{A}'_0, \mathscr{A}'_1,..., \mathscr{A}'_k$ symbolically by

(4.22) $$\mathscr{A}'_t = \mathscr{A}_t \cup (a^*+\mathscr{B}), \qquad \mathscr{B}' = \mathscr{B} \cap (\mathscr{A}_t - a^*),$$
$$\mathscr{A}'_s = \mathscr{A}_s \quad (s = 0, 1,..., t-1, t+1,..., k-1).]$$

It is important to note that $0 \in \mathscr{B}'$ (as is the case since, in forming $\mathscr{B}'$, only positive elements were removed from $\mathscr{B}$). We note also that

(4.23) $$a^*+\mathscr{B}' \subset \mathscr{A}_t.$$

The sequence $\mathscr{B}' = \mathscr{A}'_k$ certainly possesses property (iii), for the complement $\mathscr{B}''$ of $\mathscr{B}'$ with respect to $\mathscr{B}$ is non-empty. To verify (ii) it suffices to prove that $\mathscr{A}'_t+\mathscr{B}' \subset \mathscr{A}_t+\mathscr{B}$, and this is evident on noting that any element of $(a^*+\mathscr{B}'')+\mathscr{B}'$ may be written, by (4.23), in the form
$$(a^*+b'')+b' = (a^*+b')+b'' = a_t+b''.$$

It remains to confirm (i). If $a^* = 0$, $\underline{\mathscr{V}}'$ and $\underline{\mathscr{V}}$ are identical (since $\mathscr{A}'_0, \mathscr{A}'_1,..., \mathscr{A}'_k$ differ from $\mathscr{A}_0, \mathscr{A}_1,..., \mathscr{A}_k$ only in that some of the elements of $\mathscr{A}_k$ have been transferred to $\mathscr{A}_t$). In this case, therefore, the statements (4.14) and (i) are the same.

We may thus assume that $a^* > 0$. In this case $\underline{\mathscr{V}}'$ and $\underline{\mathscr{V}}$ differ in that the elements of $\mathscr{B}''$ have been translated by a distance a^* and those translates of elements of $\mathscr{B}''$ which fall outside $[0, n]$ are taken to have disappeared. Considering statement (i) for a particular m, the translation of $\mathscr{B}''$ causes only those b'' in the interval $[m-a^*+1, m]$ to disappear from $[\theta, m]$. Thus, to confirm (i) it will suffice to prove that, for each m, the interval $[\theta, m]$ remains $\underline{\mathscr{V}}$-good even if all the positive elements of $\mathscr{B}$ in the interval $[m-a^*+1, m]$ are removed from $\mathscr{B}$; we note that, when proving this, we shall no longer be concerned with the translation of $\mathscr{B}''$.

Let b_1 be the least positive b in the interval $[m-a^*+1, m]$ (if there is no such b, there is nothing to prove). We write $m = b_1+r$, so that r satisfies (4.19). The interval $[\theta, b_1-1]$ is $\underline{\mathscr{V}}$-good (when $\theta = 1$ this statement is void unless $b_1 > 1$) so that it will suffice to prove that $[b_1, m] = [b_1, b_1+r]$ is $\underline{\mathscr{V}}_1$-good (this, by (4.16), is equivalent to saying that the interval remains $\underline{\mathscr{V}}$-good even when all the positive elements are removed from $\mathscr{B}$). But $[0, r]$ is $\underline{\mathscr{V}}_1$-good, by Lemma 1, and hence $[b_1, b_1+r]$ is $\underline{\mathscr{V}}_1$-good by (II). This establishes (i), and so completes the proof of Theorem 9.

5. Bases and non-basic essential components

Theorem 5 tells us that if $\mathscr{A}$ is a sequence of positive density (less than 1) and $\mathscr{B}$ is any basis, then $\sigma(\mathscr{A}+\mathscr{B}) > \sigma\mathscr{A}$. It would be very interesting to extend this result in some (necessarily limited) sense to sequences $\mathscr{A}$ which have zero asymptotic density;† for example, by proving that there exists a sequence $\mathscr{A}$ of zero asymptotic density such that $\sigma(\mathscr{A}+\mathscr{B}) > 0$ for every basis $\mathscr{B}$. (It will be proved later that corresponding to every positive integer h there exists a basis of order h whose counting number up to n is $O(n^{1/h})$; it follows that if such a sequence $\mathscr{A}$ exists, it would have to satisfy $A(n) \gg n^{1-1/h}$ for every natural h.) A sequence which springs to mind here is $\mathscr{P}$ (consisting of 0, 1 and the primes), particularly as there is a certain amount of supporting evidence due to Romanoff.[22] Romanoff proved that if $\mathscr{Q}^{(k)}$ denotes the sequence

$$0, 1, 2^k, 3^k, 4^k, \ldots, r^k, \ldots,$$

then $\sigma(\mathscr{P}+\mathscr{Q}^{(k)}) > 0$ for every fixed $k = 2, 3, \ldots$, and he discovered the still more remarkable result that if $\mathscr{R}$ is the very thin (non-basic) sequence $0, 1, 2, 2^2, 2^3, \ldots$ then $\sigma(\mathscr{P}+\mathscr{R}) > 0$ also.‡ This led Stöhr[23] to the plausible conjecture that $\sigma(\mathscr{P}+\mathscr{B}) > 0$ for every basis $\mathscr{B}$; but the conjecture was recently disproved by Erdös[24] who constructed, in a very simple way, a basis $\mathscr{B}'$ of order 2 for which $\sigma(\mathscr{P}+\mathscr{B}') = 0$.

† This problem would lose its point when applied to Schnirelmann density. For the sequence 0, 2, 3, 4,... has Schnirelmann density 0, yet the result of adding it to any basis is $\mathscr{Z}_0$. Such trivial examples are excluded when $n^{-1}A(n)$ is small for large n.

‡ Cf. Theorem 3 of Chapter III.

(22) N. Romanoff, 'Über einige Sätze der additiven Zahlentheorie', *Math. Annalen* **109** (1934) 668–78.

(23) A. Stöhr, 'Gelöste und ungelöste Fragen über Basen der natürlichen Zahlenreihe' II, *J. reine angew. Math.* **194** (1955) 111–40; see § 11.

(24) P. Erdös, 'Einige Bemerkungen zur Arbeit von A. Stöhr' (ref. (23)), ibid. **197** (1957) 216–9.

A conjecture (also due to Stöhr), which still remains open at the present time, is that there exists a sequence $\mathscr{A}$ (other than $\mathscr{P}$) of zero asymptotic density such that $\sigma(\mathscr{A}+\mathscr{B}^{(2)}) > 0$ for every basis $\mathscr{B}^{(2)}$ of order 2.

A not dissimilar question which might be asked here is whether, given any two bases of order 2, their sum has positive density. The answer is that this is far from true—indeed, if k is any positive integer, we shall see that there exist bases $\mathscr{B}_0^{(2)}, \mathscr{B}_1^{(2)}, \ldots, \mathscr{B}_{k-1}^{(2)}$, of order 2, whose sum has zero asymptotic density. This result is a special case of the following more general theorem of Stöhr.[23]

THEOREM 10 (Stöhr). *Let h, k be positive integers with $h \geqslant 2$. Then there exist bases $\mathscr{B}_0, \mathscr{B}_1, \ldots, \mathscr{B}_{k-1}$, each of order h, with sum*

$$\mathscr{S} = \sum_{i=0}^{k-1} \mathscr{B}_i,$$

such that $(h-1)\mathscr{S}$ has zero asymptotic density. More precisely, if $\mathscr{T}$ denotes the sequence $(h-1)\mathscr{S}$, there exist positive numbers K and δ, each depending only on h and k, such that

$$T(n) < Kn^{1-\delta} \tag{5.1}$$

for all n.

We shall apply Theorem 10 in connexion with another question suggested by Theorem 5. Theorem 5 states in effect that every basis is an essential component, and one may ask whether the converse is true; is every essential component a basis? This question was settled in 1942 by Linnik[25] who found a non-basic essential component by means of an extremely complicated construction. In 1956 Stöhr and Wirsing[26] gave a much simpler example which is based on an ingenious combination of Theorems 5 and 10; we shall end this section with an account of their construction.

Proof. The proof of Theorem 10 will be carried out in a number of steps, of which Lemma 3 constitutes a result of independent interest.

Let $g \geqslant 2$ be a (fixed) natural number. Every non-negative integer z is uniquely representable in the form

$$z = {\sum_{\nu \in \mathscr{Z}_0}}' e_\nu g^\nu, \tag{5.2}$$

(25) Ju. V. Linnik, 'On Erdös's theorem on the addition of numerical sequences', *Mat. Sb.* N.S. **10** (**52**) (1942) 67–78.

(26) A. Stöhr and E. Wirsing, 'Beispiele von wesentlichen Komponenten, die keine Basen sind', *J. reine angew. Math.* **196** (1956) 96–8.

where the dash signifies that

$$(5.3) \qquad 0 \leqslant e_\nu \leqslant g-1 \quad \text{for every } \nu;$$

naturally $\sum'$ contains only finitely many terms.

On regarding the e_ν in (5.2) as independent variables, one is led to expect that if, for all ν belonging to some not too thin sequence $\mathscr{Y}$, the corresponding e_ν were confined to a proper subset of $[0, g-1]$, then the sequence of z representable by (5.2) would be thin. The following represents only a very special result of this kind, but is all we shall require for application later.

Lemma 2. *If $\mathscr{Y}$ is a congruence class modulo q, and $\mathscr{L}$ is the sequence consisting of all those z whose expansion* (5.2) *is such that*

$$(5.4) \qquad 0 \leqslant e_y \leqslant g-2 \quad \textit{for all } y \in \mathscr{Y},$$

then

$$(5.5) \qquad L(n) < K'n^{1-\delta'} \quad \textit{for every } n,$$

where K' and δ' are positive numbers depending only on g and q.

Proof. Since (5.4) reduces the number of possibilities for each e_y by the factor $(1-g^{-1})$, we have

$$(5.6) \qquad L(g^{uq}) \leqslant (1-g^{-1})^u g^{uq}$$

for any positive integer u. We define δ' by

$$1-g^{-1} = g^{-\delta' q},$$

so that δ' depends only on g, q, and $0 < \delta' \leqslant 1$.

Corresponding to any $n > 1$, there exists $r \geqslant 0$ such that

$$g^{rq} < n \leqslant g^{(r+1)q}.$$

Then, by (5.6),

$$L(n) \leqslant L(g^{(r+1)q}) \leqslant (g^{1-\delta'})^{(r+1)q} < g^q n^{1-\delta'},$$

and (5.5) is proved with $K' = g^q$.

For any sequence $\mathscr{I}$, we define the sequence $\mathscr{G}(\mathscr{I})$ to consist of all those non-negative integers x representable in the form

$$(5.7) \qquad x = \sum_{\nu \in \mathscr{I}}{}' e_\nu g^\nu.$$

The operator $\mathscr{G}$ obviously has the following two properties:

(I) $\mathscr{G}(\mathscr{Z}_0) = \mathscr{Z}_0$.

(II) If $\mathscr{X} \cap \mathscr{Y}$ is empty, then $\mathscr{G}(\mathscr{X}) + \mathscr{G}(\mathscr{Y}) = \mathscr{G}(\mathscr{X} \cup \mathscr{Y})$.

The following lemma provides a powerful tool for constructing bases economically, and is used below to construct each of the sequences $\mathscr{B}_i$ in Theorem 10:

LEMMA 3. *If $\mathscr{Z}_0$ is subdivided into h mutually exclusive sequences $\mathscr{I}^{(0)}$, $\mathscr{I}^{(1)}, \dots, \mathscr{I}^{(h-1)}$, then the sequence*

$$\mathscr{B} = \bigcup_{l=0}^{h-1} \mathscr{G}(\mathscr{I}^{(l)}) \tag{5.8}$$

is a basis of order h.

Proof. The truth of this lemma is an immediate consequence of (I) and (II), as will be clear from (5.9) below:

$$h\mathscr{B} \supset \sum_{l=0}^{h-1} \mathscr{G}(\mathscr{I}^{(l)}) = \mathscr{G}\Big(\bigcup_{l=0}^{h-1} \mathscr{I}^{(l)}\Big) = \mathscr{G}(\mathscr{Z}_0) = \mathscr{Z}_0. \tag{5.9}$$

Suppose, for the moment, that a basis $\mathscr{B}$ has been defined by means of an application of Lemma 3, so that $\mathscr{B}$ is given by (5.8) where the $\mathscr{I}^{(l)}$ are a subdivision of $\mathscr{Z}_0$. We write for $m = 0, 1, \dots, h-1$,

$$\mathscr{J}^{(m)} = \bigcup_{\substack{l=0 \\ l\neq m}}^{h-1} \mathscr{G}(\mathscr{I}^{(l)}). \tag{5.10}$$

Then,

$$(h-1)\mathscr{B} = \bigcup_{m=0}^{h-1} (h-1)\mathscr{J}^{(m)}. \tag{5.11}$$

To see this, we note that on the one hand an element of $(h-1)\mathscr{B}$ is representable as the sum of only $h-1$ elements of $\mathscr{B}$ and therefore lies in $(h-1)\mathscr{J}^{(m)}$ for at least one m; and on the other, $\mathscr{J}^{(m)} \subset \mathscr{B}$ for every m.

Let i run through the values $0, 1, 2, \dots, k-1$. For each i, let

$$\mathscr{I}_i^{(0)}, \mathscr{I}_i^{(1)}, \dots, \mathscr{I}_i^{(h-1)} \tag{5.12}$$

be a subdivision of $\mathscr{Z}_0$ into h mutually exclusive sequences, and let $\mathscr{B}_i$ be the basis defined by

$$\mathscr{B}_i = \bigcup_{l=0}^{h-1} \mathscr{G}(\mathscr{I}_i^{(l)}). \tag{5.13}$$

We write

$$\mathscr{S} = \sum_{i=0}^{k-1} \mathscr{B}_i \tag{5.14}$$

and

$$\mathscr{T} = (h-1)\mathscr{S}; \tag{5.15}$$

we shall prove that the subdivisions (5.12) of $\mathscr{Z}_0$ can be chosen so as to ensure the truth of Theorem 10, i.e. the truth of (5.1). We shall be led to the appropriate choice of these subdivisions, and of g, by first investigating the circumstances under which the sequence $\mathscr{T}$ is thin.

By analogy with (5.10), we write

$$\mathscr{J}_i^{(m)} = \bigcup_{\substack{l=0 \\ l \neq m}}^{h-1} \mathscr{G}(\mathscr{J}_i^{(l)}) \quad \text{for } i = 0, 1, \ldots, k-1, \tag{5.16}$$

so that, by (5.11),

$$(h-1)\mathscr{B}_i = \bigcup_{m=0}^{h-1} (h-1)\mathscr{J}_i^{(m)} \quad \text{for } i = 0, 1, \ldots, k-1. \tag{5.17}$$

Now the sequence $\mathscr{T}$ consists, by (5.15), (5.14), and (5.17), of all numbers t of the form

$$t = f_0 + f_1 + \cdots + f_{k-1}$$

where $f_i \in (h-1)\mathscr{J}_i^{(m_i)}$ for some m_i. Hence

$$\mathscr{T} = \bigcup_{m_0=0}^{h-1} \bigcup_{m_1=0}^{h-1} \cdots \bigcup_{m_{k-1}=0}^{h-1} \sum_{i=0}^{k-1} (h-1)\mathscr{J}_i^{(m_i)}. \tag{5.18}$$

For a fixed set of values $m_0, m_1, \ldots, m_{k-1}$, we consider the sequence

$$\mathscr{C} = \sum_{i=0}^{k-1} (h-1)\mathscr{J}_i^{(m_i)}. \tag{5.19}$$

Since $\mathscr{T}$ is, by (5.18), the union of a finite number of sequences of type (5.19), we seek sufficient conditions for $\mathscr{C}$ to be thin. The following lemma will enable us to provide such conditions.

LEMMA 4. *Suppose there exists a sequence* $\mathscr{Y}$ *with the property*

$$\mathscr{Y} \subset \mathscr{J}_i^{(m_i)} \quad \textit{for } i = 0, 1, \ldots, k-1. \tag{5.20}$$

Then, in the expansion

$$c = \sum_{\nu \in \mathscr{Z}_0}' e_\nu g^\nu \tag{5.21}$$

of an element c *of* $\mathscr{C}$,

$$e_y \leqslant k(h-1)-1 \quad \textit{for every } y \in \mathscr{Y}. \tag{5.22}$$

As was observed earlier, (5.22) implies that $\mathscr{C}$ is thin, provided that

$$k(h-1) < g \tag{5.23}$$

and that the sequence $\mathscr{Y}$ is not itself too thin. In particular, we have by Lemma 2 that:

COROLLARY. *If the sequence* $\mathscr{Y}$ *of Lemma* 4 *is (the non-negative part of) a congruence class* (mod q), *and* (5.23) *is satisfied,*

$$C(n) < K'n^{1-\delta'} \quad \textit{for every } n. \tag{5.24}$$

Proof of Lemma 4. Let c and y be fixed elements of $\mathscr{C}$ and $\mathscr{Y}$ respectively. Define d by

$$d \equiv c \pmod{g^{y+1}}, \qquad 0 \leqslant d < g^{y+1}, \tag{5.25}$$

so that, if (5.21) is the expansion of c,

$$d = \sum_{\nu=0}^{y}{}' e_\nu g^\nu, \tag{5.26}$$

and, in particular,

$$e_y g^y \leqslant d. \tag{5.27}$$

By using (5.20) and the representation (5.19) of $\mathscr{C}$ we shall now obtain an effective upper bound for d.

By (5.19), we can choose a representation

$$c = f_0 + f_1 + \dots + f_{k-1} \tag{5.28}$$

of c with $f_i \in (h-1)\mathscr{I}_i^{(m_i)}$. Now, by (5.16), there exists, corresponding to each element x_i of $\mathscr{I}_i^{(m_i)}$, an $l \neq m_i$ such that $x_i \in \mathscr{G}(\mathscr{I}_i^{(l)})$; so that x_i has an expansion of type (5.7) with $\mathscr{I} = \mathscr{I}_i^{(l)}$ $(l \neq m_i)$. We may thus write

$$f_i = x_i^{(1)} + x_i^{(2)} + \dots + x_i^{(h-1)}, \tag{5.29}$$

where

$$x_i^{(r)} = \sum_{\nu \in \mathscr{I}_i^{(l)}}{}' e_\nu^* g^\nu, \quad \text{with } l \neq m_i; \tag{5.30}$$

here l and the coefficients e_ν^* will, of course, depend on i and r, but we require only the fact that there is no term $e_y^* g^y$, a circumstance assured by (5.20) and the restriction $l \neq m_i$ (since the h sequences (5.12) are mutually exclusive).

Since there is no term in (5.30) corresponding to $\nu = y$ we have by (5.28), (5.29), and (5.30) that

$$d \leqslant \sum_{i=0}^{k-1} \sum_{r=1}^{h-1} \sum_{\nu=0}^{y-1} (g-1)g^\nu = k(h-1)(g^y-1). \tag{5.31}$$

Comparison of (5.27) and (5.31) yields $e_y < k(h-1)$, so that (5.22) is proved.

We now choose g to satisfy (5.23); let

$$g = hk.$$

To prove (5.1) we need to find subdivisions (5.12) such that, corresponding to each of the h^k possible sets of values $m_0, m_1, \dots, m_{k-1}$ there exists a congruence class $\mathscr{Y}$ (which may depend on the m's) satisfying (5.20). By the corollary to Lemma 4, (5.24) will then be true for each $\mathscr{C}$ defined by (5.19), and since, by (5.18), $\mathscr{T}$ is the union of a finite number of such sequences $\mathscr{C}$, (5.1) will follow.

In seeking such subdivisions (5.12), it is natural (for a suitable fixed q) to restrict each of $\mathscr{I}_i^{(l)}$ to be a union of congruence classes modulo q. For with this restriction, any set of $\mathscr{I}$'s having an element in common automatically have a congruence class in common. Representing each congruence class by its least (non-negative) element, we see that we need only prove the following result.

LEMMA 5. *If q is suitably chosen, then for $i = 0, 1, \ldots, k-1$ there exist subdivisions of the set of numbers*

$$0, 1, 2, \ldots, q-1 \tag{5.32}$$

into h mutually exclusive classes

$$\mathscr{I}_i^{(0)}, \mathscr{I}_i^{(1)}, \ldots, \mathscr{I}_i^{(h-1)} \tag{5.33}$$

such that for any set of values $m_0, m_1, \ldots, m_{k-1}$, the classes

$$\mathscr{I}_0^{(m_0)}, \mathscr{I}_1^{(m_1)}, \ldots, \mathscr{I}_{k-1}^{(m_{k-1})} \tag{5.34}$$

have an element in common.

Proof. The following construction immediately suggests itself (and is, in fact, the most economical). Take $q = h^k$ and, representing each element of (5.32) in the form

$$\sum_{i=0}^{k-1} j_i h^i \quad (0 \leqslant j_i \leqslant h-1),$$

place in $\mathscr{I}_i^{(l)}$ those elements of (5.32) for which $j_i = l$. Then the classes (5.33) are mutually exclusive, and the classes (5.34) have the element

$$\sum_{i=0}^{k-1} m_i h^i$$

in common.

In view of Lemma 5 and the remarks preceding it, the proof of Theorem 10 is now complete.

We now come to the Stöhr–Wirsing proof of Linnik's theorem. (For references, see the discussion below Theorem 10.)

THEOREM 11 (Linnik). *There exists an essential component which is not a basis.*

Proof (Stöhr-Wirsing). We begin by proving that there exist bases

$$\mathscr{S}_2, \mathscr{S}_3, \mathscr{S}_4, \ldots \tag{5.35}$$

of orders 2, 3, 4,... respectively, with the following two properties:

(I) $(h-1)\mathscr{S}_h$ has zero asymptotic density for $h = 2, 3, 4, \ldots$.

(II) If $\mathscr{A}$ is any sequence whose Schnirelmann density $\sigma\mathscr{A} = \alpha$ satisfies $0 < \alpha < 1$, then $\sigma(\mathscr{A}+\mathscr{S}_h)-\sigma\mathscr{A} \geqslant \phi(\alpha)$, where $\phi(\alpha)$ is a *positive* quantity depending only on α (and *not* on h).

Roughly speaking, properties (I) and (II) tell us that, as h increases, the effectiveness of $\mathscr{S}_h$ as a basis diminishes whilst its effectiveness as an essential component does not. These two contrasting features of the sequence (5.35) will enable us, by what might be described as a diagonal argument, to construct a sequence $\mathscr{W}$ which, while no longer a basis, yet remains an essential component.

To establish the existence of the bases (5.35) we turn to Theorem 10. The sequence† $\mathscr{S} = \mathscr{S}_{h,k}$ considered there is such that $(h-1)\mathscr{S}_{h,k}$ has zero (asymptotic) density. Furthermore, $\mathscr{S}_{h,k}$, being the sum of bases of order h, is itself a basis of order h; and is therefore, by Theorem 5, an essential component. This in itself is of no value to us, for the amount by which the Schnirelmann density of a sequence is increased when $\mathscr{S}_{h,k}$ is added to it, may depend on h. We shall, however, prove the following result.

LEMMA 6. *Corresponding to each $h = 2, 3,\ldots$, there exists $k' = k'(h)$ such that, for every sequence $\mathscr{A}$,*

(5.36) $$\sigma(\mathscr{A}+\mathscr{S}_{h,k'}) \geqslant \sigma\mathscr{A}+\tfrac{1}{2}(1-\sigma\mathscr{A})\sigma\mathscr{A}.$$

Proof. We write

(5.37) $$\psi_h(\alpha) = \alpha+\frac{1}{2h}(1-\alpha)\alpha$$

and note that this function is increasing in the interval $[0,1]$. Now Theorem 5 tells us that if $\sigma\mathscr{A}' \geqslant \alpha' \geqslant 0$ and $\mathscr{B}$ is a basis of order h,

(5.38) $$\sigma(\mathscr{A}'+\mathscr{B}) \geqslant \psi_h(\alpha').$$

We define the function $\psi_h^{(r)}(\alpha)$ inductively by

(5.39) $$\psi_h^{(0)}(\alpha) = \alpha, \qquad \psi_h^{(r+1)}(\alpha) = \psi_h\{\psi_h^{(r)}(\alpha)\} \quad \text{for } r = 0, 1, 2,\ldots;$$

so that, for $r \geqslant 1$, $\psi_h^{(r)}(\alpha)$ is the rth iterate of $\psi_h(\alpha)$.

To obtain a lower bound for

$$\sigma(\mathscr{A}+\mathscr{S}_{h,k}) = \sigma(\mathscr{A}+\mathscr{B}_0+\mathscr{B}_1+\ldots+\mathscr{B}_{k-1})$$

in terms of $\sigma\mathscr{A} = \alpha$ we apply (5.38) k times in succession; at the rth step we take $\mathscr{A}' = \mathscr{A}+\mathscr{B}_0+\mathscr{B}_1+\ldots+\mathscr{B}_{r-2}$, $\mathscr{B} = \mathscr{B}_{r-1}$ and obtain

$$\sigma(\mathscr{A}'+\mathscr{B}_{r-1}) \geqslant \psi_h^{(r)}(\alpha).$$

The final result of this iterative process is

(5.40) $$\sigma(\mathscr{A}+\mathscr{S}_{h,k}) \geqslant \psi_h^{(k)}(\alpha),$$

and to prove the lemma it now suffices to show that there exists $k' = k'(h)$ such that

(5.41) $$\psi_h^{(k')}(\alpha) \geqslant \alpha+\tfrac{1}{2}(1-\alpha)\alpha \quad \text{for } 0 \leqslant \alpha \leqslant 1.$$

† Strictly speaking, $\mathscr{S}$ depends on $\mathscr{B}_0, \mathscr{B}_1,\ldots, \mathscr{B}_{k-1}$, but only its dependence on h and k will be relevant here.

To prove (5.41) we keep α and h fixed, and write

$$\psi_h^{(r)}(\alpha) = f(r).$$

By (5.37) and (5.39), $f(0) = \alpha$ and

$$f(r+1) = f(r)+\frac{1}{2h}f(r)\{1-f(r)\} \quad \text{for } r = 0,\ 1,\ 2,\dots. \tag{5.42}$$

We shall prove (5.41) with $k' = 2h$, by showing that

$$f(2h) \geqslant f(0)+\tfrac{1}{2}\{1-f(0)\}f(0) \tag{5.43}$$

provided that $0 \leqslant f(0) \leqslant 1$.

As was remarked earlier, $\alpha+(1/2h)(1-\alpha)\alpha$ is an increasing function in the interval $[0, 1]$. Thus $f(r)$ has 1 as an upper bound and is a (not necessarily strictly) increasing function of r. By (5.42), we have

$$f(2h) = f(0)+\sum_{r=0}^{2h-1}\{f(r+1)-f(r)\} = f(0)+\frac{1}{2h}\sum_{r=0}^{2h-1}f(r)\{1-f(r)\}.$$

Hence, if

$$f(r)\{1-f(r)\} > \tfrac{1}{2}f(0)\{1-f(0)\} \quad \text{for } 0 < r < 2h,$$

(5.43) is satisfied. We may suppose therefore that there exists an r^*, $0 < r^* < 2h$, such that

$$f(r^*)\{1-f(r^*)\} \leqslant \tfrac{1}{2}f(0)\{1-f(0)\}.$$

But then

$$\begin{aligned}\tfrac{1}{2}f(0)\{1-f(0)\} &\leqslant |f(0)\{1-f(0)\}-f(r^*)\{1-f(r^*)\}|\\ &= \{f(r^*)-f(0)\}|f(r^*)+f(0)-1|\\ &\leqslant f(r^*)-f(0),\end{aligned}$$

and (5.43) is again true because $f(r^*) \leqslant f(2h)$. This completes the proof of the lemma.

The right-hand side of (5.36) depends only on $\sigma\mathscr{A}$, and if, for $h = 2, 3,\dots$, we take $\mathscr{S}_h = \mathscr{S}_{h,k'}$ (with any definite choice of $\mathscr{B}_0, \mathscr{B}_1,\dots, \mathscr{B}_{k'-1}$), we have established the existence of a sequence (5.35) of bases $\mathscr{S}_h$ satisfying (I) and (II) (with $\phi(\alpha) = \frac{1}{2}\alpha(1-\alpha)$). We are ready now to embark on the diagonal argument mentioned in the summary of the proof of Theorem 11. We define the sequence $\mathscr{W}$ sectionally as follows:

Let $$0 = m_1 < m_2 < m_3 < \dots$$

be an infinite sequence of integers, increasing sufficiently rapidly in a manner to be specified later. This sequence defines a sub-division of $\mathscr{Z}_0$ into intervals

$$I_h = [m_{h-1}, m_h-1] \quad (h = 2, 3,\dots).$$

For each h we place into I_h those elements of $m_{h-1}+\mathscr{S}_h$ which fall into this interval (we may picture the basis $\mathscr{S}_h$ translated so that its zero element coincides with the first element of I_h, and then cut off at the end of the interval). We define $\mathscr{W}$ to be the entire sequence so formed; or, written symbolically, $\mathscr{W}$ is defined by the relations

$$\mathscr{W} \cap I_h = (m_{h-1}+\mathscr{S}_h) \cap I_h \quad (h = 2, 3,...).$$

We prove first that $\mathscr{W}$ is an essential component.

Let $\mathscr{A}$ be any sequence of *positive* integers with Schnirelmann density $\sigma\mathscr{A} = \alpha$ satisfying $0 < \alpha < 1$. The integers of $\mathscr{A}+\mathscr{W}$ in $[m_{h-1}+1, m_h]$ (that is to say, in the 'interval' $1+I_h$) will certainly include all the integers of the sequence

$$\mathscr{A}+(m_{h-1}+\mathscr{S}_h) = m_{h-1}+(\mathscr{A}+\mathscr{S}_h)$$

in this interval. But this latter set of integers is simply the translation (by a distance m_{h-1}) of the elements of $\mathscr{A}+\mathscr{S}_h$ in the interval $[1, m_h-m_{h-1}]$. It follows, from (II), that in $[m_{h-1}+1, m_h]$ and also in those of its sub-intervals having the same starting-point (i.e. sub-intervals of type $[m_{h-1}+1, n]$) at least a proportion $\alpha+\phi(\alpha)$ of the elements belong to $\mathscr{A}+\mathscr{W}$. The same is now true of any interval $[1, n]$, for we can always subdivide it into intervals

$$[m_1+1, m_2], \quad [m_2+1, m_3], \quad ..., \quad [m_{t-2}+1, m_{t-1}], \quad [m_{t-1}+1, n]$$

where $m_{t-1} < n \leqslant m_t$. This proves that $\mathscr{W}$ is an essential component.

It remains to prove that $\mathscr{W}$ is not a basis provided that the sequence $\{m_h\}$ increases sufficiently rapidly. In particular, it suffices to prove that if m_h is sufficiently large as a function of m_{h-1}, the sequence $(h-1)\mathscr{W}$ does not contain all the integers of the interval $[0, m_h-1]$; for it then follows at once, for each $h = 2, 3,...$, that $\mathscr{W}$ is not a basis of order $h-1$, and hence not a basis of finite order.

We denote the interval† $[0, m_h-1]$ by J_h and $[0, m_h]$ by $J_h^{(+1)}$. We consider an element w of $\mathscr{W} \cap J_h$. Such an element lies in either J_{h-1} or I_h, and therefore in at least one of J_{h-1} and $m_{h-1}+\mathscr{S}_h$. Since $0 \in \mathscr{S}_h$ and $m_{h-1} \in J_{h-1}^{(+1)}$, it follows that

$$w \in J_{h-1}^{(+1)}+\mathscr{S}_h.$$

It will therefore suffice to show that J_h is not entirely covered by the sequence

$$(h-1)(J_{h-1}^{(+1)}+\mathscr{S}_h) = (h-1)J_{h-1}^{(+1)}+(h-1)\mathscr{S}_h.$$

But if m_h is chosen to be sufficiently large as a function of m_{h-1}, this follows from (I); for the sequence $(h-1)J_{h-1}^{(+1)}$ is finite and depends only on h and m_{h-1}. Thus $\mathscr{W}$ is not a basis, and the proof of Theorem 11 is complete.

† As was remarked earlier, an interval is here understood to mean the set of all integers in the interval.

The sequence $\mathscr{W}$ constructed above provides the simplest known example of a non-basic essential component. Linnik's example is far more complicated but is also, in a way, far more striking. For Linnik constructed an essential component $\mathscr{V}$ which is exceedingly thin; so thin, indeed, that it satisfies $V(n) = O(n^{\epsilon})$ for every ϵ and thus could not (on trivial combinatorial grounds) be a basis. Linnik's highly ingenious method is based on the use of exponential sums. Many of the very considerable complications in his method arise from the need to establish the existence of a thin sequence $\mathscr{S}$ such that the associated exponential sums

$$F(N,\alpha) = \sum_{\substack{n=1\\ n\in\mathscr{S}}}^{N} e^{2\pi i\alpha n} \tag{5.44}$$

have certain properties.

Linnik actually constructed such a sequence $\mathscr{S}$ in terms of the power sequences $\mathscr{Q}^{(k)} = \{x^k\}$ with various k. Very profound and difficult results are known about the exponential sums associated with these sequences, and Linnik applied such results† to establish the required properties of the exponential sums (5.44).

Erdös and Roth have recently‡ found that, by the use of probability methods, it is possible to prove the existence of a sequence $\mathscr{S}$ of the required kind, without the use of any deep results (other than the central limit theorem), and that Linnik's method can be much simplified in this way.

'Thin' bases

We remarked at the end of section 2 that a basis $\mathscr{B}$ of order h cannot be too thin, in the sense that its counting number $B(n)$ must satisfy $\{B(n)+1\}^h \geqslant n+1$. Thus $B(n)$ must satisfy

$$\limsup_{n\to\infty} n^{-1/h}B(n) \geqslant 1. \tag{5.45}$$

This result can, however, be improved. At the present time, the most precise inequality of type (5.45) known, due to Stöhr, is

$$\limsup_{n\to\infty} n^{-1/h}B(n) \geqslant (h!)^{1/h}/\Gamma(1+h^{-1}); \tag{5.46}$$

the right-hand side of this relation is asymptotic to

$$0{\cdot}3678...\,h \quad \text{as } h\to\infty.$$

† The results are required to be uniform in k, and this adds to the complications.

‡ Unpublished.

(Stöhr's survey[6] discusses various types of lower estimates for $B(n)$, and we quote only (5.46) as this is of the type most suitable for comparison with (5.48) below.)

On the other hand, as was also remarked earlier, Lemma 3 provides a powerful tool for the construction of 'thin' bases. The following construction already leads to a basis $\mathscr{B}$ whose counting number $B(n)$ is, for each n, as small as possible to within a constant factor.

In Lemma 3, for each $l = 0, 1,..., h-1$, take $\mathscr{I}^{(l)}$ to be the congruence class† $l \pmod h$, and let $g = 2$. Writing $\mathscr{G}_l$ for $\mathscr{G}(\mathscr{I}^{(l)})$, it follows from Lemma 3 that

$$\mathscr{B} = \bigcup_{l=0}^{h-1} \mathscr{G}_l$$

is a basis of order h.

Corresponding to any $n > 1$, there exist integers t and r $(0 \leqslant r < h)$, such that

$$2^{ht+r-1} < n \leqslant 2^{ht+r}.$$

Then
$$B(n) \leqslant B(2^{ht+r}) = \sum_{l=0}^{h-1} G_l(2^{ht+r}),$$

and it requires only a straightforward calculation to verify that

$$G_l(2^{ht+r}) = \begin{cases} 2^{t+1}-1 & (l < r); \\ 2^t & (l = r); \\ 2^t-1 & (l > r). \end{cases}$$

Hence

$$B(2^{ht+r}) = r(2^{t+1}-1)+2^t+(h-1-r)(2^t-1) = (h+r)2^t+1-h,$$

so that
$$B(n) \leqslant (r+h)\left(\frac{n}{2^{r-1}}\right)^{1/h}.$$

We have thus arrived at the following result.

For every natural number h there exists a basis $\mathscr{B}$ of order h such that

$$\limsup_{n\to\infty} \frac{B(n)}{n^{1/h}} \leqslant \max_{0 \leqslant r \leqslant h-1} \{(r+h)2^{-(r-1)/h}\}. \tag{5.47}$$

Although the right-hand side of (5.47) has the correct order of magnitude, a more precise result was obtained by Stöhr[27] (see also Raikov[28]);

† Here $\mathscr{I}^{(l)}$ is really only the non-negative part of the congruence class, consisting of all non-negative integers $\equiv l \pmod h$.

(27) A. Stöhr, 'Eine Basis h-ter Ordnung für die Menge aller natürlichen Zahlen', *Math. Z.* **42** (1937) 739–43; 'Anzahlabschätzung einer bekannten Basis h-ter Ordnung', ibid. **47** (1942) 778–87; **48** (1942–3) 792.

(28) D. Raikov, 'Über die Basen der natürlichen Zahlenreihe', *Mat. Sb.* N.S. **2 (44)** (1937) 595–7.

he (and Raikov) used the sequence $\mathscr{B}$ constructed above but, in place of (5.47), proved by more careful analysis that

$$\limsup_{n\to\infty} \frac{B(n)}{n^{1/h}} = \max_{r=1,2,\dots,h} (r+h)\left(\frac{1-2^{-h}}{2^{r-1}}\right)^{1/h}. \tag{5.48}$$

The right-hand side of this relation, as of (5.47), is asymptotic to

$$\frac{2}{e\log 2}h < 1{\cdot}062h, \quad \text{as } h \to \infty.$$

The result (5.48) should be compared with (5.46) above.

We conclude this section by describing some striking results of Cassels,[29] although we shall prove only a special case of one of his theorems. We have seen that the 'thinnest' bases $\mathscr{B}$ of order h have counting numbers $B(n)$ lying between two constant multiples of $n^{1/h}$. This raises the question† whether there exists a basis of order h whose counting number is approximately equal to a constant multiple of $n^{1/h}$. Cassels has answered this question by way of Theorem B below and has obtained a good deal of additional information.

Theorem B (Cassels[29]). *For each $h = 2, 3,\dots$, there exists a basis $\mathscr{B}$, of order h, whose nth positive term b_n is such that*

$$\lambda_h = \lambda_h(\mathscr{B}) = \lim_{n\to\infty} \frac{b_n}{n^h} \tag{5.49}$$

exists and is positive. In fact, there exists a basis $\mathscr{B}'$, of order h, for which

$$b'_n = \mu_h n^h + O(n^{h-1}), \tag{5.50}$$

where $\mu_h = \mu_h(\mathscr{B}')$ is a positive number independent of n.

Cassels also proves the deeper Theorem A below. In fact, Cassels deduces Theorem B from Theorem A, although he mentions that he had proved Theorem B some years earlier as a deduction from a somewhat weaker form of Theorem A.

Theorem A (Cassels[29]). *For each $h = 2, 3,\dots$, there exists a basis $\mathscr{B}$ of order h whose n-th positive term b_n is such that*

$$\liminf_{n\to\infty} \frac{b_{n+1}-b_n}{n^{h-1}} > 0.$$

In the general case, Cassels was interested mainly in the qualitative nature of the above results. In the special case $h = 2$, however, he also included a quantitative investigation in which the bases were so

† Cf. Stöhr (6), (23).

(29) J. W. S. Cassels, 'Über Basen der natürlichen Zahlenreihe', *Abh. math. Semin. Univ. Hamburg*, **21** (1957) 247–57.

constructed as to make the constants λ_2 and μ_2 (and certain other constants) as large as his method would allow. We shall concentrate our attention entirely on the case $h = 2$, which is related to questions considered elsewhere in this book.† Although we do not include proofs of the quantitative results, we shall prove the case $h = 2$ of Theorem B, which is restated as Theorem 12 below.

Cassels' method, in the case $h = 2$, hinges on the following lemma.

LEMMA A (Cassels). *Let $q_1, q_2,\ldots$, where $q_1 = 1$, be a sequence of natural numbers such that*

$$(5.51) \qquad (q_j, q_{j+1}) = (q_j, q_{j+2}) = 1$$

for $j = 1, 2,\ldots$, and

$$(5.52) \qquad q_{j-1}(q_j+q_{j+1})+q_j(q_{j+1}+q_{j+2}) \leqslant q_{j+1}(q_{j+2}+q_{j+3})$$

for $j = 2, 3,\ldots$. For each $j = 1, 2,\ldots$, let

$$(5.53) \qquad Q_j = \sum_{i=2}^{j} q_{i-1}(q_i+q_{i+1}),$$

where the empty sum for Q_1 is interpreted to be 0. Furthermore, let $\mathscr{A}$ be the sequence consisting of all those integers a representable in the form

$$(5.54) \qquad a = Q_j+rq_j, \quad \text{where } 0 \leqslant r < q_{j+1}+q_{j+2}$$

and j is a natural number.

Then the sequence $\mathscr{A}$ is a basis of order 2.

As was pointed out by Cassels, the hypotheses (5.51) and (5.52) are satisfied when the sequence $\{q_j\}$ is taken to be the Fibonacci sequence (cf. the proof of Theorem 12) although this choice does not constitute the most effective application of the lemma from the quantitative point of view. We shall formulate the proof of Theorem 12 in terms of the Fibonacci sequence from the outset, and Lemma A will not be used explicitly (although a special case of the lemma is embodied in the proof). Nevertheless, to avoid concealing the motivation of Cassels' method, we say a few words concerning the proof of this lemma.

The proof is based on the fact that if u, v are fixed natural numbers which are relatively prime and x, y range independently over the integers of sufficiently long intervals, then all the integers in a suitably chosen interval are representable in the form $ux+vy$. (Explicit results

† The reader should compare Theorem 12 below with Theorem 4 of Chapter III (§ 2). Although the latter is concerned with sequences of pseudo-squares satisfying $a_n = (1+o(1))n^2$, corresponding results for sequences $\mathscr{A}$ satisfying $a_n = (1+o(1))\beta n^2$ (cf. (5.55) below) can be obtained in exactly the same way. Note that $\beta < 1$ in Theorem 12, in view of the remark immediately preceding the theorem.

of this general nature are easily established.) If $\mathscr{A}$ is defined as in the lemma, it is clear that, for each j, $2\mathscr{A}$ contains all integers of the form

$$Q_j+Q_{j+1}+(q_j x'+q_{j+1}y')$$

where x', y' range over the intervals

$$0 \leqslant x' < q_{j+1}+q_{j+2}, \qquad 0 \leqslant y' < q_{j+2}+q_{j+3},$$

and also contains all integers of the form

$$Q_j+Q_{j+2}+(q_j x''+q_{j+2}y'')$$

where x'', y'' range over the intervals

$$0 \leqslant x'' < q_{j+1}+q_{j+2}, \qquad 0 \leqslant y'' < q_{j+3}+q_{j+4}.$$

In view of (5.51), it can be shown by means of considerations of the type described above that $2\mathscr{A}$ contains all integers in a certain system of intervals; and the premisses of the lemma are so constructed as to ensure that this system of intervals in fact entirely covers $\mathscr{Z}_0$. It is, of course, quite accidental that the Fibonacci sequence turns out to be useful in this connexion.

For the quantitative results, we refer the reader to Cassels' paper. Here we mention only that these results include the construction of a 2-basis satisfying (5.49) with $\lambda_2 = \frac{1}{27}$, and the existence of an h-basis $\mathscr{B}$ satisfying (5.49) is shown to imply the existence of an h-basis $\mathscr{B}'$ satisfying (5.50) corresponding to every number μ_h satisfying $\mu_h < \lambda_h$. On the other hand, Cassels shows by means of a very simple combinatorial argument that

$$\liminf_{n\to\infty} \frac{b_n}{n^2} \leqslant \tfrac{1}{8}\pi \quad \text{for every 2-basis } \mathscr{B}.$$

THEOREM 12 (Cassels). *There exists a basis $\mathscr{B}$ of order 2 whose n-th positive element b_n satisfies*

(5.55) $$b_n = \beta n^2+O(n);$$

β being a positive constant independent of n.

Proof. Let $\mathscr{F}$ denote the Fibonacci sequence

$$f_1, f_2, f_3, \ldots,$$

determined by

(5.56) $$f_1 = 1, \quad f_2 = 2, \quad f_i = f_{i-1}+f_{i-2} \quad (i = 3, 4, 5, \ldots).$$

It is well known (see Appendix, § 1) that

(5.57) $$(f_i, f_{i+1}) = (f_i, f_{i+2}) = 1$$

and

(5.58) $$\lim_{i\to\infty} \frac{f_{i-1}}{f_i} = \tfrac{1}{2}(\sqrt{5}-1) = \rho,$$

say. Define

(5.59) $$g_1 = 0, \qquad g_j = \sum_{i=2}^{j} f_{i-1}f_{i+2} \quad (j = 2,3,\ldots),$$

and let $\mathscr{A}$ be the sequence given by

(5.60) $$a = g_j + rf_j, \qquad 0 \leqslant r < f_{j+3} \quad (j = 1,2,\ldots).$$

Then g_1, g_2,... is a subsequence of $\mathscr{A}$, and since

(5.61) $$g_j + f_j f_{j+3} = g_{j+1} \quad (j = 1,2,\ldots),$$

it is clear that

(5.62) $$g_j + rf_j \in \mathscr{A} \quad \text{for } 0 \leqslant r \leqslant f_{j+3} \quad (j = 1,2,\ldots).$$

We shall prove that $\mathscr{A}$ is a basis of order 2.

We have to show that every natural number m can be written in the form

$$m = a + a' \quad (a \in \mathscr{A},\, a' \in \mathscr{A}).$$

By (5.62) with $j = 1$, the numbers 0, 1, 2, 3, 4, 5 belong to $\mathscr{A}$; so that, since $g_2 = 5$, the sum sequence $2\mathscr{A}$ contains all the integers 0, 1, 2,..., $2g_2$. If $m > 2g_2$, we proceed as follows.

Suppose first that, for some particular $j \geqslant 2$,

(I) $$2g_j < m \leqslant g_j + g_{j+1} - f_{j-1}(f_j - 1).$$

Any number of the form

$$g_j - sf_{j-1} = g_{j-1} + (f_{j+2} - s)f_{j-1}, \qquad 0 \leqslant s \leqslant f_{j+2},$$

is, by (5.62), an element of $\mathscr{A}$, and we shall show now that integers r, s can be found to satisfy

(5.63) $$0 \leqslant r \leqslant f_{j+3}, \qquad 0 \leqslant s \leqslant f_{j+2}$$

and

(5.64) $$m = (g_j + rf_j) + (g_j - sf_{j-1});$$

it will follow at once that $m \in 2\mathscr{A}$. Since $(f_{j-1}, f_j) = 1$, there exists a (unique) number s such that

$$sf_{j-1} \equiv 2g_j - m \pmod{f_j}, \qquad 0 \leqslant s < f_j.$$

Then

$$m - 2g_j + sf_{j-1} = rf_j,$$

and since $m > 2g_j$, by (I), and $s \geqslant 0$, r is clearly positive; also by (I), the left-hand side does not exceed

$$g_{j+1} - g_j + f_{j-1}(f_j - 1) - f_{j-1}(f_j - 1) = f_j f_{j+3},$$

so that $r \leqslant f_{j+3}$. Thus the representation of m in the form (5.64), subject to (5.63), is established, and m is seen to lie in $2\mathscr{A}$.

Suppose next that, for some particular $j \geqslant 2$,

(II) $$g_j + g_{j+1} - f_{j-1}(f_j - 1) < m \leqslant 2g_{j+1}.$$

We shall again prove that $m \in 2\mathscr{A}$, this time by showing that integers r, s can be found to satisfy the two conditions

(5.65) $$m = (g_{j+1} + rf_{j+1}) + (g_j - sf_{j-1})$$

and

(5.66) $$0 \leqslant r \leqslant f_{j+4}, \qquad 0 \leqslant s \leqslant f_{j+2}.$$

We proceed as before. Since $(f_{j-1}, f_{j+1}) = 1$, a (unique) integer s exists satisfying

$$sf_{j-1} \equiv g_j + g_{j+1} - m \pmod{f_{j+1}},$$

and

$$f_j < s \leqslant f_j + f_{j+1} = f_{j+2}.$$

Then

$$m - g_j - g_{j+1} + sf_{j-1} = rf_{j+1}$$

where, by (II), $\quad rf_{j+1} > f_j f_{j-1} - f_{j-1}(f_j - 1) = f_{j-1}$

so that r is positive; furthermore, again using (II), we have

$$\begin{aligned} rf_{j+1} &\leqslant g_{j+1} - g_j + f_{j-1} f_{j+2} = f_j f_{j+3} + f_{j-1} f_{j+2} \\ &< f_j(f_{j+3} + f_{j+2}) = f_j f_{j+4} < f_{j+1} f_{j+4}, \end{aligned}$$

so that $r < f_{j+4}$. Thus m is representable in the form (5.65) subject to (5.66), and therefore m is an element of $2\mathscr{A}$.

We have proved that $\mathscr{A}$ is a basis of order 2. To describe the growth of $\mathscr{A}$, we define

(5.67) $$N_0 = 0 \quad \text{and} \quad N_j = \sum_{i=1}^{j} f_{i+3} \quad (j = 1, 2, \ldots).$$

Then if $j \geqslant 1$ and $\quad N_{j-1} \leqslant n \leqslant N_j,$

we see from (5.60) and (5.61) that the nth positive element of $\mathscr{A}$ is given by

(5.68) $$a_n = g_j + (n - N_{j-1}) f_j.$$

The sequence $\mathscr{A}$ is not distributed with sufficient regularity to permit a conclusion of type (5.55) for a_n; indeed, it can be shown by the kind of argument used below that both $\limsup_{n\to\infty} a_n/n^2$ and $\liminf_{n\to\infty} a_n/n^2$ are finite and positive, but *different*. However, we shall see that it is possible to embed $\mathscr{A}$ in a sequence $\mathscr{B}$ (which is then automatically a basis of order 2) possessing property (5.55). We shall now prove (5.69) below, which contains all the information concerning the distribution of terms in the sequence $\mathscr{A}$ that we require in the sequel.

(5.69) $$\liminf_{n\to\infty} \frac{a_{n+1} - a_n}{a_n^{\frac{1}{2}}} = \rho^2.$$

For $N_{j-1} \leqslant n < N_j$,

$$\frac{(a_{n+1} - a_n)^2}{a_n} = \frac{f_j^2}{g_j + (n - N_{j-1}) f_j} \geqslant \frac{(a_{N_j} - a_{N_j - 1})^2}{a_{N_j - 1}},$$

and we shall prove that

(5.70) $$\lim_{j\to\infty} \frac{(a_{N_j} - a_{N_j - 1})^2}{a_{N_j - 1}} = \rho^4;$$

(5.69) is then an immediate consequence of (5.70). By (5.67), (5.68), and (5.61),

$$a_{N_j-1} = g_j+(N_j-N_{j-1}-1)f_j = g_{j+1}-f_j,$$

so that, since $f_j/g_{j+1} \to 0$ as $j \to \infty$, (5.70) is equivalent to

$$\lim_{j\to\infty} \frac{f_j^2}{g_{j+1}} = \rho^4;$$

and it follows from (5.58) that this relation, in turn, is equivalent to

$$\lim_{j\to\infty} \frac{f_j f_{j+3}}{g_{j+1}} = \rho. \tag{5.71}$$

It suffices therefore to prove (5.71).

We require the well-known fact (see Appendix, § 1) that the jth term of $\mathscr{F}$ is given by the formula

$$f_j = A\rho^{-j}+B(-\rho)^j \quad (j = 1, 2, 3,...),$$

where A and B are non-zero constants. We also need to recall that

$$\rho^2+\rho = 1 \quad \text{and} \quad 0 < \rho < 1.$$

A simple computation gives

$$\begin{aligned} f_j f_{j+3} &= A^2\rho^{-2j-3}+4(-1)^j AB+B^2(-\rho)^{2j+3} \\ &= A^2\rho^{-2j-3}+O(1), \end{aligned} \tag{5.72}$$

and, in view of (5.59), g_{j+1} is therefore given by

$$g_{j+1} = A^2 \sum_{r=1}^{j} \rho^{-2r-3}+O(j) = \frac{A^2}{1-\rho^2}\rho^{-2j-3}+O(j).$$

Since $1-\rho^2 = \rho$, and $j\rho^j \to 0$ as $j \to \infty$, (5.71) follows at once from (5.72), and this completes the proof of (5.69).

We enter upon the final stage of the proof of Theorem 12. Suppose that α is a number satisfying

$$0 < \alpha < \rho^2; \tag{5.73}$$

since $\rho^2 = \frac{1}{2}(3-\sqrt{5})$, we could, for instance, take $\alpha = \frac{3}{8}$. We shall now construct a sequence $\mathscr{B}$ containing $\mathscr{A}$, whose nth positive term b_n satisfies

$$b_n = (\tfrac{1}{2}\alpha)^2 n^2+O(n). \tag{5.74}$$

Since $2\mathscr{B} \supset 2\mathscr{A} = \mathscr{Z}_0$, Theorem 12 follows, with $\beta = \frac{1}{4}\alpha^2$.

Let $\mathscr{C}$ be a sequence of natural numbers

$$0 = c_0 < c_1 < c_2 < ...$$

such that

$$c_n = (\tfrac{1}{2}\alpha)^2 n^2+O(1); \tag{5.75}$$

evidently such a sequence can be constructed, and we observe that

$$\lim_{n\to\infty} \frac{c_{n+1}-c_n}{c_n^{\frac{1}{2}}} = \alpha. \tag{5.76}$$

As a preliminary step in the construction of $\mathscr{B}$ we need to show that there exists an absolute constant M such that for every $m \geqslant M$, the interval

$$c_m < x \leqslant c_{m+1}$$

contains at most one element of $\mathscr{A}$. Since $\alpha < \rho^2$, a positive number ϵ_0 can be found such that

$$0 < \frac{\alpha^2+\epsilon_0}{\rho^4-\epsilon_0} < 1. \tag{5.77}$$

Then, by (5.69), there exists $n_0 = n_0(\epsilon_0)$ such that

$$(a_{n+1}-a_n)^2 > (\rho^4-\epsilon_0)a_n \quad (n \geqslant n_0);$$

and by (5.76) there exists $m_0 = m_0(\epsilon_0)$ such that

$$(c_{m+1}-c_m)^2 < (\alpha^2+\epsilon_0)c_m \quad (m \geqslant m_0).$$

Suppose, if possible, that there exists an integer $m_1 > m_0$ such that

$$c_{m_1} < a_n < a_{n+1} \leqslant c_{m_1+1}$$

for some $n \geqslant n_0$. Then $a_{n+1}-a_n < c_{m_1+1}-c_{m_1}$, and it follows that

$$a_n < \left(\frac{\alpha^2+\epsilon_0}{\rho^4-\epsilon_0}\right)c_{m_1},$$

which, by (5.77), is contrary to the supposition $a_n > c_{m_1}$. Thus we have proved that any consecutive pair c_m, c_{m+1} of elements of $\mathscr{C}$ is separated by at most one member of $\mathscr{A}$ provided only that m is sufficiently large.

We are now in a position to construct $\mathscr{B}$. Let N be the integer determined by

$$a_N \leqslant c_M < a_{N+1},$$

and define $b_n = a_n$ for $1 \leqslant n \leqslant N$. For $n > N$, let b_n denote the element of $\mathscr{A}$ in the interval

$$c_{n-N+M-1} < x \leqslant c_{n-N+M}$$

if such an element exists (we have seen that there is at most one such element), and otherwise let $b_n = c_{n-N+M}$. Thus, in any case,

$$c_{n-N+M-1} < b_n \leqslant c_{n-N+M} \quad (n > N),$$

and (5.74) follows at once from (5.75).

The proof of Theorem 12 is now complete.

6. Asymptotic analogues and p-adic analogues

As we have now seen, Schnirelmann density is certainly a type of density appropriate to the problems considered in the preceding sections. Nevertheless, the most natural type of sequence density is asymptotic density, and it is desirable to obtain analogous results in terms of this type also; or better, in terms of lower asymptotic density, since the asymptotic density of a sequence may not exist. Extensions of preceding results to other kinds of density, and in other contexts (general abelian groups, for instance), have given rise to a vast literature, but for these the reader is referred to H. Ostmann's[30] comprehensive survey.

In the special case when each of the sequences under consideration coincides (from some point onwards) with a union of congruence classes to a fixed modulus g, we may interpret addition modulo g, and, in place of the asymptotic density of each sequence, consider the number of congruence classes in the corresponding union. The resulting g-adic analogues (and, in particular, the p-adic analogues, which arise when g is a prime p) of the various questions concerning density in the addition of sequences are of particular importance and the subject of separate study.

In this section we shall give analogues of Theorems 5′ and 11 in terms of (lower) asymptotic density. The exact asymptotic analogue of Theorem 7 (and of its special case, Theorem 3′) is false for trivial reasons. However, the situations in which the exact analogue of Theorem 7 is false can be isolated, and this task will be carried out in sections 7–10.

Finally, we shall consider the g-adic analogue of Theorem 3′ (suitably modified), although this matter also will be carried further in the later sections.

If $\mathscr{A}$ is a subsequence of $\mathscr{Z}$ and $A(n)$ is the counting number of the positive part of $\mathscr{A}$, the *lower asymptotic density* $\underline{\mathbf{d}}\mathscr{A}$ of $\mathscr{A}$ is defined by

$$\underline{\mathbf{d}}\mathscr{A} = \liminf_{n\to\infty} \frac{A(n)}{n}.$$

An essential difference between Schnirelmann density and asymptotic density is that the latter is invariant under translation; thus

$$\underline{\mathbf{d}}(t+\mathscr{A}) = \underline{\mathbf{d}}\mathscr{A} \quad \text{for any integer } t, \tag{6.1}$$

a fact which will frequently be used in the sequel. Moreover, comparison of the two kinds of density leads at once to the following two obvious comments.

LEMMA 7. *For any sequence* $\mathscr{A}$,

(i) $\sigma\mathscr{A} \leqslant \underline{\mathbf{d}}\mathscr{A}$;

(ii) *if* $\underline{\mathbf{d}}\mathscr{A} > 0$ *and* $1 \in \mathscr{A}$, *then* $\sigma\mathscr{A} > 0$.

(30) H. Ostmann, 'Additive Zahlentheorie' I, II (Ergebnisse Series, Springer, 1956).

Definition 2*. $\mathscr{B}$ is said to be an *asymptotic basis* of order h if all sufficiently large positive integers lie in $h\mathscr{B}$; in other words, if there exists a natural number n_0 such that $n \in h\mathscr{B}$ whenever $n \geqslant n_0$.

We define the *average asymptotic order* $\mathbf{h}^*$ of the asymptotic basis $\mathscr{B}$ by

$$\mathbf{h}^* = \limsup_{n\to\infty} \frac{1}{n} \sum_{m=1}^{n} h(m)$$

where, for each $m \geqslant n_0$, $h(m)$ is the least number of elements of $\mathscr{B}$ with sum m, and $h(m) = 0$ for $m < n_0$.

It is clear that every basis is an asymptotic basis and, conversely, that an asymptotic basis which contains 0 and 1 is a basis. To illustrate the consequences of this remark, the sequence $\mathscr{W}$ which was constructed in the proof of Theorem 11 is not an asymptotic basis; for it is not a basis and yet contains 0 and 1. Thus the difference between the two kinds of bases is slight, and it is hardly surprising that Theorems 5′ and 11 possess exact 'asymptotic' analogues, the proofs of which raise no new difficulties. We have the following analogue of Theorem 5′.

THEOREM 13 (Rohrbach[31]). *If $\mathscr{B}$ is an asymptotic basis of average asymptotic order* $\mathbf{h}^*$, *then for every sequence* $\mathscr{A}$,

$$\underline{\mathbf{d}}(\mathscr{A}+\mathscr{B}) \geqslant \underline{\mathbf{d}}\mathscr{A} + \frac{1}{2\mathbf{h}^*}(1-\underline{\mathbf{d}}\mathscr{A})\,\underline{\mathbf{d}}\mathscr{A}. \tag{6.2}$$

Proof. It requires only minor modifications in the proof of Theorem 5′ to obtain Theorem 13. We indicate the changes, retaining the notation of the former proof.

We recall that the proof of Theorem 5′ was based on the inequalities (the first of these appeared as (3.5))

$$\text{(i)} \qquad \bar{D}_n^* \leqslant h^*\{C(n)-A(n)\},$$

$$\text{(ii)} \qquad \bar{D}_n^* \geqslant \tfrac{1}{2}n\left\{\sigma\mathscr{A} - \left(\frac{A(n)}{n}\right)^2\right\}.$$

For an arbitrary $\epsilon > 0$, let $n_0 = n_0(\epsilon)$ be chosen sufficiently large for the following statements to be true:

$$A(n) > (\underline{\mathbf{d}}\mathscr{A}-\epsilon)n \quad \text{for } n \geqslant n_0,$$

$$h(m) \geqslant 1 \quad \text{for } m \geqslant n_0,$$

$$\sum_{m=n_0}^{n} h(m) < (\mathbf{h}^*+\epsilon)n \quad \text{for } n \geqslant n_0.$$

(31) H. Rohrbach, 'Einige neuere Untersuchungen über die Dichte in der additiven Zahlentheorie', *Jber. dtsch. Mat.-Verein.* 48 (1938) 199–236.

It requires only automatic changes in the proof to obtain in place of (i),

$$\frac{1}{n}\sum_{m=n_0}^{n} \bar{D}_n(m) < (\mathbf{h}^*+\epsilon)\{C(n)-A(n)\}. \tag{6.3}$$

As regards a modified form of (ii), we have for $n \geqslant 2n_0$,

$$\begin{aligned}\frac{1}{n}\sum_{m=n_0}^{n} \bar{D}_n(m) &= \frac{1}{n}\sum_{m=n_0}^{n} A(n-m)-\frac{1}{n}\sum_{m=n_0}^{n} D_n(m)\\ &\geqslant \frac{1}{n}(\underline{\mathbf{d}}\mathscr{A}-\epsilon)\sum_{m=n_0}^{n-n_0}(n-m)-\frac{1}{n}\sum_{m=n_0}^{n} D_n(m)\\ &\geqslant \tfrac{1}{2}(\underline{\mathbf{d}}\mathscr{A}-\epsilon)(n-2n_0+1)-\frac{1}{2n}(A(n)-1)A(n)\end{aligned}$$

so that

$$\frac{1}{n}\sum_{m=n_0}^{n} \bar{D}_n(m) \geqslant \tfrac{1}{2}n\left\{\underline{\mathbf{d}}\mathscr{A}-\epsilon-\left(\frac{A(n)}{n}\right)^2\right\}-n_0\,\underline{\mathbf{d}}\mathscr{A}. \tag{6.4}$$

The derivation of (6.2) from (6.3) and (6.4) presents no difficulty.

In the case of Theorem 11, it is actually possible to give the analogue in a strengthened form. In the statement of this analogue (Theorem 14 below) we shall make use of the following definition.

Definition 7. A sequence $\mathscr{W}$ is said to be an *asymptotic essential component* if

$$\underline{\mathbf{d}}(\mathscr{W}+\mathscr{A}) > \underline{\mathbf{d}}\mathscr{A}$$

for every sequence $\mathscr{A}$ with $0 < \underline{\mathbf{d}}\mathscr{A} < 1$.

THEOREM 14 (Stöhr–Wirsing).† *There exists an asymptotic essential component which is not an asymptotic basis. In fact, there exists a sequence $\mathscr{W}_1$ which is not an asymptotic basis but has the property that for any sequence $\mathscr{A}$ with $\underline{\mathbf{d}}\mathscr{A} > 0$,*

$$\underline{\mathbf{d}}(\mathscr{A}+\mathscr{W}_1) = 1.$$

Proof. We recall that the sequence $\mathscr{W}$ in Theorem 11 was constructed from the bases $\mathscr{S}_h$ of (5.35). We construct the sequence $\mathscr{W}_1$ from suitable bases $\mathscr{S}_h$ in exactly the same way, but redefine the bases $\mathscr{S}_h = \mathscr{S}_{h,k'}$ themselves by choosing the integer $k' = k'(h)$ in Lemma 6 so large that, in place of (5.36),

$$\sigma(\mathscr{A}+\mathscr{S}_{h,k'}) \geqslant 1-\exp(-h\sigma\mathscr{A}) \tag{6.5}$$

for every sequence $\mathscr{A}$. To show that this is possible, we prove that $k'(h) = 2h^2$ is an appropriate choice. To do this we need only verify

† Stöhr–Wirsing; see ref. (26).

(retaining the notation of the proof of Lemma 6) that

$$f(2h^2) \geqslant 1-\exp\{-hf(0)\}. \tag{6.6}$$

Now, by (5.42),

$$1-f(r+1) = \{1-f(r)\}\left\{1-\frac{f(r)}{2h}\right\}$$

$$= \{1-f(0)\}\prod_{i=0}^{r}\left\{1-\frac{f(i)}{2h}\right\} \quad \text{for } r = 0, 1, 2,...,$$

and since $f(r)$ is a monotonically increasing function of r, we have

$$1-f(r+1) \leqslant \left\{1-\frac{f(0)}{2h}\right\}^{r+1}.$$

Hence
$$1-f(2h^2) \leqslant \left\{1-\frac{f(0)}{2h}\right\}^{2h^2},$$

and (6.6) follows since

$$\left\{1-\frac{f(0)}{2h}\right\}^{2h} \leqslant \exp\{-f(0)\}.$$

In accordance with an earlier remark, $\mathscr{W}_1$ is not an asymptotic basis; for it contains 0, 1 and yet is not a basis. It remains to show that $\underline{\mathbf{d}}(\mathscr{A}+\mathscr{W}_1) = 1$ whenever $\underline{\mathbf{d}}\mathscr{A} > 0$.

The following simple argument, which will be in frequent use for the rest of the chapter, shows that there is no loss of generality in assuming that $1 \in \mathscr{A}$. A translation of $\mathscr{A}$ leaves both $\underline{\mathbf{d}}\mathscr{A}$ and $\underline{\mathbf{d}}(\mathscr{A}+\mathscr{W}_1)$ invariant; thus we may replace $\mathscr{A}$ by $t+\mathscr{A}$, where the integer t is chosen to satisfy $1 \in t+\mathscr{A}$.

We have redefined $\mathscr{S}_h = \mathscr{S}_{h,k'}$ so that (6.5) replaces (5.36). Thus (cf. the proof of Theorem 11) the interval $1+I_h$, and those of its sub-intervals having the same starting-point, contain at least a proportion $1-\exp(-h\sigma\mathscr{A})$ of elements belonging to $\mathscr{A}+\mathscr{W}_1$. If $\underline{\mathbf{d}}\mathscr{A} > 0$ and $1 \in \mathscr{A}$, then $\sigma\mathscr{A} > 0$ so that $\exp(-h\sigma\mathscr{A}) \to 0$ as $h \to \infty$. We may thus conclude that $\underline{\mathbf{d}}(\mathscr{A}+\mathscr{W}_1) = 1$ provided $\underline{\mathbf{d}}\mathscr{A} > 0$.

The following example makes it clear that, in marked contrast, there is no exact analogue of Theorem 7 (or of Theorem 3′).

Let $h_0, h_1,..., h_k$ be non-negative numbers and g be a natural number such that $\sum_{r=0}^{k} (h_r+1) < g$. For $r = 0, 1,..., k$, let $\mathscr{A}_r$ coincide, from some point onwards, with the union of the congruence classes $0, 1,..., h_r$ (mod g). Then

$$\lim_{n\to\infty} n^{-1} \sum_{r=0}^{k} A_r(n) = g^{-1} \sum_{r=0}^{k} (h_r+1),$$

whereas $\sum_{r=0}^{k} \mathscr{A}_r$ consists of the residue classes $0, 1,..., h_0+h_1+...+h_k$, so that

$$\underline{\mathbf{d}} \sum_{r=0}^{k} \mathscr{A}_r = g^{-1}\Big(1+\sum_{r=0}^{k} h_r\Big).$$

Thus

$$\underline{\mathbf{d}} \sum_{r=0}^{k} \mathscr{A}_r = \Big\{\liminf_{n\to\infty} n^{-1} \sum_{r=0}^{k} A_r(n)\Big\} - \frac{k}{g}. \tag{6.7}$$

Since $k/g > 0$ and the first term on the right of (6.7) is less than 1, the above provides a counter-example† to the exact analogue of Theorem 7 (and, in particular, of Theorem 3′).

In the above example each of the sequences $\mathscr{A}_r$ is essentially the union of congruence classes modulo g. It will be shown in the following sections that the exact analogue of Theorem 7 can, indeed, fail only in this or a closely related situation. Meanwhile we obtain, in this 'degenerate' case, a weakened analogue of Theorem 3′,‡ namely Theorem 15, below. This theorem, which is the best possible of its type and has many applications in number theory, dates back, in essence, to Cauchy.

The condition (6.8) in the statement of Theorem 15 is rather restrictive; some such condition is necessary to ensure the conclusion of the theorem, but the more general situation will be investigated in later sections (cf. remarks following Theorem 16).

It is clear that, when considering the degenerate case described above, we may replace each sequence by the union of congruence classes with which it coincides from some point onward. (For the remainder of this section, we deviate from our system of notation, in that $\mathscr{A}$, $\mathscr{B}$ need not denote subsequences of $\mathscr{Z}$. When $\mathscr{A}$ is a union of congruence classes, we retain the convention that only positive elements are counted in $A(n)$, so that the definition of asymptotic density remains unaffected.) The relationship of Theorem 15 to the degenerate case in the addition of sequences is relevant to the investigations of §§ 7–10. For this reason, we state the result in terms of asymptotic density. At the same time, the theorem is, of course, a statement about sets of residues modulo g.

† The simplest such counter-example is provided by the case $h_0 = h_1 = ... = h_k = 0$.

‡ A weakened analogue of Theorem 7 could easily be deduced from Theorem 15 by induction, but we shall, in any case, obtain more general results in later sections.

THEOREM 15 (Cauchy–Davenport–Chowla).† *Let* $\mathscr{A}$ *be the union of* r *incongruent residue classes* $a_1, a_2, \ldots, a_r \pmod g$ *and* $\mathscr{B}$ *the union of* s *incongruent residue classes* $0, b_1, b_2, \ldots, b_{s-1} \pmod g$, *where*

$$(b_i, g) = 1 \quad \textit{for } i = 1, 2, \ldots, s-1. \tag{6.8}$$

Then

$$\mathbf{d}(\mathscr{A}+\mathscr{B}) \geqslant \min\left(1, \mathbf{d}\mathscr{A}+\mathbf{d}\mathscr{B}-\frac{1}{g}\right); \tag{6.9}$$

in other words, $\mathscr{A}+\mathscr{B}$ *is the union of at least* $\min(g, r+s-1)$ *incongruent residue classes modulo* g.

Proof. There is no loss of generality in assuming

$$r+s-1 \leqslant g. \tag{6.10}$$

For suppose the theorem were proved subject to this restriction; then for $\mathscr{A}$, $\mathscr{B}$ with $r+s-1 > g$, we simply discard $r+s-1-g$ of the congruence classes forming $\mathscr{A}$ (so that (6.10) then holds with equality) and apply the restricted form of the theorem to obtain $\mathscr{A}+\mathscr{B} = \mathscr{Z}$.

The proof of the theorem (subject to (6.10)) now runs parallel to that of Theorem 3′, although the details become much simpler. We suppose the result to be false, and from among those pairs $\mathscr{A}$, $\mathscr{B}$ satisfying the premisses of the theorem (and (6.10)) for which (6.9) is false (and thus $r \geqslant 2$, $s \geqslant 2$), we choose a pair for which s is minimal. We obtain a contradiction in the following manner (cf. the proof of Theorem 3′).

We shall construct, by removing certain non-zero congruence classes $\pmod g$ from $\mathscr{B}$ and placing certain additional congruence classes $\pmod g$ in $\mathscr{A}$, a pair of new sequences $\mathscr{A}'$, $\mathscr{B}'$ with the following properties:

(i) $r'+s' = r+s$,

(ii) $\mathscr{A}'+\mathscr{B}' \subset \mathscr{A}+\mathscr{B}$,

(iii) $s' < s$.

The new pair $\mathscr{A}'$, $\mathscr{B}'$ clearly satisfy the premisses of the theorem (and (6.10)); but $\mathscr{A}'$, $\mathscr{B}'$ do not satisfy (6.9) in view of (i), (ii), and (6.10), so that (iii) contradicts the minimal property of $\mathscr{B}$.

As in the proof of Theorem 3′, there exists an element a^* of $\mathscr{A}$ such that

$$a^*+\mathscr{B} \text{ is not entirely contained in } \mathscr{A}, \tag{6.11}$$

† (32) A. L. Cauchy, 'Recherches sur les nombres', *J. Éc. polytech.* **9** (1813) 99–116.
(33) H. Davenport, 'On the addition of residue classes', *J. Lond. math. Soc.* **10** (1935) 30–2; 'A historical note', ibid. **22** (1947) 100–1.
(34) I. Chowla, 'A theorem on the addition of residue classes', *Proc. Indian Acad. Sci.* **2** (1935) 242–3.

although for a different reason. If such an a^* did not exist, then to each element a_j of $\mathscr{A}$ there would correspond another element a_k such that $b_1+a_j \equiv a_k \pmod g$. This would imply

$$\sum_{j=1}^{r} (b_1+a_j) \equiv \sum_{k=1}^{r} a_k \pmod g$$

and hence $rb_1 \equiv 0 \pmod g$. But this is not possible, since $(b_1, g) = 1$ and $0 < r < g$ in view of (6.10) and the fact that $r \geqslant 2$, $s \geqslant 2$.

Let a^* be chosen to satisfy (6.11), and let $\mathscr{B}''$ be the union of all those congruence classes $b_i \pmod g$ for which

$$a^*+b_i \notin \mathscr{A}.$$

As in the proof of Theorem 3′, we form the new pair $\mathscr{A}'$, $\mathscr{B}'$ by removing the elements of $\mathscr{B}''$ (which does not contain 0) from $\mathscr{B}$ and adjoining the sequence $a^*+\mathscr{B}''$ to $\mathscr{A}$.

The relations (ii) and (iii) are satisfied for exactly the same reasons as the corresponding relations in the case of Theorem 3′. Furthermore, (i) (the justification of which was the main source of difficulty in the proof of Theorem 3′) is here obvious. The theorem is thus established.

The following variant of Theorem 15 is sometimes useful.

THEOREM 15′ (Kemperman and Scherk[35]). *Let $\mathscr{A}$ be the union of r incongruent residue classes $0, a_1, a_2,..., a_{r-1} \pmod g$ and $\mathscr{B}$ the union of s incongruent residue classes $0, b_1, b_2,..., b_{s-1} \pmod g$. Suppose that 0 has no representation of the form a_i+b_j, except for the trivial representation corresponding to $a_i \equiv b_j \equiv 0$. Then $\mathscr{A}+\mathscr{B}$ is the union of at least* $\min(g, r+s-1)$ *distinct residue classes modulo g.*

The proof is exactly the same as that of Theorem 15 except in one particular; namely the proof of the existence of the element a^* (cf. the proof of Theorem 15). Here we need only observe that if a^* does not exist then $b_1+\mathscr{A}$ $(b_1 \not\equiv 0)$ coincides with $\mathscr{A}$ modulo g and therefore the element 0 (which lies in $\mathscr{A}$) has a representation b_1+a_i, contrary to hypothesis.

It will be seen later that Theorems 15 and 15′ are only special cases of Kneser's Theorem 16 (in § 7).

The original Cauchy–Davenport theorem is the case $g = p$ (where p denotes a prime) of Theorem 15. This case of the theorem may be restated as follows.

(35) J. H. B. Kemperman and P. Scherk, 'Complexes in abelian groups', *Can. J. Math.* **6** (1954) 230–7.
See also Problem 4466 set by L. Moser, *Am. math. Mon.* **62** (1955) 46–7.

Let each of $\mathscr{A}$, $\mathscr{B}$ *be a set of incongruent residues modulo* p. *Then, interpreting addition modulo* p,

$$|\mathscr{A}+\mathscr{B}| \geqslant \min(p, |\mathscr{A}|+|\mathscr{B}|-1), \tag{6.12}$$

where $|\mathscr{S}|$ *denotes the number of elements in* $\mathscr{S}$.

Vosper[36] succeeded in classifying those pairs $\mathscr{A}$, $\mathscr{B}$ of sets of residues for which equality holds in (6.12). It is easily verified that equality holds in each of the cases (i)–(iv) below, and Vosper proved that apart from these cases, (6.12) holds with strict inequality.

(i) $|\mathscr{A}|+|\mathscr{B}| > p$;

(ii) $|\mathscr{A}| = 1$ or $|\mathscr{B}| = 1$;

(iii) there exists a number d such that $\mathscr{A}$ is the complement of $d-\mathscr{B}$ with respect to the complete set of residues modulo p;

(iv) $\mathscr{A}$ and $\mathscr{B}$ are representable as arithmetic progressions with the same common difference. (We say that an arithmetic progression represents a set $\mathscr{A}$ of residues modulo p, if it provides a set of representatives modulo p for $\mathscr{A}$.)

Kemperman[37] has generalized the work of Vosper. Theorem 15 is not a suitable starting point for such a generalization for, as we shall see in the next section, this theorem is rather artificial when g is composite (because condition (6.8), which was required for Chowla's proof, does not correspond to any intrinsic feature of the situation).

In essence, Kemperman carries out the (much more difficult) analysis required for an investigation analogous to that of Vosper, in relation to Theorem 16; he does so in the more general context of additive abelian groups. As is to be expected, his results are complicated, and we do not quote them here.

7. Kneser's theorem

In this and the next three sections we shall give an account of Kneser's remarkable achievement[38] in this field—his formulation and proof of the 'asymptotic' analogue of Theorem 7 (Dyson's generalization of Mann's theorem). The problem before us is, given a system $\mathfrak{A}$ of $k+1$ integer sequences $\mathscr{A}_0, \mathscr{A}_1, \ldots, \mathscr{A}_k$, to obtain a lower bound for

(36) A. G. Vosper, 'The critical pairs of sub-sets of a group of prime order', *J. Lond. math. Soc.* **31** (1956) 200–5; Addendum (containing a simpler proof), ibid. 280–2.

(37) J. H. B. Kemperman, 'On small sumsets in an abelian group', *Acta math.* **103** (1960) 63–88.

(38) M. Kneser, 'Abschätzungen der asymptotischen Dichte von Summenmengen', *Math. Z.* **58** (1953) 459–84. For earlier, partial, results see Kneser's acknowledgements and ref. (30).

$\underline{\mathbf{d}}(\mathscr{A}_0+\mathscr{A}_1+...+\mathscr{A}_k)$ in terms of

$$\liminf_{n\to\infty}\frac{1}{n}\sum_{\nu=0}^{k} A_\nu(n). \tag{7.1}$$

Although sequences of non-negative integers are under consideration, it will be convenient to permit $\mathscr{A}_0$, $\mathscr{A}_1$,..., $\mathscr{A}_k$ to contain negative elements (cf. § 2). In view of the fact that only positive elements are counted in $A_\nu(n)$, this relaxation makes no essential difference. Hereafter 'integer sequence' will mean 'subsequence of $\mathscr{Z}$', that is, a strictly increasing sequence of integers (which may, of course, contain any finite number of negative elements). The only deviation from this usage will occur in connexion with Definition 10, and is discussed immediately below that definition.

We begin with some preparatory remarks and definitions. We first reintroduce the notion of a strong union, which was defined in § 4 (Definition 4); with the slight difference that here the sequences $\mathscr{A}_\nu$ need not† contain 0. Accordingly, we shall understand by the strong union

$$\underline{\mathscr{V}}(\mathfrak{A}) = \underline{\mathscr{V}}(\mathscr{A}_0, \mathscr{A}_1,..., \mathscr{A}_k) = \bigvee_{\nu=0}^{k} \mathscr{A}_\nu \tag{7.2}$$

the aggregate of the elements of $\mathscr{A}_0$, $\mathscr{A}_1$,..., $\mathscr{A}_k$, each element being counted according to its multiplicity; except that 0, if it occurs at all, is to be counted only once. As in Definition 5, we denote by

$$\underline{V}(n) = \underline{V}(\mathfrak{A}; n),$$

the counting number of the positive part of $\underline{\mathscr{V}}(\mathfrak{A})$. In this notation, (7.1) becomes $\liminf_{n\to\infty} n^{-1}\underline{V}(n)$ and, by analogy with the definition of $\underline{\mathbf{d}}\mathscr{A}$, we write

$$\underline{\mathbf{d}}\underline{\mathscr{V}}(\mathfrak{A}) = \liminf_{n\to\infty}\frac{1}{n}\underline{V}(\mathfrak{A}; n). \tag{7.3}$$

We denote the sum of the sequences of the system $\mathfrak{A}$ by

$$\mathscr{C} = \mathscr{C}(\mathfrak{A}) = \mathscr{C}(\mathscr{A}_0, \mathscr{A}_1,..., \mathscr{A}_k),$$

so that $\mathscr{C}(\mathfrak{A})$ is defined by

$$\mathscr{C}(\mathfrak{A}) = \sum_{\nu=0}^{k} \mathscr{A}_\nu. \tag{7.4}$$

We remark at the outset that for our purpose, namely to obtain a lower bound for $\underline{\mathbf{d}}\mathscr{C}(\mathfrak{A})$ in terms of $\underline{\mathbf{d}}\underline{\mathscr{V}}(\mathfrak{A})$, we are always entitled to

† Actually, where the multiplicity of the 0 element is material, the sequences will all contain 0.

replace the system $\mathfrak{A}$ by any system $\mathfrak{A}'$ which satisfies both the following conditions:

(i) $\mathscr{A}_\nu \subset \mathscr{A}'_\nu$ for $\nu = 0, 1, 2,..., k$;
(ii) $\mathscr{C}(\mathfrak{A})$ and $\mathscr{C}(\mathfrak{A}')$ are identical from some point onward.

Definition 8. If two systems $\mathfrak{A}$ and $\mathfrak{A}'$ stand in the relation described by (i) and (ii), we say that $\mathfrak{A}'$ is a *worse* system than $\mathfrak{A}$.

Definition 9. If two sequences $\mathscr{A}$ and $\mathscr{B}$ are identical from some point onward, we say that they are *equivalent* and write

$$\mathscr{A} \sim \mathscr{B}.$$

Thus condition (ii) above states that $\mathscr{C}(\mathfrak{A})$ and $\mathscr{C}(\mathfrak{A}')$ are equivalent.

It will be useful to have on record the following obvious consequence of (6.1).

Lemma 8. *For any given set $t_0, t_1,..., t_k$ of integers define the new system $\mathfrak{A}'$ of sequences $\mathscr{A}'_0, \mathscr{A}'_1,..., \mathscr{A}'_k$ by*

$$\mathscr{A}'_\nu = \mathscr{A}_\nu - t_\nu \quad (\nu = 0, 1,..., k).$$

Then $$\underline{\mathbf{d}}\mathscr{C}(\mathfrak{A}') = \underline{\mathbf{d}}\mathscr{C}(\mathfrak{A}) \quad \textit{and} \quad \underline{\mathbf{d}}\mathscr{V}(\mathfrak{A}') = \underline{\mathbf{d}}\mathscr{V}(\mathfrak{A}).$$

Definition 10. If for some g, each $\mathscr{A}_\nu$ of the system $\mathfrak{A}$ is a union of (entire) congruence classes modulo g, we say that the system $\mathfrak{A}$ is *degenerate modulo g*. If we say merely that the system $\mathfrak{A}$ is *degenerate*, we mean that it is degenerate modulo g, for some g.

It has already been pointed out, near the beginning of § 6, that addition of sequences, each coinciding from some point onwards with a union of congruence classes modulo g, may for our purposes be interpreted as addition modulo g. In the latter form of addition, each 'congruence class' is regarded as a single entity and it is natural to think in terms of *entire* congruence classes. Accordingly, we use Definitions 9 and 10 to relate addition of the former kind to addition modulo g; even though this entails the use of the word 'sequence' in connexion with the $\mathscr{A}_\nu$ occurring in Definition 10.

As we have previously used the word 'degenerate' for descriptive purposes, we stress that from now on 'degenerate system' will have the strict technical meaning of Definition 10. We shall also use the term 'degenerate sequence modulo g' to describe a union of congruence classes modulo g.

In the preceding section, we constructed a system $\mathfrak{A}$ satisfying (6.7). It follows that the inequality $\underline{\mathbf{d}}\mathscr{C}(\mathfrak{A}) \geqslant \min(1, \underline{\mathbf{d}}\mathscr{V}(\mathfrak{A}))$, suggested by direct analogy to Dyson's theorem is false; and that for systems $\mathfrak{A}$ which are degenerate† modulo g we cannot hope to prove more than

$$\mathbf{d}\mathscr{C}(\mathfrak{A}) \geqslant \mathbf{d}\mathscr{V}(\mathfrak{A}) - \frac{k}{g}. \tag{7.5}$$

It is clear, however, that even this inequality does not hold, in general, unless g is minimal in some sense. For if $\mathfrak{A}$ is degenerate modulo g it is also degenerate modulo mg for any integer $m > 1$; and if $\mathfrak{A}$ is chosen to satisfy

$$\mathbf{d}\mathscr{C}(\mathfrak{A}) = \mathbf{d}\mathscr{V}(\mathfrak{A}) - \frac{k}{g},$$

then (7.5) becomes false when g is replaced by mg. It will transpire that the appropriate minimal condition for g is that none of the systems $\mathfrak{A}'$ worse than $\mathfrak{A}$ (including $\mathfrak{A}$ itself) shall be degenerate to a modulus less than g. We shall, in fact, prove the following generalization of Theorems 15 and 15′.

THEOREM 16 (Kneser). *If the system $\mathfrak{A}$ is degenerate modulo g, there exists a (minimal) divisor g' of g and a system $\mathfrak{A}'$, degenerate modulo g', such that $\mathfrak{A}'$ is worse‡ than $\mathfrak{A}$ and*

$$\mathbf{d}\mathscr{C}(\mathfrak{A}') \geqslant \mathbf{d}\mathscr{V}(\mathfrak{A}') - \frac{k}{g'}. \tag{7.6}$$

In particular, (7.5) *is true if g has the above-mentioned minimal property.*

The connexion between Theorem 15 and Theorem 16 (with $k = 1$) is not at once obvious, especially as condition (6.8) implies (6.9) even when g is not minimal. We write (7.6) (with $k = 1$ and in the notation of Theorem 15) in the form

$$\mathbf{d}(\mathscr{A}+\mathscr{B}) \geqslant \mathbf{d}\mathscr{A}+\mathbf{d}\mathscr{B}-\frac{1}{g}+(\mathbf{d}\mathscr{A}'-\mathbf{d}\mathscr{A})+\left(\mathbf{d}\mathscr{B}'-\mathbf{d}\mathscr{B}+\frac{1}{g}-\frac{1}{g'}\right)$$

(cf. (7.7) below). Since $0 \in \mathscr{B}$, the sequence $\mathscr{B}'$ contains the entire congruence class 0 (mod g') so that, if $\mathscr{B}$ satisfies (6.8),

$$\mathbf{d}\mathscr{B}'-\mathbf{d}\mathscr{B} \geqslant \frac{1}{g'}-\frac{1}{g}$$

provided $g' > 1$ (the case $g' = 1$ corresponds to $\mathbf{d}(\mathscr{A}+\mathscr{B}) = 1$). Also, $\mathbf{d}\mathscr{A}' \geqslant \mathbf{d}\mathscr{A}$ because $\mathfrak{A}'$ is worse than $\mathfrak{A}$. Hence, given condition (6.8), the conclusion (6.9) of Theorem 15 follows from Theorem 16 (with

† In view of the special role played by the zero elements of the sequences in a strong union, it is essential to note the distinction between the zero element of a sequence and the zero element modulo g. The zero element modulo g actually corresponds to a sequence of asymptotic density g^{-1}.

‡ Here, of course, condition (ii) implies that $\mathscr{C}(\mathfrak{A}') = \mathscr{C}(\mathfrak{A})$.

$k = 1$). It is clear from this analysis that the condition (6.8) in Theorem 15 is rather artificial.

Theorem 15′ arises as a special case of Theorem 16 in much the same way. Suppose that there exists a system $\mathfrak{A}'$ worse than $\mathfrak{A}$ which is degenerate modulo g', where g' is a proper divisor of g. Then, since each of $\mathscr{A}$, $\mathscr{B}$ contains the zero element modulo g, each of $\mathscr{A}'$, $\mathscr{B}'$ contains the zero element modulo g', and hence contains all the residues

$$0,\, g',\, 2g', \ldots,\, (d-1)g' \pmod{g},$$

where $d = g/g'$. Since the zero element has no non-trivial representation of the form $a+b$ (with $a \in \mathscr{A}$, $b \in \mathscr{B}$) modulo g, it follows that for each $r = 1, 2, \ldots, d-1$, at most one of the statements

$$rg' \in \mathscr{A}, \qquad (d-r)g' \in \mathscr{B}$$

can be true. It follows that

$$\mathbf{d}\mathscr{A}' + \mathbf{d}\mathscr{B}' - \mathbf{d}\mathscr{A} - \mathbf{d}\mathscr{B} \geqslant (d-1)\frac{1}{g} = \frac{1}{g'} - \frac{1}{g},$$

and on substituting this in the lower estimate for $\mathbf{d}(\mathscr{A}+\mathscr{B})$ displayed above, we obtain the desired result.

Theorem 16 solves our problem in the case when $\mathfrak{A}$ is degenerate. It is clear, however, that a degenerate system $\mathfrak{A}$ may (and usually will) become non-degenerate when some of the elements are removed from one or more of the $\mathscr{A}_\nu$. Indeed, this will happen when a single element is removed; an operation which clearly has no effect on our problem. It is reasonable, therefore, to include in the degenerate case systems $\mathfrak{A}$ which are not themselves degenerate but corresponding to which there is a worse system which is degenerate. Accordingly, we regard our problem as solved if a system worse than $\mathfrak{A}$ is degenerate. In this case Theorem 16 yields (7.6) for some worse system $\mathfrak{A}'$ (degenerate modulo g'), and we have, of course, that $\underline{\mathbf{d}}\mathscr{C}(\mathfrak{A}) = \mathbf{d}\mathscr{C}(\mathfrak{A}')$, $\underline{\mathbf{d}}\underline{\mathscr{V}}(\mathfrak{A}) \leqslant \mathbf{d}\underline{\mathscr{V}}(\mathfrak{A}')$ for all systems $\mathfrak{A}'$ worse than $\mathfrak{A}$. We may write the resulting inequality in the form

$$\underline{\mathbf{d}}\mathscr{C}(\mathfrak{A}) \geqslant \underline{\mathbf{d}}\underline{\mathscr{V}}(\mathfrak{A}) - \frac{k}{g'} + \{\mathbf{d}\underline{\mathscr{V}}(\mathfrak{A}') - \underline{\mathbf{d}}\underline{\mathscr{V}}(\mathfrak{A})\}, \tag{7.7}$$

where the last term (in brackets) on the right is non-negative. We observe that (7.7) is best possible in the sense that if

$$\mathbf{d}\mathscr{C}(\mathfrak{A}') = \mathbf{d}\underline{\mathscr{V}}(\mathfrak{A}') - k/g'$$

(as was the case in the counter-example given in the previous section), then equality holds in (7.7).

For the remaining systems, those not covered by the preceding discussion, Kneser proved the following remarkable theorem.

THEOREM 17 (Kneser). *If no system worse than $\mathfrak{A}$ (including $\mathfrak{A}$ itself) is degenerate, then*

$$\underline{\mathbf{d}}\mathscr{C}(\mathfrak{A}) \geqslant \underline{\mathbf{d}}\mathscr{V}(\mathfrak{A}). \tag{7.8}$$

If (7.8) is false for a given system $\mathfrak{A}$, there exists, by Theorem 17, a natural number g' and a system $\mathfrak{A}'$ degenerate modulo g', such that $\mathfrak{A}'$ is worse than $\mathfrak{A}$. It then follows from Theorem 16 that there exists a natural number g'' and a system $\mathfrak{A}''$ degenerate modulo g'', such that $\mathfrak{A}''$ is worse than $\mathfrak{A}'$ (hence also worse than $\mathfrak{A}$) and

$$\mathbf{d}\mathscr{C}(\mathfrak{A}'') \geqslant \mathbf{d}\underline{\mathscr{V}}(\mathfrak{A}'')-k/g''.$$

We may therefore combine Theorems 16 and 17 to give:

THEOREM 18 (Kneser). *For any given system $\mathfrak{A}$, either $\underline{\mathbf{d}}\mathscr{C}(\mathfrak{A}) \geqslant \underline{\mathbf{d}}\mathscr{V}(\mathfrak{A})$ or there exists a natural number g' and a system $\mathfrak{A}'$ degenerate modulo g', such that $\mathfrak{A}'$ is worse than $\mathfrak{A}$ and*

$$\mathbf{d}\mathscr{C}(\mathfrak{A}') \geqslant \mathbf{d}\underline{\mathscr{V}}(\mathfrak{A}')-\frac{k}{g'}. \tag{7.9}$$

We shall in fact prove Theorems 16 and 17 by way of the equivalent Theorems 16′ and 17′ below. Before stating these we introduce some further notation and several auxiliary results.

Definition 11. For a given sequence $\mathscr{A}$ and a given natural number g, we denote by $\mathscr{A}^{(g)}$ the sequence of all numbers of the form $a+gz$, $a \in \mathscr{A}$, $z \in \mathscr{Z}$.

Thus $\mathscr{A}^{(g)}$ is the minimal sequence, containing $\mathscr{A}$, which is degenerate modulo g.

LEMMA 9. $\{\mathscr{A}^{(g)}\}^{(h)} = \mathscr{A}^{(d)}$ *where* $d = (g,h)$.

Proof. The sequence $\{\mathscr{A}^{(g)}\}^{(h)}$ consists of all integers of the form $a+gz+hz'$ where z, z' run independently through $\mathscr{Z}$, and the module of integers of the form $gz+hz'$ consists of all multiples of d.

LEMMA 10. *If* $\mathscr{C} = \sum_{\nu=0}^{k} \mathscr{A}_\nu$, *then*

$$\mathscr{C}^{(g)} = \sum_{\nu=0}^{k} \mathscr{A}_\nu^{(g)}.$$

LEMMA 11. *If* $\mathscr{A}^{(g_1)} = \mathscr{A}^{(g_2)} = \ldots = \mathscr{A}^{(g_r)}$, *then each of these sequences is equal to* $A^{(d)}$, *where* $d = (g_1, g_2,\ldots, g_r)$.

Proof. If $r = 2$ the result follows from Lemma 9 since, if $\mathscr{A}^{(g_1)} = \mathscr{A}^{(g_2)}$, $\{\mathscr{A}^{(g_1)}\}^{(g_2)} = \{\mathscr{A}^{(g_2)}\}^{(g_2)} = \mathscr{A}^{(g_2)}$ and similarly $\{\mathscr{A}^{(g_2)}\}^{(g_1)} = \mathscr{A}^{(g_1)}$. We proceed by induction on r. Let $d' = (g_1, g_2,\ldots, g_{r-1})$ so that $d = (d', g_r)$, and assume that each of $\mathscr{A}^{(g_1)}$, $\mathscr{A}^{(g_2)}$,..., $\mathscr{A}^{(g_{r-1})}$ is equal to $\mathscr{A}^{(d')}$. If then

$\mathscr{A}^{(d')} = \mathscr{A}^{(g_r)}$, we again apply the case $r = 2$ obtaining $\mathscr{A}^{(d')} = \mathscr{A}^{(g_r)} = \mathscr{A}^{(d)}$. Hence the result.

Henceforward it will be convenient, when there is no danger of confusion, to write $\mathscr{C}$ for $\mathscr{C}(\mathfrak{A})$, $\underline{\mathscr{V}}$ for $\underline{\mathscr{V}}(\mathfrak{A})$, $\mathscr{C}'$ for $\mathscr{C}(\mathfrak{A}')$, $\underline{\mathscr{V}}'$ for $\underline{\mathscr{V}}(\mathfrak{A}')$, etc.

We shall now state equivalent forms of Theorems 16 and 17.

THEOREM 16′ (Kneser). *If the system $\mathfrak{A}$ is degenerate modulo g, then there exists a divisor g' of g such that $\mathscr{C}^{(g')} = \mathscr{C}$ and*

$$\mathbf{d}\left(\sum_{\nu=0}^{k} \mathscr{A}_\nu^{(g')}\right) \geqslant \mathbf{d}\left(\bigvee_{\nu=0}^{k} \mathscr{A}_\nu^{(g')}\right) - \frac{k}{g'}.$$

THEOREM 17′ (Kneser). *If the system $\mathfrak{A}$ satisfies $\underline{\mathbf{d}}\mathscr{C} < \underline{\mathbf{d}}\underline{\mathscr{V}}$, there exists a natural number g such that $\mathscr{C} \sim \mathscr{C}^{(g)}$.*

It follows from Lemma 10, that Theorems 16′ and 17′ imply Theorems 16 and 17. For in Theorem 16′ the system $\mathscr{A}_0^{(g')}, \mathscr{A}_1^{(g')}, \ldots, \mathscr{A}_k^{(g')}$ is worse than $\mathfrak{A}$; and in Theorem 17′ the system $\mathscr{A}_0^{(g)}, \mathscr{A}_1^{(g)}, \ldots, \mathscr{A}_k^{(g)}$ is worse than $\mathfrak{A}$. Conversely, Theorems 16 and 17 imply Theorems 16′ and 17′. For, there exists, by Theorem 16, a divisor g' of g and a system $\mathfrak{A}'$, degenerate modulo g' and worse than $\mathfrak{A}$, such that (7.6) is true. Here $\mathfrak{A}'$ is also worse than the system $\mathscr{A}_0^{(g')}, \mathscr{A}_1^{(g')}, \ldots, \mathscr{A}_k^{(g')}$ and, in particular, $\mathscr{C} \sim \mathscr{C}^{(g')}$. Moreover, $\mathscr{C} = \mathscr{C}^{(g')}$, for if $\mathscr{C}^{(g')}$ were to contain an element not in $\mathscr{C}$ it would contain an entire congruence class modulo g not in $\mathscr{C}$, contrary to $\mathscr{C}^{(g')} \sim \mathscr{C}$. Also, it follows from (7.6) that

$$\mathbf{d}\mathscr{C}^{(g')} \geqslant \mathbf{d}\left(\bigvee_{\nu=0}^{k} \mathscr{A}_\nu^{(g')}\right) - \frac{k}{g'},$$

completing the deduction of Theorem 16′. Further, there exists, by Theorem 17, a system $\mathfrak{A}'$, identical with or worse than $\mathfrak{A}$, which is degenerate to some modulus g. Then $\mathscr{C} \subset \mathscr{C}^{(g)} \subset \mathscr{C}'$ by Lemma 10, and $\mathscr{C} \sim \mathscr{C}'$. Hence $\mathscr{C} \sim \mathscr{C}^{(g)}$ and Theorem 17′ follows.

The proofs of Theorems 16′ and 17′ are difficult and complicated in structure. Both will be seen later to follow from Theorem 19 below, which embodies in its proof the principal ideas of Kneser's argument.

THEOREM 19. *If the system $\mathfrak{A}$ is such that $0 \in \bigcap_{\nu=0}^{k} \mathscr{A}_\nu$ and $\underline{\mathbf{d}}\mathscr{C} < \underline{\mathbf{d}}\underline{\mathscr{V}}$, there exists a subsequence $\mathscr{E}$, containing 0, of $\mathscr{C}$, and a natural number g, such that*

(7.10) $$\mathscr{E} \sim \mathscr{E}^{(g)}$$

and

(7.11) $$\mathbf{d}\mathscr{E} \geqslant \underline{\mathbf{d}}\underline{\mathscr{V}} - \frac{k}{g}.$$

The proof of Theorem 19 occupies the next two sections. If we were to concentrate on Theorem 16′ alone and deal solely with degenerate systems, the proof of Theorem 19 would become appreciably simpler; the argument then would be more along the lines followed in the proof of Theorem 15 (the Cauchy–Davenport–Chowla theorem).

8. Kneser's theorem (continued)—the τ-transformations

The proof of Theorem 19 requires the systematic use of transformations of the kind we have already met in the proofs of Mann's and Dyson's theorems. We introduce some additional notation for the description of such transformations.

Let $\mathfrak{A}$ be a given system of sequences $\mathscr{A}_0, \mathscr{A}_1, \ldots, \mathscr{A}_k$. Although the final results we shall obtain are symmetrical in $\mathscr{A}_0, \mathscr{A}_1, \ldots, \mathscr{A}_k$, the method of proof will single out $\mathscr{A}_0$ for special treatment. To begin with, let $\mathscr{U} = \mathscr{U}(\mathfrak{A})$ denote the sequence formed by the union $\bigcup_{\nu=1}^{n} \mathscr{A}_\nu$ of all sequences of $\mathfrak{A}$ other than the first.

Next, if a_0 is an element of $\mathscr{A}_0$ and $\mathscr{A}_l$ is one of the sequences $\mathscr{A}_1, \mathscr{A}_2, \ldots, \mathscr{A}_k$, we define the new system $\mathfrak{A}^\tau$, where $\tau = \tau(a_0, l)$, by

(8.1) $$\mathscr{A}_0^\tau = \mathscr{A}_0 \cup (a_0 + \mathscr{A}_l), \qquad \mathscr{A}_l^\tau = \mathscr{A}_l \cap (\mathscr{A}_0 - a_0),$$
$$\mathscr{A}_\nu^\tau = \mathscr{A}_\nu \quad (1 \leqslant \nu \leqslant k, \ \nu \neq l),$$

and write

$$\mathscr{C}^\tau = \mathscr{C}(\mathfrak{A}^\tau) = \sum_{\nu=0}^{k} \mathscr{A}_\nu^\tau, \qquad \mathscr{U}^\tau = \mathscr{U}(\mathfrak{A}^\tau) = \bigcup_{\nu=1}^{k} \mathscr{A}_\nu^\tau,$$

$$\underline{\mathscr{V}}^\tau = \underline{\mathscr{V}}(\mathfrak{A}^\tau) = \bigvee_{\nu=0}^{k} \mathscr{A}_\nu^\tau.$$

We recall, especially by reference to (4.22), that such a transformation played a vital part in the proof of Theorem 9, as well as in the proof of Theorem 3′ (with $\mathscr{A}_0 = \mathscr{A}$, $\mathscr{A}_l = \mathscr{B}$, and $a_0 = a^*$). The transformation τ has the following properties, of which much use will be made:

(8.2) $\mathscr{A}_0 \subset \mathscr{A}_0^\tau, \quad \mathscr{A}_\nu^\tau \subset \mathscr{A}_\nu \quad (\nu = 1, 2, \ldots, k)$ and hence $\mathscr{U}^\tau \subset \mathscr{U}$;

(8.3) $$\mathscr{C}^\tau \subset \mathscr{C};$$

(8.4) $$a_0 + \mathscr{A}_l^\tau \subset \mathscr{A}_0 \subset \mathscr{A}_0^\tau;$$

(8.5) if $0 \in \mathscr{A}_\nu$, then also $0 \in \mathscr{A}_\nu^\tau$ (where ν is any one of $0, 1, \ldots, k$);

(8.6) $$\mathbf{d}\underline{\mathscr{V}}^\tau = \mathbf{d}\underline{\mathscr{V}}.$$

Of these, (8.2), (8.4), and (8.5) are obvious. To verify (8.3), it suffices to show that $\mathscr{A}_0^\tau + \mathscr{A}_l^\tau \subset \mathscr{A}_0 + \mathscr{A}_l$. Now this is certainly true, for the sum $(a_0+a_l)+a_l^\tau$ of an element a_0+a_l of $\mathscr{A}_0^\tau$ and an element a_l^τ of $\mathscr{A}_l^\tau$

may be written in the form $(a_0+a_l^\tau)+a_l$ and therefore belongs to $\mathscr{A}_0+\mathscr{A}_l$ by condition (8.4). To prove (8.6) we observe that the systems $\mathfrak{A}$ and $\mathfrak{A}^\tau$ differ in that a certain subset $\mathscr{A}_l''$ has been removed from $\mathscr{A}_l$, leaving $\mathscr{A}_l^\tau$; and $a_0+\mathscr{A}_l''$ has been included in $\mathscr{A}_0$, thus making up $\mathscr{A}_0^\tau$ (it is helpful here to refer back to the relevant parts of the proofs of Theorems 9 or 3′). Hence the counting numbers of $\underline{\mathscr{V}}$ and $\underline{\mathscr{V}}^\tau$ corresponding to any given interval $[1,x]$ differ at most by the number of integers (of $\mathscr{A}_l''$) lying in an interval of length $|a_0|$; that is, they differ by at most $O(1)$.

The transformation τ depends on the choice of a particular element a_0 of $\mathscr{A}_0$ and a particular positive integer l. This dependence will rarely need to be made explicit but should be borne in mind. Thus, when we write

$$\mathfrak{A}^{\tau_1\tau_2} = (\mathfrak{A}^{\tau_1})^{\tau_2}$$

we shall mean that $\mathfrak{A}^{\tau_1\tau_2}$ is derived from $\mathfrak{A}$ by first applying τ_1 to $\mathfrak{A}$ by means of some element a_0 of $\mathscr{A}_0$ and some sequence $\mathscr{A}_l$ chosen from among $\mathscr{A}_1, \mathscr{A}_2,\ldots, \mathscr{A}_k$; and then by applying τ_2 to $\mathfrak{A}^{\tau_1}$ by means of some element $a_0^{\tau_1}$ of $\mathscr{A}_0^{\tau_1}$ and some sequence $\mathscr{A}_{l'}^{\tau_1}$ from among $\mathscr{A}_1^{\tau_1}, \mathscr{A}_2^{\tau_1},\ldots, \mathscr{A}_k^{\tau_1}$. We reserve the letter T (with or without suffixes) for a (finite) succession of τ-transformations, and write $\mathfrak{A}^T$ for $\mathfrak{A}^{\tau_1\tau_2\ldots\tau_r}$ if $\mathfrak{A}^T$ is the result of applying τ_1 to $\mathfrak{A}$, τ_2 to $\mathfrak{A}^{\tau_1},\ldots, \tau_r$ to $\mathfrak{A}^{\tau_1\tau_2\ldots\tau_{r-1}}$. We shall refer to $\mathfrak{A}^T$ as a system *derived* from $\mathfrak{A}$, or as a *derivation* of $\mathfrak{A}$. Evidently a derivation $(\mathfrak{A}^T)^{T'}$ of $\mathfrak{A}^T$ is at the same time the derivation $\mathfrak{A}^{TT'}$ of $\mathfrak{A}$.

For any derived system $\mathfrak{A}^T$, comprising the sequences $\mathscr{A}_0^T, \mathscr{A}_1^T,\ldots, \mathscr{A}_k^T$, we have

(8.7) $$\mathscr{A}_0 \subset \mathscr{A}_0^T, \qquad \mathscr{U}^T \subset \mathscr{U};$$

(8.8) $$\mathscr{C}^T \subset \mathscr{C};$$

(8.9) if $0 \in \mathscr{A}_\nu$, then also $0 \in \mathscr{A}_\nu^T$ (where ν is any one of 0, 1,..., k);

(8.10) $$\mathbf{d}\underline{\mathscr{V}}^T = \mathbf{d}\underline{\mathscr{V}}.$$

All these are obvious consequences of (8.2), (8.3), (8.5), and (8.6). Also, property (8.4) (in conjunction with (8.2) and (8.7)) leads to the following result.

LEMMA 12. *If $\mathscr{F}$ is a finite subset of $\mathscr{A}_0$ and $0 \in \bigcap_{\nu=1}^{k} \mathscr{A}_\nu$, there exists a derivation $\mathfrak{A}^T$ of $\mathfrak{A}$ such that*

(8.11) $$\mathscr{F}+\mathscr{U}^T \subset \mathscr{A}_0^T.$$

Proof. Suppose that $\mathscr{F} = \{a_0^{(1)}, a_0^{(2)},\ldots, a_0^{(n)}\}$. If τ_1 is the transformation applied to $\mathfrak{A}$ by means of $a_0^{(1)}$ and $\mathscr{A}_1$, $a_0^{(1)}+\mathscr{A}_1^{\tau_1} \subset \mathscr{A}_0^{\tau_1}$ by (8.4). Similarly if $\tau_2 = \tau_2(a_0^{(2)}, 1)$, effected by means of $a_0^{(2)}$ and $\mathscr{A}_1^{T_1}$, is applied to $\mathfrak{A}^{\tau_1}$, $a_0^{(2)} + \mathscr{A}_1^{\tau_1\tau_2} \subset \mathscr{A}_0^{\tau_1\tau_2}$. By (8.2), $a_0^{(1)} + \mathscr{A}_1^{\tau_1\tau_2} \subset a_0^{(1)} + \mathscr{A}_1^{\tau_1} \subset \mathscr{A}_0^{\tau_1} \subset \mathscr{A}_0^{\tau_1\tau_2}$, and hence

$$\{a_0^{(1)}, a_0^{(2)}\} + \mathscr{A}_1^{\tau_1\tau_2} \subset \mathscr{A}_0^{\tau_1\tau_2}.$$

At the rth stage of this process, we show in the same way that if $\tau_r = \tau_r(a_0^{(r)}, 1)$ is applied to $\mathfrak{A}^{\tau_1\tau_2\dots\tau_{r-1}}$, then

$$a_0^{(r)} + \mathscr{A}_1^{\tau_1\tau_2\dots\tau_r} \quad \text{and} \quad \{a_0^{(1)}, a_0^{(2)}, \dots, a_0^{(r-1)}\} + \mathscr{A}_1^{\tau_1\tau_2\dots\tau_r}$$

each lie in $\mathscr{A}_0^{\tau_1\tau_2\dots\tau_r}$, and hence

$$\{a_0^{(1)}, a_0^{(2)}, \dots, a_0^{(r)}\} + \mathscr{A}_1^{\tau_1\tau_2\dots\tau_r} \subset \mathscr{A}_0^{\tau_1\tau_2\dots\tau_r}.$$

Writing T_1 for the succession $\tau_1 \tau_2 \dots \tau_n$, we thus arrive after n stages at a derivation $\mathfrak{A}^{T_1}$ of $\mathfrak{A}$ for which

$$\mathscr{F} + \mathscr{A}_1^{T_1} \subset \mathscr{A}_0^{T_1}.$$

Next, (8.7) permits us to regard $\mathscr{F}$ as a subset of $\mathscr{A}_0^{T_1}$. We therefore apply the above procedure to $\mathfrak{A}^{T_1}$ instead of $\mathfrak{A}$, this time with each τ-transformation characterized by $l = 2$ (instead of $l = 1$), and arrive at a derivation $(\mathfrak{A}^{T_1})^{T_2} = \mathfrak{A}^{T_1T_2}$ for which

$$\mathscr{F} + \mathscr{A}_2^{T_1T_2} \subset \mathscr{A}_0^{T_1T_2};$$

but $\mathscr{F} + \mathscr{A}_1^{T_1T_2} \subset \mathscr{F} + \mathscr{A}_1^{T_1} \subset \mathscr{A}_0^{T_1} \subset \mathscr{A}_0^{T_1T_2}$ by successive applications of (8.2), and therefore

$$\mathscr{F} + (\mathscr{A}_1^{T_1T_2} \cup \mathscr{A}_2^{T_1T_2}) \subset \mathscr{A}_0^{T_1T_2}.$$

The procedure for arriving at (8.11) is now clear.

It is clear from (8.8), (8.9), and (8.10) that *every* derivation of $\mathfrak{A}$ satisfies the hypotheses of Theorem 19, and that if the conclusions of Theorem 19 hold for *some* derivation of $\mathfrak{A}$, they hold also for the system $\mathfrak{A}$. Thus:

LEMMA 13. *The statement constituting Theorem* 19 *is true if the corresponding statement, with* $\mathfrak{A}$ *replaced by some derivation of* $\mathfrak{A}$, *is true.*

We shall make much use of this circumstance, the course we shall follow being to find, among the class of derivations of $\mathfrak{A}$, one for which the conclusions of Theorem 19 are relatively simple to establish. Indeed, we shall show in this fashion that Theorem 19 can be made to depend on the following Theorem 20, whose proof below also provides an illustration of the value of τ-transformations.

THEOREM 20. *If each sequence of* $\mathfrak{A}$ *contains* 0, *if* $\mathscr{A}_0$ *contains m consecutive integers, and if*

$$\underline{\mathrm{d}}\mathscr{C} < \frac{m}{m+k}\,\underline{\mathrm{d}}\mathscr{V},$$

then $\mathscr{C} \sim \mathscr{Z}$.

Proof. Suppose that $\mathscr{A}_0$ contains the m consecutive integers $a_0, a_0+1, a_0+2, \ldots, a_0+m-1$. We define the system $\mathfrak{A}'$ by writing $\mathscr{A}_0' = \mathscr{A}_0 - a_0$ and $\mathscr{A}_\nu' = \mathscr{A}_\nu$ ($\nu = 1, 2, \ldots, k$). Then each sequence of $\mathfrak{A}'$ contains 0 and, by Lemma 8, $\underline{\mathbf{d}}\mathscr{C}' = \underline{\mathbf{d}}\mathscr{C}$, $\underline{\mathbf{d}}\mathscr{V}' = \underline{\mathbf{d}}\mathscr{V}$. Hence the result is invariant under the translation of $\mathscr{A}_0$ considered above, and we may suppose without loss of generality that $\{0, 1, 2, \ldots, m-1\} \subset \mathscr{A}_0$.

We may suppose even that

$$\{0, 1, 2, \ldots, m-1\} + \mathscr{U} \subset \mathscr{A}_0. \tag{8.12}$$

For by Lemma 12, (8.12) is certainly true for some derivation of $\mathfrak{A}$, and the hypotheses of the theorem hold for every derivation of $\mathfrak{A}$; and if we can prove, for some T, that $\mathscr{C}^T \sim \mathscr{Z}$, it follows from (8.8) that also $\mathscr{C} \sim \mathscr{Z}$.

Let γ be a positive number satisfying

$$\gamma < \underline{\mathbf{d}}\mathscr{V} \tag{8.13}$$

and write

$$\gamma = \frac{m+k}{m} \quad \text{if} \quad \frac{m}{m+k}\underline{\mathbf{d}}\mathscr{V} > 1. \tag{8.14}$$

We have then, in any case,

$$\frac{m}{m+k}\gamma \leqslant 1. \tag{8.15}$$

Inequality (8.13) implies the existence of a positive integer $x_0 = x_0(\gamma)$ such that

$$\sum_{\nu=0}^{k} A_\nu(x) \geqslant \gamma x \qquad (x \geqslant x_0), \tag{8.16}$$

and we choose x_0 to be the least positive integer for which (8.16) is true. Our choice of x_0 implies that

$$\sum_{\nu=0}^{k} A_\nu(x_0-1) \leqslant \gamma(x_0-1),$$

and if we subtract this inequality from (8.16) and replace x by $x+x_0$, we arrive at

$$\sum_{\nu=0}^{k} A_\nu(x_0, x_0+x) \geqslant \gamma(x+1) \quad (x = 0, 1, 2, \ldots); \tag{8.17}$$

in other words, $[x_0, x_0+x]$ is $\mathscr{V}$-good with respect to γ for $x = 0, 1, 2, \ldots$. (See Definition 6 in § 4.)

On putting $x = 0$ (in 8.17) we see that $x_0 \in \mathscr{V}$; thus $x_0 \in \mathscr{A}_0$ or $x_0 \in \mathscr{U}$. But, by (8.12), $\mathscr{U} \subset \mathscr{A}_0$ and hence $x_0 \in \mathscr{A}_0$. For each $\nu = 1, 2, \ldots, k$, let x_ν be the least element of $\mathscr{A}_\nu$ not less than x_0, and suppose without loss of generality that the sequences of $\mathfrak{A}$ are numbered in such a way that $x_1 \leqslant x_2 \leqslant ... \leqslant x_k$. Define the new system $\mathfrak{A}'$ of sequences

$$\mathscr{A}'_\nu = (\mathscr{A}_\nu - x_\nu) \cap \mathscr{Z}_0 \quad (\nu = 0, 1, 2,..., k)$$

so that each new sequence $\mathscr{A}'_\nu$ contains 0 and consists otherwise of natural numbers. We shall prove that

(8.18) $[0, x]$ is $\underline{\mathscr{V}}'$-good with respect to $\dfrac{m}{m+k}\gamma$ for $x = 0, 1, 2,....$

By the case $\theta = 0$ of Theorem 9 it will then follow that

(8.19) $[0, x]$ is $\mathscr{C}'$-good with respect to $\dfrac{m}{m+k}\gamma$ for $x = 0, 1, 2,...,$

and the theorem will be seen to be an almost immediate consequence of (8.19).

By the definition of x_ν the interval $[x_0, x_\nu]$ contains exactly one element of $\mathscr{A}_\nu$, so that

$$A'_\nu(x) = A_\nu(x_\nu+1, x_\nu+x) = A_\nu(x_0, x_\nu+x)-1 \geqslant A_\nu(x_0, x_0+x)-1.$$

It follows from (8.17) that

$$1+\sum_{\nu=0}^{k} A'_\nu(x) \geqslant \gamma(x+1)-k \geqslant \frac{m}{m+k}\gamma(x+1)$$

if $x \geqslant (m+k)/\gamma-1$. This is equivalent to saying that $[0, x]$ is $\underline{\mathscr{V}}'$-good with respect to $\{m/(m+k)\}\gamma$ provided that $x \geqslant (m+k)/\gamma-1$; to complete the proof of (8.18) it suffices to show that this statement is true also if $x < (m+k)/\gamma-1$.

If $0 \leqslant x < x_1-x_0$, $[x_0, x_0+x]$ contains no element of $\mathscr{U}$ so that, by (8.17), $[0, x]$ is even $\mathscr{A}'_0$-good with respect to γ.

For the remaining values of x we use the fact that, by (8.12), $x_1+\{0, 1, 2,..., m-1\} \subset \mathscr{A}_0$. First, if $x_1-x_0 \leqslant x < x_1-x_0+m$, we write† $[0, x] = [0, x_1-x_0-1]+[x_1-x_0, x]$; the first of these intervals has just been shown to be $\mathscr{A}'_0$-good with respect to γ, and the second is entirely contained in $\mathscr{A}'_0$ and therefore certainly $\mathscr{A}'_0$-good with respect to $\{m/(m+k)\}\gamma$. Thus the desired result follows again because the property of being $\mathscr{A}'_0$-good is additive. Next, suppose that‡

$$x_1-x_0+m \leqslant x < (m+k)/\gamma-1;$$

here it suffices to observe that $[0, x]$ contains m elements of $\mathscr{A}'_0$ and that

$$m > \frac{m}{m+k}\gamma(x+1).$$

† Here, as on earlier occasions, we understand by $[a, b]$ the set of all integers in the closed interval $[a, b]$.

‡ If this interval is empty, the proof of (8.18) is complete.

It remains to show that (8.19) leads to the conclusion of Theorem 20. We may suppose that $\{m/(m+k)\}\underline{\mathbf{d}}\mathscr{V} > 1$. For if $\{m/(m+k)\}\underline{\mathbf{d}}\mathscr{V} \leqslant 1$, (8.19) implies that $\underline{\mathbf{d}}\mathscr{C} = \underline{\mathbf{d}}\mathscr{C}' \geqslant \{m/(m+k)\}\gamma$ for any γ less than $\underline{\mathbf{d}}\mathscr{V}$, and therefore

$$\underline{\mathbf{d}}\mathscr{C} \geqslant \frac{m}{m+k}\underline{\mathbf{d}}\mathscr{V},$$

contrary to hypothesis.

Suppose then that $\{m/(m+k)\}\underline{\mathbf{d}}\mathscr{V} > 1$. By (8.14), $\{m/(m+k)\}\gamma = 1$, and (8.19) now tells us that, for each $x = 0, 1, 2,\ldots$ all the integers of $[0, x]$ lie in $\mathscr{C}'$; in other words, $\mathscr{C}' \sim \mathscr{Z}_0$. But

$$\mathscr{C} = \sum_{\nu=0}^{k} \mathscr{A}_\nu \supset \sum_{\nu=0}^{k} (\mathscr{A}'_\nu + x_\nu) = \mathscr{C}' + \sum_{\nu=0}^{k} x_\nu,$$

so that $\mathscr{C} \sim \mathscr{Z}$.

Corollary. *Let the system* $\mathfrak{A}$ *be such that* $0 \in \bigcap_{\nu=0}^{k} \mathscr{A}_\nu$ *and*

$$\underline{\mathbf{d}}\mathscr{C} < \underline{\mathbf{d}}\mathscr{V}.$$

Then if there exist an infinite sequence of natural numbers $m_1, m_2,\ldots$ *such that* (i) $m_i \to \infty$ *as* $i \to \infty$, *and a sequence of corresponding derivations* $\mathfrak{A}^{T_1}, \mathfrak{A}^{T_2},\ldots$ *of* $\mathfrak{A}$ *such that* (ii) $\mathscr{A}_0^{T_i}$ *contains* m_i *consecutive integers* $(i = 1, 2,\ldots)$, *it follows that* $\mathscr{C} \sim \mathscr{Z}$.

Proof. Since $m_i/(m_i+k) \to 1-0$ as $i \to \infty$ and $\underline{\mathbf{d}}\mathscr{C} < \underline{\mathbf{d}}\mathscr{V}$, there exists an integer r such that

$$\underline{\mathbf{d}}\mathscr{C} < \frac{m_r}{m_r+k}\underline{\mathbf{d}}\mathscr{V},$$

and the result follows from (8.8), (8.10) and Theorem 20.

9. Kneser's theorem (continued): proof of Theorem 19—sequence functions associated with the derivations of a system

We remarked earlier that our aim was to discover a derivation of $\mathfrak{A}$ for which Theorem 19 would, in some sense, reduce to Theorem 20. In this connexion it will prove useful to introduce the following two sequence functions.

Definition 12. A given sequence $\mathscr{S}$ determines uniquely the set of positive differences of pairs of elements of $\mathscr{S}$. We define $f(\mathscr{S})$ to be the least element† of this set of positive differences. Symbolically,

$$f(\mathscr{S}) = \min_{\substack{s,s' \in \mathscr{S} \\ s \neq s'}} |s-s'|.$$

† In all applications $\mathscr{S}$ will contain the element 0. If $\mathscr{S}$ contains no other element, we interpret $f(\mathscr{S})$ and $g(\mathscr{S})$ (see Definition 13 below) as being infinite. The motivation for this interpretation will become apparent later.

Definition 13. For a given sequence $\mathscr{S}$, let $g(\mathscr{S})$ denote the highest common factor of the elements of $\mathscr{S}$.

Then $g(\mathscr{S})$ is clearly a divisor of $f(\mathscr{S})$ and, in particular,

$$g(\mathscr{S}) \leqslant f(\mathscr{S}). \tag{9.1}$$

The functions $f(\mathscr{U}^T)$, $g(\mathscr{U}^T)$ corresponding to the derivations $\mathfrak{A}^T$ of $\mathfrak{A}$ play a vital part in the subsequent argument.

LEMMA 14. *If each sequence of the system* $\mathfrak{A}$ *contains* 0, *then for every derivation* $\mathfrak{A}^T$,

$$\underline{\mathbf{d}}\mathscr{C} \geqslant \underline{\mathbf{d}}\mathscr{V} - \frac{k}{f(\mathscr{U}^T)} \tag{9.2}$$

and

$$\underline{\mathbf{d}}\mathscr{C}^T \geqslant \underline{\mathbf{d}}\mathscr{V}^T - \frac{k}{g(\mathscr{U}^T)}. \tag{9.3}$$

Proof. For any derivation $\mathfrak{A}^T$, $0 \in \mathscr{U}^T$ by (8.9). Put $u_0 = 0$ and let u_r be the rth element of $\mathscr{U}^T$, with the ordering according to magnitude. Then $u_r = \sum_{n=1}^{r} (u_n - u_{n-1}) \geqslant rf(\mathscr{U}^T)$ and, taking u_r to be the largest element of $\mathscr{U}^T$ not exceeding x, we obtain for each $\nu = 1, 2, \ldots, k$,

$$A_\nu^T(x) \leqslant U^T(x) \leqslant \frac{x}{f(\mathscr{U}^T)}.$$

Hence

$$\underline{\mathbf{d}}\mathscr{C}^T \geqslant \underline{\mathbf{d}}\mathscr{A}_0^T \geqslant \liminf_{x\to\infty} \frac{1}{x}\left\{A_0^T(x) + A_1^T(x) + \ldots + A_k^T(x) - \frac{kx}{f(\mathscr{U}^T)}\right\}$$
$$= \underline{\mathbf{d}}\mathscr{V}^T - \frac{k}{f(\mathscr{U}^T)},$$

so that (9.2) follows from (8.8) and (8.10), and (9.3) from (9.1).

If the set of numbers $f(\mathscr{U}^T)$ corresponding to all possible derivations $\mathfrak{A}^T$ is unbounded, (9.2) implies that $\underline{\mathbf{d}}\mathscr{C} \geqslant \underline{\mathbf{d}}\mathscr{V}$. But this is not consistent with the hypotheses of Theorem 19, so that we may assume from now on that†:

$$\begin{cases} \text{(I) the set } \{f(\mathscr{U}^T)\} \text{ is bounded,} \\ \text{(II) the set } \{g(\mathscr{U}^T)\} \text{ is bounded,} \end{cases} \tag{9.4}$$

(where (II) follows from (I) and (9.1)). Let

$$g = \max_{\mathfrak{A}^T} g(\mathscr{U}^T). \tag{9.5}$$

We shall prove Theorem 19 with g as given by (9.5) and with $\mathscr{E} = \mathscr{C}^T$,

† If $\mathscr{U}^T$ contains no non-zero element, we again have $\underline{\mathbf{d}}\mathscr{C} \geqslant \underline{\mathbf{d}}\mathscr{V}$ contrary to hypothesis; we may therefore exclude this possibility, and suppose that $f(\mathscr{U}^T)$, $g(\mathscr{U}^T)$ are finite. (See footnote to Definition 12.)

where $\mathfrak{A}^T$ is a certain derivation of $\mathfrak{A}$ for which $g(\mathscr{U}^T) = g$. By (8.8), (8.9), (8.10) and Lemma 14, the following statement implies Theorem 19.

If each sequence of $\mathfrak{A}$ *contains* 0, *if* $\underline{\mathbf{d}}\mathscr{C} < \underline{\mathbf{d}}\mathscr{V}$ *and if the derivations of* $\mathfrak{A}$ *have the property* (II), *there exists a derivation* $\mathfrak{A}^T$ *with* $g(\mathscr{U}^T) = g$ *such that* $\mathscr{C}^T \sim (\mathscr{C}^T)^{(g)}$.

Actually, we shall express this in a still more convenient form. For this purpose, we require the lemma below.

Lemma 15. *Under successive* T*-transformations,* $f(\mathscr{U})$ *and* $g(\mathscr{U})$ *are increasing functions.*

Proof. If $\mathfrak{A}^T$ is a derivation of $\mathfrak{A}$, we have, by (8.7), that $\mathscr{U}^T \subset \mathscr{U}$, and the lemma follows from the Definitions 12 and 13.

Let $\mathfrak{A}^T$ be any derivation with maximal $g(\mathscr{U}^T)$, that is, with $g(\mathscr{U}^T) = g$. It follows from Lemma 15 that if $(\mathfrak{A}^T)^{T'}$ is a derivation of $\mathfrak{A}^T$, and hence also of $\mathfrak{A}$, $g(\mathscr{U}^{TT'}) = g(\mathscr{U}^T) = g$. In other words, the g-functions associated with all systems derived from $\mathfrak{A}^T$ have the same value g. We conclude, by analogy to Lemma 13, that to prove Theorem 19 it suffices to prove the following theorem.

Theorem 21. *If each sequence of* $\mathfrak{A}$ *contains* 0, *if* $\underline{\mathbf{d}}\mathscr{C} < \underline{\mathbf{d}}\mathscr{V}$ *and if* $g(\mathscr{U}^T) = g(\mathscr{U}) = g$ *for every derivation* $\mathfrak{A}^T$, *then* $\mathscr{C} \sim \mathscr{C}^{(g)}$.

As a first step in the proof of Theorem 21 we shall show, by means of a simple mapping argument, that Theorem 21 is equivalent to the following special case, expressed in Theorem 22 below.

Theorem 22. *Let* $\mathfrak{A}$ *be a system with the following properties*: (i) *every sequence of* $\mathfrak{A}$ *contains* 0; (ii) $\underline{\mathbf{d}}\mathscr{C} < \underline{\mathbf{d}}\mathscr{V}$; (iii) $g(\mathscr{U}^T) = g(\mathscr{U}) = g$ *for every derivation* $\mathfrak{A}^T$; *and* (iv) $\mathscr{A}_0$ *contains* g *consecutive integers. Then* $\mathscr{C} \sim \mathscr{Z}$.

The difference between Theorems 21 and 22 is that in the latter the system $\mathfrak{A}$ possesses the additional property (iv); we note that if $\mathscr{A}_0$ possesses g consecutive integers, $\mathscr{C}^{(g)} = \mathscr{Z}$.

Proof that Theorems 21 *and* 22 *are equivalent*

Let n be a natural number and let $s_0, s_1, \ldots, s_j$ be integers, incongruent modulo n. We define the sequence

$$\mathscr{T}^{(n)}(s_0, s_1, \ldots, s_j)$$

to consist of all integers of the form

$$s_\mu + tn, \tag{9.6}$$

where μ runs through the values $\mu = 0, 1, \ldots, j$ and t runs through $\mathscr{Z}$.

Suppose now that

$$\mathscr{T}^{(n)}(r_0, r_1, \ldots, r_{h-1}), \qquad \mathscr{T}^{(n^*)}(r_0^*, r_1^*, \ldots, r_{h-1}^*)$$

are both defined (so that the r and the r^* are incongruent modulo n and n^* respectively) and that

(9.7) $$r_0 = r_0^* = 0.$$

We define the fundamental mapping Ω of $\mathscr{T}^{(n)}(r_0, r_1,..., r_{h-1})$ onto $\mathscr{T}^{(n^*)}(r_0^*, r_1^*,..., r_{h-1}^*)$ by

(9.8) $$\Omega(r_\mu+tn) = r_\mu^*+tn^*.$$

In particular, $\Omega r_\mu = r_\mu^*$ and (in view of (9.7)) $\Omega n = n^*$, so that we may adopt, without inconsistency, the notation

$$\Omega x = x^*.$$

Moreover, if x runs through the elements of a subsequence $\mathscr{X}$ of $\mathscr{T}^{(n)}(r_0, r_1,..., r_{h-1})$, we denote by $\mathscr{X}^*$ the sequence of corresponding elements x^*.

It is important to note that the inverse mapping Ω^{-1} is the fundamental map of $\mathscr{T}^{(n^*)}(r_0^*, r_1^*,..., r_{h-1}^*)$ onto $\mathscr{T}^{(n)}(r_0, r_1,..., r_{h-1})$.

The fundamental mapping Ω has the following obvious properties:

(9.9) $$\frac{x}{n} = \frac{x^*}{n^*}+O(1) \quad \text{if } x \text{ is the only variable;}$$

(9.10) if $\mathscr{X} \subset \mathscr{T}^{(n)}(0)$, then $n^{-1}g(\mathscr{X}) = (n^*)^{-1}g(\mathscr{X}^*)$;

(9.11) if a^* is defined and $x \in \mathscr{T}^{(n)}(0)$,

$$(a+x)^* = a^*+x^*;$$

(9.12) if a and x both lie in $\mathscr{T}^{(n)}(r_\mu)$,

$$(x-a)^* = x^*-a^*.$$

Suppose now that a system $\mathfrak{A}$ of sequences $\mathscr{A}_0, \mathscr{A}_1,..., \mathscr{A}_k$ satisfies

(9.13) $$\mathscr{A}_0 \subset \mathscr{T}^{(n)}(r_0, r_1,..., r_{h-1}), \qquad \mathscr{U} = \mathscr{U}(\mathfrak{A}) \subset \mathscr{T}^{(n)}(0).$$

We consider the system $\mathfrak{A}^\tau$ arising from the transformation $\tau = \tau(a_0, l)$, defined by (8.1). As we shall see, we may infer from (9.11) and (9.12) that the result $(\mathfrak{A}^\tau)^*$ of applying the mapping Ω to $\mathfrak{A}^\tau$ is simply the system $(\mathfrak{A}^*)^{\tau^*}$ which arises on applying the transformation $\tau^* = \tau(a_0^*, l)$ to $\mathfrak{A}^*$. For by (9.11)

$$(a_0+\mathscr{A}_l)^* = a_0^*+\mathscr{A}_l^*;$$

and by (9.12)

$$\{\mathscr{A}_l \cap (\mathscr{A}_0-a_0)\}^* = \mathscr{A}_l^* \cap (\mathscr{A}_0^*-a_0^*),$$

since every element x of $\mathscr{A}_0$ satisfying $x-a_0 \in \mathscr{A}_l \subset \mathscr{T}^{(n)}(0)$ must satisfy $x \in \mathscr{T}^{(n)}(r_\mu)$ where μ is defined by $a_0 \in \mathscr{T}^{(n)}(r_\mu)$.

It follows that Ω maps every derived system $\mathfrak{A}^T$ of $\mathfrak{A}$ onto a derivation $(\mathfrak{A}^*)^{T^*}$ of $\mathfrak{A}^*$. Further, since $\sum_{\nu=1}^{k} \mathscr{A}_\nu \subset \mathscr{T}^{(n)}(0)$, we have by (9.11) that

$$\mathscr{C}^* = \Bigl(\mathscr{A}_0 + \sum_{\nu=1}^{k} \mathscr{A}_\nu\Bigr)^* = \mathscr{A}_0^* + \Bigl(\sum_{\nu=1}^{k} \mathscr{A}_\nu\Bigr)^* = \sum_{\nu=0}^{k} \mathscr{A}_\nu^*,$$

and it is obvious that

$$\underline{\mathscr{V}}^* = \bigvee_{\nu=0}^{k} \mathscr{A}_\nu^*, \qquad \mathscr{U}^* = \bigcup_{\nu=1}^{k} \mathscr{A}_\nu^*.$$

It follows from (9.9) and (9.10) that, for given $\mathfrak{A}$, $\mathfrak{A}^T$, each of the quantities

$$n\,\underline{\underline{\mathbf{d}}}\mathscr{C}, \quad n\,\underline{\underline{\mathbf{d}}}\underline{\mathscr{V}}, \quad \frac{1}{n} g(\mathscr{U}^T)$$

is invariant under Ω.

Now suppose that a given system $\mathfrak{A}$ satisfies the premisses of Theorem 21. Let h be the maximal number of elements that can be chosen from $\mathscr{A}_0$ so as to be incongruent modulo g. It is then possible to choose elements $r_0, r_1,\ldots, r_{h-1}$ of $\mathscr{A}_0$ so that

$$\mathscr{A}_0 \subset \mathscr{T}^{(g)}(r_0, r_1,\ldots, r_{h-1})$$

(actually $\mathscr{T}^{(g)}(r_0,\ldots, r_{h-1}) = \mathscr{A}_0^{(g)}$) and, since $0 \in \mathscr{A}_0$, we may take $r_0 = 0$. Furthermore, the premisses of Theorem 21 ensure that $\mathscr{U}^T \subset \mathscr{T}^{(g)}(0)$ always.

We now take Ω to be the fundamental mapping of $\mathscr{T}^{(g)}(r_0, r_1,\ldots, r_{h-1})$ onto $\mathscr{T}^{(h)}(0, 1,\ldots, h-1)$. The sequence $\Omega\mathscr{A}_0$ contains the h consecutive integers $0, 1,\ldots, h-1$ and hence $\Omega\mathfrak{A}$ satisfies the premisses of Theorem 22 with h in place of g. If then the conclusion of Theorem 22 holds for $\Omega\mathfrak{A}$, Theorem 21 for the system $\mathfrak{A}$ follows on applying the fundamental mapping Ω^{-1}.

Finally we shall deduce Theorem 22 from the corollary to Theorem 20.

Proof of Theorem 22. We shall require the following lemma.

LEMMA 16. *Let the system $\mathfrak{A}$ satisfy conditions* (i) *and* (iii) *of Theorem* 22 *and condition* (I) *of* (9.4). *Then, if $\mathscr{F}$ is a finite subset of $\mathscr{A}_0$, there exists an integer y and a derivation $\mathfrak{A}^T$ of $\mathfrak{A}$ such that*

$$(y+\mathscr{F}) \cup (y+g+\mathscr{F}) \subset \mathscr{A}_0^T. \tag{9.14}$$

Before proving this lemma we show how Theorem 22 follows from it. Let $\mathscr{S}$ denote the set $\{a_0, a_0+1,\ldots, a_0+g-1\}$ of consecutive integers of $\mathscr{A}_0$. Applying the lemma with $\mathscr{F} = \mathscr{S}$, (9.14) states that $\mathscr{A}_0^T$ contains the set

$$\mathscr{S}^T = \{a_0+y, a_0+y+1,\ldots, a_0+y+g-1, a_0+y+g,\ldots, a_0+y+2g-1\}$$

of $2g$ consecutive integers; the same result, when applied to $\mathfrak{A}^T$ and $\mathscr{S}^T$ in place of $\mathfrak{A}$ and $\mathscr{S}$, then implies the existence of a further derivation $\mathfrak{A}^{TT_1}$ for which $\mathscr{A}_0^{TT_1}$ contains $3g$ consecutive integers, and so on. Hence the hypotheses of the corollary to Theorem 20 hold with $m_i = ig$ $(i = 1, 2, 3,...)$, and Theorem 22 follows at once from this corollary.

Proof of Lemma 16. We recall that all the derivations of $\mathfrak{A}$ possess property I of (9.4), and we denote by f the greatest value taken by $f(\mathscr{U}^T)$. We shall narrow our attention to the class of all those derivations $\mathfrak{A}^T$ of $\mathfrak{A}$ for which $f(\mathscr{U}^T) = f$. It follows at once from the maximal property of f and from Lemma 15 that this class is closed with respect to the operation of derivation.

We select from this same class of derivations one system, call it $\mathfrak{A}^{T_1}$, for which the number of residue classes modulo f with representatives in $\mathscr{U}^{T_1}$ is minimal; let $\rho_1, \rho_2,..., \rho_r \pmod{f}$ be these residue classes. Then every derivation $(\mathfrak{A}^{T_1})^T = \mathfrak{A}^{T_1T}$ has, by (8.7), the property that $\mathscr{U}^{T_1T}$ is distributed among these (and no fewer) residue classes modulo f. We have thus arrived at a subclass Σ of derivations of $\mathfrak{A}$ such that, if $\mathfrak{A}^T \in \Sigma$,

(*a*) $g(\mathscr{U}^T) = g(\mathscr{U}) = g$;

(*b*) $f(\mathscr{U}^T) = f$ and, in particular, $\mathscr{U}^T$ contains two elements whose difference is f;

(*c*) $\mathscr{U}^T$ is distributed among the residue classes $\rho_i \pmod{f}$ $(i = 1,2,...,r)$, and each class $\rho_i \pmod{f}$ contains at least one element of $\mathscr{U}^T$;

(*d*) Σ is closed under derivation.

By (*a*), g can be expressed as a linear combination of elements of $\mathscr{U}^T$, and hence, by (*c*), there exist integers $n_1, n_2,..., n_r$ such that $g \equiv \sum_{i=1}^{r} n_i \rho_i$ $\pmod{f}$; we may assume that the n_i are non-negative. It will be more convenient to write this congruence in the form

$$g \equiv \sum_{j=1}^{s} \sigma_j \pmod{f} \tag{9.15}$$

where $s = n_1+n_2+...+n_r$ and the set $\sigma_1, \sigma_2,..., \sigma_s$ consists of the numbers $\rho_1, \rho_2,..., \rho_r$, each ρ_i appearing with multiplicity n_i $(i = 1, 2,..., r)$.

We are now in a position to prove (9.14). Let $\mathfrak{A}^T$ be any member of Σ. Since $\mathscr{F} \subset \mathscr{A}_0$, $\mathscr{F} \subset \mathscr{A}_0^T$. By Lemma 12 (applied to $\mathfrak{A}^T$) and (*d*), there exists a derivation $\mathfrak{A}^{T_1}$ such that $\mathscr{F}+\mathscr{U}^{T_1} \subset \mathscr{A}_0^{T_1}$ and $\mathfrak{A}^{T_1} \in \Sigma$. Now, by (*c*), $\mathscr{U}^{T_1}$ contains an element $x_1 \equiv \sigma_1 \pmod{f}$ so that, in particular, $\mathscr{F}+x_1 \subset \mathscr{A}_0^{T_1}$. Applying the same procedure with $\mathscr{F}_1 = x_1+\mathscr{F}$ and $\mathfrak{A}^{T_1}$ in place of $\mathscr{F}$ and $\mathfrak{A}^T$, we are led to a derivation $\mathfrak{A}^{T_2}$, belonging to Σ, such that

$$(x_1+x_2)+\mathscr{F} = x_2+\mathscr{F}_1 \subset \mathscr{A}_0^{T_2}, \qquad x_2 \equiv \sigma_2 \,(\mathrm{mod}\, f).$$

Continuing in this way we arrive at a derivation $\mathfrak{A}^{T_s}$ in Σ such that

$$(x_1+x_2+...+x_s)+\mathscr{F} \subset \mathscr{A}_0^{T_s}, \qquad x_j \equiv \sigma_j \,(\mathrm{mod}\, f) \quad (j = 1,2,..., s).$$

But by (9.15), $x_1+x_2+...+x_s = g+nf$ for some integral n, and we know that $\mathscr{F} \subset \mathscr{A}_0 \subset \mathscr{A}_0^{T_s}$; hence

$$\mathscr{F} \cup (\mathscr{F}+g+nf) \subset \mathscr{A}_0^{T_s}. \tag{9.16}$$

If $n = 0$, (9.16) implies (9.14) with $y = 0$. If $n \neq 0$, we proceed as follows:

We apply Lemma 12 to $\mathfrak{A}^{T_s}$ and obtain a derivation $\mathfrak{A}^{T'}$ such that

$$\{\mathscr{F} \cup (\mathscr{F}+g+nf)\}+\mathscr{U}^{T'} \subset \mathscr{A}_0^{T'}.$$

By (d), $\mathfrak{A}^{T'} \in \Sigma$ and hence, by (b), $\mathscr{U}^{T'}$ contains two elements whose difference is f, say u' and $u'+f$. It follows that

$$(\mathscr{F}+u') \cup (\mathscr{F}+u'+g+nf) \subset \mathscr{A}_0^{T'} \tag{9.17}$$

and

$$(\mathscr{F}+u'+f) \cup (\mathscr{F}+u'+g+(n+1)f) \subset \mathscr{A}_0^{T'}; \tag{9.18}$$

if $n < 0$, we put $\mathscr{F}' = \mathscr{F}+u'$, so that (using both (9.17) and (9.18))

$$\mathscr{F}' \cup (\mathscr{F}'+g+(n+1)f) \subset \mathscr{A}_0^{T'},$$

and if $n > 0$ we put $\mathscr{F}' = \mathscr{F}+u'+f$ so that (again using both (9.17) and (9.18))

$$\mathscr{F}' \cup (\mathscr{F}'+g+(n-1)f) \subset \mathscr{A}_0^{T'}.$$

In either case we have found a relation of type (9.16) with $\mathscr{F}'$ in place of $\mathscr{F}$, T' in place of T_s, and the coefficient of f diminished by 1 in absolute value. If then we repeat this procedure $|n|$ times we must arrive at a system $\mathfrak{A}^{T''}$ in Σ, a set $\mathscr{F}''$ and an integer y such that $\mathscr{F}'' = \mathscr{F}+y$ and

$$\mathscr{F}'' \cup (\mathscr{F}''+g) \subset \mathscr{A}_0^{T''}.$$

This completes the proof of (9.14), the lemma, and of Theorem 22.

To sum up, we have proved above that Theorem 22 implies Theorem 21 and earlier that Theorem 21 implies Theorem 19. Hence the proof of Theorem 19 is now complete.

10. Kneser's theorem (continued): proofs of Theorems 16′ and 17′

The hypothesis

$$0 \in \bigcap_{\nu=0}^{k} \mathscr{A}_\nu$$

of Theorem 19 played a vital part in the effective use of τ-transformations. However, we shall show now that Theorem 19 has an immediate and powerful application to systems for which this hypothesis is not necessarily valid.

THEOREM 19′. *If the system* $\mathfrak{A}$ *is such that* $\underline{\mathbf{d}}\mathscr{C} < \underline{\mathbf{d}}\mathscr{V}$, *there exists, corresponding to each element c of* $\mathscr{C}$, *a subsequence* $\mathscr{E}_c$ *of* $\mathscr{C}$ *which contains c, and a natural number* g_c, *such that*

$$\mathscr{E}_c^{(g_c)} \sim \mathscr{E}_c$$

and

$$\mathbf{d}\mathscr{E}_c \geqslant \underline{\mathbf{d}}\mathscr{V} - \frac{k}{g_c}. \tag{10.1}$$

Proof. A given element c of $\mathscr{C}$ can be expressed in the form

$$c = a_0 + a_1 + \ldots + a_k, \qquad a_\nu \in \mathscr{A}_\nu \quad (\nu = 0, 1, \ldots, k).$$

Let $\mathfrak{A}'$ be the system of sequences $\mathscr{A}'_0, \mathscr{A}'_1, \ldots, \mathscr{A}'_k$ defined by

$$\mathscr{A}'_\nu = \mathscr{A}_\nu - a_\nu \quad (\nu = 0, 1, \ldots, k),$$

so that

$$\mathscr{C}' = \mathscr{C} - c, \qquad 0 \in \bigcap_{\nu=0}^{k} \mathscr{A}'_\nu$$

and, by Lemma 8,

$$\underline{\mathbf{d}}\mathscr{C}' = \underline{\mathbf{d}}\mathscr{C} \quad \text{and} \quad \underline{\mathbf{d}}\mathscr{V}' = \underline{\mathbf{d}}\mathscr{V}.$$

Then Theorem 19 is applicable to $\mathfrak{A}'$, so that there exists a subsequence $\mathscr{E}'$ of $\mathscr{C}'$ which contains 0, and a natural number $g = g_c$, such that $\mathscr{E}' \sim \mathscr{E}'^{(g_c)}$ and

$$\mathbf{d}\mathscr{E}' \geqslant \underline{\mathbf{d}}\mathscr{V}' - \frac{k}{g_c}.$$

It is clear that the sequence $\mathscr{E}_c = \mathscr{E}' + c$ fulfils the requirements of Theorem 19′.

We are now able to prove Theorem 17′; namely, that *if the system* $\mathfrak{A}$ *is such that* $\underline{\mathbf{d}}\mathscr{C} < \underline{\mathbf{d}}\mathscr{V}$, *there exists a natural number g such that* $\mathscr{C} \sim \mathscr{C}^{(g)}$.

We first observe that the sequence of numbers g_c, whose existence was established in Theorem 19′, is bounded. For if it is not, inequality (10.1) holds for arbitrarily large g_c and therefore, combined with the fact that $\mathscr{C} \supset \mathscr{E}_c$ for all c, implies that $\underline{\mathbf{d}}\mathscr{C} \geqslant \underline{\mathbf{d}}\mathscr{V}$.

Let g be the least common multiple of the finite set of values taken by g_c. By Theorem 19′, $\mathscr{C}$ contains, with each element c, all integers of the form $c + tg_c$ from some point onward, and since $g_c \mid g$, $\mathscr{C}$ contains also all sufficiently large integers of the form $c + ug$. By considering a finite set of representatives of the congruence classes of $\mathscr{C}^{(g)}$, chosen from $\mathscr{C} \cap \mathscr{C}^{(g)}$, we see that this is equivalent to saying that $\mathscr{C} \sim \mathscr{C}^{(g)}$.

It remains to prove Theorem 16′, which concerns the degenerate systems and is a generalization of the Cauchy–Davenport–Chowla Theorem 15. Our proof is by induction on k. The most difficult part of the proof is to show that the result is true for $k = 1$; we shall find that the inductive part of the argument gives little trouble.

To prove Theorem 16′ for $k = 1$ we require a generalization of the case $k = 1$ of Theorem 19′. We require it only in the case when $\mathfrak{A}$ is degenerate, but with only very little more trouble we prove Theorem 19″ below for an unrestricted pair of integer sequences $\mathscr{A}$ and $\mathscr{B}$.

THEOREM 19″. *If $\mathscr{C}$ is the sum of the two sequences $\mathscr{A}$ and $\mathscr{B}$ and if $\underline{\mathbf{d}}\mathscr{C} < \underline{\mathbf{d}}(\mathscr{A} \vee \mathscr{B})$, there exists, corresponding to each finite sub-set $\{c_1, c_2, \ldots, c_n\}$ of $\mathscr{C}$, a subsequence $\mathscr{E}$ of $\mathscr{C}$ which contains $c_1, c_2, \ldots, c_n$, and a natural number g, such that*

$$\mathscr{E}^{(g)} \sim \mathscr{E}$$

and

$$\mathbf{d}\mathscr{E} \geqslant \underline{\mathbf{d}}(\mathscr{A} \vee \mathscr{B}) - \frac{1}{g}. \tag{10.2}$$

Proof. The proof is by induction on n. If $n = 1$ the theorem reduces to Theorem 19′ with $k = 1$, and is therefore established. Suppose then that $n > 1$, and assume that the result is true for the sub-set $\{c_1, c_2, \ldots, c_{n-1}\}$. Then there exists a sequence $\mathscr{E}_1$ and a natural number l such that

$$\{c_1, c_2, \ldots, c_{n-1}\} \subset \mathscr{E}_1 \subset \mathscr{C}, \qquad \mathscr{E}_1 \sim \mathscr{E}_1^{(l)}, \qquad \mathbf{d}\mathscr{E}_1 \geqslant \underline{\mathbf{d}}(\mathscr{A} \vee \mathscr{B}) - \frac{1}{l},$$

and by Theorem 19′ there exists a sequence $\mathscr{E}_2$ and a natural number m such that

$$c_n \in \mathscr{E}_2 \subset \mathscr{C}, \qquad \mathscr{E}_2 \sim \mathscr{E}_2^{(m)}, \qquad \mathbf{d}\mathscr{E}_2 \geqslant \underline{\mathbf{d}}(\mathscr{A} \vee \mathscr{B}) - \frac{1}{m}.$$

Let $\mathscr{E} = \mathscr{E}_1 \cup \mathscr{E}_2$; then certainly $\{c_1, c_2, \ldots, c_n\} \subset \mathscr{E} \subset \mathscr{C}$.

First, if $\mathscr{E}_1 \subset \mathscr{E}_2^{(m)}$, we choose $g = m$ and note that

$$\mathscr{E}^{(g)} = (\mathscr{E}_1 \cup \mathscr{E}_2)^{(m)} = \mathscr{E}_2^{(m)} \sim \mathscr{E}_2 \subset \mathscr{E};$$

since also $\mathscr{E} \subset \mathscr{E}^{(g)}$ it follows that

$$\mathscr{E}^{(g)} \sim \mathscr{E} \quad \text{and} \quad \mathbf{d}\mathscr{E} \geqslant \mathbf{d}\mathscr{E}_2 \geqslant \underline{\mathbf{d}}(\mathscr{A} \vee \mathscr{B}) - 1/g.$$

Again, if $\mathscr{E}_2 \subset \mathscr{E}_1^{(l)}$ we choose $g = l$ and arrive at (10.2) by the same argument.

Hence we may suppose that $\mathscr{E}_1$ contains an element not in $\mathscr{E}_2^{(m)}$ and $\mathscr{E}_2$ contains an element not in $\mathscr{E}_1^{(l)}$, so that $\mathscr{E}_1^{(l)} \cap \mathscr{E}_2^{(m)}$ is a proper subset of each of $\mathscr{E}_1^{(l)}$ and $\mathscr{E}_2^{(m)}$. We choose g to be the lowest common multiple of l and m; then

$$\mathscr{E} \subset \mathscr{E}^{(g)} = (\mathscr{E}_1 \cup \mathscr{E}_2)^{(g)} \subset (\mathscr{E}_1^{(l)} \cup \mathscr{E}_2^{(m)}) \sim (\mathscr{E}_1 \cup \mathscr{E}_2) = \mathscr{E},$$

so that $\mathscr{E}^{(g)} \sim \mathscr{E}$, and it remains to prove (10.2).

We write $\mathscr{X}$ for $\mathscr{E}_1^{(l)}$, $\mathscr{Y}$ for $\mathscr{E}_2^{(m)}$, and D for $\underline{\mathbf{d}}(\mathscr{A} \vee \mathscr{B})$, and we observe that

(i) $\mathscr{E} \sim (\mathscr{X} \cup \mathscr{Y})$,

(ii) $\mathbf{d}\mathscr{X} \geqslant D-1/l, \quad \mathbf{d}\mathscr{Y} \geqslant D-1/m$,

(iii) $\mathscr{X} \cap \mathscr{Y}$ is a proper subset of $\mathscr{X}$ and of $\mathscr{Y}$.

By (i), it suffices to prove that, for any D, (ii) and (iii) imply

$$\mathbf{d}(\mathscr{X} \cup \mathscr{Y}) \geqslant D-\frac{1}{g}. \tag{10.3}$$

Let $\mathscr{X}$ be the union of the complete residue classes $x_1, x_2,..., x_r \pmod{l}$ and $\mathscr{Y}$ the union of the complete residue classes $y_1, y_2,..., y_s \pmod{m}$. Then

$$1 \leqslant r \leqslant l-1, \quad 1 \leqslant s \leqslant m-1 \tag{10.4}$$

by (iii), and

$$\mathbf{d}\mathscr{X} = r/l, \quad \mathbf{d}\mathscr{Y} = s/m. \tag{10.5}$$

It is clear that

$$\mathbf{d}(\mathscr{X} \cup \mathscr{Y}) = \mathbf{d}\mathscr{X}+\mathbf{d}\mathscr{Y}-\mathbf{d}(\mathscr{X} \cap \mathscr{Y}), \tag{10.6}$$

and if $\mathbf{d}(\mathscr{X} \cap \mathscr{Y}) = 0$ we have, by (ii) and (10.5), that

$$\mathbf{d}(\mathscr{X} \cup \mathscr{Y}) \geqslant \max\left(D-\frac{1}{l}+\frac{s}{m}, D-\frac{1}{m}+\frac{r}{l}\right)$$

$$\geqslant D+\max\left(\frac{1}{m}-\frac{1}{l}, \frac{1}{l}-\frac{1}{m}\right) \geqslant D,$$

which is a stronger inequality than (10.3). Hence we may suppose that

$$\mathbf{d}(\mathscr{X} \cap \mathscr{Y}) > 0. \tag{10.7}$$

We write $d = (l, m)$ and put $l = l_1 d$, $m = m_1 d$ so that $g = l_1 m = lm_1$. We partition $\mathscr{X}$ into d mutually exclusive sets $\mathscr{X}_1, \mathscr{X}_2,..., \mathscr{X}_d$, putting class $x_i \pmod{l}$ into the set $\mathscr{X}_\mu$ if and only if $x_i \equiv \mu \pmod{d}$, and we perform a similar partition of $\mathscr{Y}$. Let r_μ be the number of classes $\pmod{l}$ of $\mathscr{X}$ in $\mathscr{X}_\mu$, and s_ν the number of classes $\pmod{m}$ of $\mathscr{Y}$ in $\mathscr{Y}_\nu$; evidently $0 \leqslant r_\mu \leqslant l_1$ and $0 \leqslant s_\nu \leqslant m_1$. If $r_\mu = l_1$ we shall say that $\mathscr{X}_\mu$ is *full*, and we shall use the same terminology with regard to $\mathscr{Y}_\nu$. From elementary congruence theory, $\mathscr{X}_\mu \cap \mathscr{Y}_\nu$ is empty unless $\mu = \nu$, and

$$\mathbf{d}(\mathscr{X}_\mu \cap \mathscr{Y}_\mu) = (r_\mu s_\mu)/g.$$

Hence

$$\mathbf{d}(\mathscr{X} \cap \mathscr{Y}) = \sum_{\mu=1}^{d} \frac{r_\mu s_\mu}{g}, \tag{10.8}$$

and therefore, by (10.5) and (10.6),

$$\text{(10.9)}\quad \mathbf{d}(\mathscr{X}\cup\mathscr{Y}) = \mathbf{d}\mathscr{X}+\frac{1}{g}\sum_{\mu=1}^{d} s_\mu(l_1-r_\mu) = \mathbf{d}\mathscr{Y}+\frac{1}{g}\sum_{\mu=1}^{d} r_\mu(m_1-s_\mu).$$

We may assume, without loss of generality, that

(iv) $\mathscr{X}_\mu$, $\mathscr{Y}_\mu$ are not both full for any μ, $1 \leqslant \mu \leqslant d$.

For if (iv) is not satisfied, suppose that $\mathscr{X}\cap\mathscr{Y}$ contains exactly t 'full' congruence classes modulo d. We form $\mathscr{X}'$, $\mathscr{Y}'$ by omitting from $\mathscr{X}$, $\mathscr{Y}$ the t pairs of full sets $\mathscr{X}_\mu$, $\mathscr{Y}_\mu$ they have in common. Then

$$\mathbf{d}\mathscr{X}-\mathbf{d}\mathscr{X}' = \mathbf{d}\mathscr{Y}-\mathbf{d}\mathscr{Y}' = t/d,$$

and $\mathscr{X}'$, $\mathscr{Y}'$ satisfy conditions† of type (ii) and (iii) (the former with $D' = D-t/d$ in place of D), and also (iv). Clearly, to prove (10.3) is equivalent to proving that $\mathbf{d}(\mathscr{X}'\cup\mathscr{Y}') \geqslant D'-1/g$.

By (10.7) and (10.8) there exists μ_0, $1 \leqslant \mu_0 \leqslant d$, such that $r_{\mu_0}s_{\mu_0} \geqslant 1$, and by (iv) either (*a*) one of $\mathscr{X}_{\mu_0}$, $\mathscr{Y}_{\mu_0}$ (but not both) is full, or (*b*) neither is full.

Taking alternative (*a*) first, suppose that $r_{\mu_0} = l_1$ and $1 \leqslant s_{\mu_0} < m_1$. By (10.9) and (ii),

$$\mathbf{d}(\mathscr{X}\cup\mathscr{Y}) \geqslant D-\frac{1}{m}+\frac{1}{g}r_{\mu_0}(m_1-s_{\mu_0}) \geqslant D-\frac{1}{m}+\frac{l_1}{g} = D,$$

which is better than (10.3). (This also settles the case $1 \leqslant r_{\mu_0} < l_1$, $s_{\mu_0} = m_1$, by symmetry.)

Taking alternative (*b*), we have that $1 \leqslant r_{\mu_0} < l_1$ and $1 \leqslant s_{\mu_0} < m_1$. By (10.9),

$$\mathbf{d}(\mathscr{X}\cup\mathscr{Y}) \geqslant D-\frac{1}{g}+\max\left(\frac{1}{g}-\frac{1}{l}+\frac{s_{\mu_0}}{g}(l_1-r_{\mu_0}),\ \frac{1}{g}-\frac{1}{m}+\frac{r_{\mu_0}}{g}(m_1-s_{\mu_0})\right),$$

and (10.3) follows if we can prove that

$$\text{(10.10)}\quad \left\{\frac{1}{g}-\frac{1}{l}+\frac{s_{\mu_0}}{g}(l_1-r_{\mu_0})\right\}+\left\{\frac{1}{g}-\frac{1}{m}+\frac{r_{\mu_0}}{g}(m_1-s_{\mu_0})\right\} \geqslant 0.$$

After multiplication by g, the left-hand side of (10.10) becomes

$$\begin{aligned} s_{\mu_0}(l_1-r_{\mu_0})&+r_{\mu_0}(m_1-s_{\mu_0})-l_1-m_1+2 \\ &= l_1(s_{\mu_0}-1)+m_1(r_{\mu_0}-1)-2r_{\mu_0}s_{\mu_0}+2 \\ &\geqslant (r_{\mu_0}+1)(s_{\mu_0}-1)+(s_{\mu_0}+1)(r_{\mu_0}-1)-2r_{\mu_0}s_{\mu_0}+2 = 0 \end{aligned}$$

so that (10.10) follows. This completes the proof of (10.3) and of Theorem 19″.

The case $k = 1$ of Theorem 16′ follows readily from Theorem 19″. Let $\mathfrak{A}$ be the system consisting of the two sequences $\mathscr{A}$ and $\mathscr{B}$ and

† It is to be shown that (10.3) is a consequence of (ii) and (iii).

suppose that $\mathfrak{A}$ is degenerate modulo g. Then $\mathscr{C} = \mathscr{C}^{(g)}$ (as usual, $\mathscr{C} = \mathscr{A}+\mathscr{B}$), and we denote by g' the least natural number possessing this property; by Lemma 11, g' is a factor of g, and by Lemma 10 $\mathscr{C}^{(g')} = \mathscr{A}^{(g')}+\mathscr{B}^{(g')}$. We shall prove that

$$(10.11) \qquad \mathbf{d}\mathscr{C}^{(g')} \geqslant \mathbf{d}(\mathscr{A}^{(g')} \vee \mathscr{B}^{(g')})-1/g',$$

and for this purpose we may assume that $\mathbf{d}\mathscr{C}^{(g')} < \mathbf{d}(\mathscr{A}^{(g')} \vee \mathscr{B}^{(g')})$; otherwise (10.11) is obviously true.

Let $c_1, c_2,..., c_n$ be a set of representatives of all the distinct residue classes (mod g') which occur in $\mathscr{C}^{(g')}$, and apply Theorem 19″ to the sequences $\mathscr{A}^{(g')}$ and $\mathscr{B}^{(g')}$. Theorem 19″ tells us that there exists a subsequence $\mathscr{E}$ of $\mathscr{C}^{(g')}$ which contains $c_1, c_2,..., c_n$—so that $\mathscr{E}^{(g')} = \mathscr{C}^{(g')}$—and a natural number g'' such that $\mathscr{E}^{(g'')} \sim \mathscr{E}$ and

$$\mathbf{d}\mathscr{E} \geqslant \mathbf{d}(\mathscr{A}^{(g')} \vee \mathscr{B}^{(g')})-1/g''.$$

Let $d = (g', g'')$. By Lemma 9,

$$\mathscr{C}^{(d)} = (\mathscr{C}^{(g')})^{(g'')} = (\mathscr{E}^{(g')})^{(g'')} = (\mathscr{E}^{(g'')})^{(g')} = \mathscr{E}^{(g')} = \mathscr{C}^{(g')},$$

so that, by the minimal property of g', $d = g'$. Hence $g'' \geqslant g'$ and therefore

$$\mathbf{d}\mathscr{C}^{(g')} = \mathbf{d}\mathscr{E}^{(g')} \geqslant \mathbf{d}\mathscr{E} \geqslant \mathbf{d}(\mathscr{A}^{(g')} \vee \mathscr{B}^{(g')})-1/g'' \geqslant \mathbf{d}(\mathscr{A}^{(g')} \vee \mathscr{B}^{(g')})-1/g',$$

proving (10.11), and with it Theorem 16′ with $k = 1$.

To complete the proof of Theorem 16′, we now proceed by induction on k; we prove the theorem for the system $\mathfrak{A}$ of $k+1$ sequences $\mathscr{A}_0, \mathscr{A}_1,...,$ $\mathscr{A}_k$, assuming it to hold for systems of k sequences. Let g^* be the least g for which $\mathscr{C}^{(g)} = \mathscr{C}$, and denote by $\mathfrak{A}^*$ the system $\mathscr{A}_0^{(g^*)}, \mathscr{A}_1^{(g^*)},..., \mathscr{A}_k^{(g^*)}$. Writing $\underline{\mathscr{V}}^* = \underline{\mathscr{V}}(\mathfrak{A}^*)$, $\mathscr{C}^* = \mathscr{C}(\mathfrak{A}^*)$, it suffices to prove

$$(10.12) \qquad \mathbf{d}\mathscr{C}^* \geqslant \mathbf{d}\underline{\mathscr{V}}^*-k/g^*;$$

for, by Lemma 11, $\mathscr{C}^{(g)} = \mathscr{C}^{(g^*)}$ implies $g^* \mid g$.

We denote by $\mathfrak{A}^*_{(-)}$ the system $\mathscr{A}_0^{(g^*)}, \mathscr{A}_1^{(g^*)},..., \mathscr{A}_{k-1}^{(g^*)}$, and interpret $\mathscr{C}^*_{(-)}$ and $\underline{\mathscr{V}}^*_{(-)}$ accordingly. We note that g^* is also the least g for which $\mathscr{C}^*_{(-)} = \mathscr{C}^{*(g)}_{(-)}$. For if $\mathscr{A}$, $\mathscr{B}$ are any two non-empty sequences,

$$(\mathscr{A}+\mathscr{B})^{(g)} = \mathscr{A}^{(g)}+\mathscr{B};$$

thus if $\mathscr{C}^*_{(-)}$ degenerates modulo a divisor of g^*, so does

$$\mathscr{C} = \mathscr{C}^* = \mathscr{C}^*_{(-)}+\mathscr{A}_k^{(g^*)}.$$

In view of the minimal property of g^*, we obtain, on applying the induction hypothesis to $\mathfrak{A}^*_{(-)}$,

$$(10.13) \qquad \mathbf{d}\mathscr{C}^*_{(-)} \geqslant \mathbf{d}\underline{\mathscr{V}}^*_{(-)}-\frac{k-1}{g^*}.$$

We now apply the case $k = 1$ of Theorem 16′ to the system consisting of the two sequences $\mathscr{C}^*_{(-)}$, $\mathscr{A}_k^{(g^*)}$. Again bearing in mind the minimal property of g^*, we obtain

$$\text{(10.14)} \qquad \mathbf{d}\mathscr{C} \geqslant \mathbf{d}(\mathscr{C}^*_{(-)} \vee \mathscr{A}_k^{(g^*)}) - \frac{1}{g^*} = \mathbf{d}\mathscr{C}^*_{(-)} + \mathbf{d}\mathscr{A}_k^{(g^*)} - \frac{1}{g^*}.$$

Since $\underline{\mathscr{V}}^* = \underline{\mathscr{V}}^*_{(-)} \vee \mathscr{A}_k^{(g^*)}$ and hence $\mathbf{d}\underline{\mathscr{V}}^* = \mathbf{d}\underline{\mathscr{V}}^*_{(-)} + \mathbf{d}\mathscr{A}_k^{(g^*)}$,

(10.13) together with (10.14) imply (10.12). This completes the proof of Theorem 16′.

11. Hanani's conjecture (*Revised May* 1965)

We conclude Chapter I by reporting on some recent progress with an interesting problem concerning pairs of 'complementary' sequences $\mathscr{A}$, $\mathscr{B}$ (that is, sequences satisfying $\mathscr{A}+\mathscr{B} \sim \mathscr{Z}_0$—see Definition 9). If $\mathscr{A}$, $\mathscr{B}$ is such a pair, and the number c is such that all natural numbers greater than c lie in $\mathscr{A}+\mathscr{B}$, it follows trivially that $A(n)B(n) \geqslant n-c$, and in particular that

$$\text{(11.1)} \qquad \limsup_{n\to\infty} \frac{A(n)B(n)}{n} \geqslant 1.$$

G. Hanani conjectured that if $\mathscr{A}$, $\mathscr{B}$ are infinite complementary sequences, then (11.1) holds with strict inequality. Clearly this need not be the case if one of $\mathscr{A}$, $\mathscr{B}$ is finite. For let k be a natural number, and consider the case of

$$\mathscr{A} = \{0, 1, 2, \ldots, k-1\}, \qquad \mathscr{B} = \{kx : x = 0, 1, 2, \ldots\};$$

then $\mathscr{A}+\mathscr{B} = \mathscr{Z}_0$, but

$$\text{(11.2)} \qquad \limsup_{n\to\infty} \frac{A(n)B(n)}{n} = 1.$$

Using this example as an underlying idea, Danzer[39] recently disproved Hanani's conjecture by constructing, in a highly ingenious manner, two infinite complementary sequences satisfying (11.2). He went on to conjecture, with Erdős, that if $\mathscr{A}$, $\mathscr{B}$ are complementary sequences for which (11.2) is true, then $\liminf_{n\to\infty}\{A(n)B(n)-n\} = \infty$; we are informed that this has been confirmed by A. Sárközi and E. Szemerédi. The precise rate of growth of $A(n)B(n)-n$ as $n \to \infty$ has yet to be settled.

[39] L. Danzer, 'Über eine Frage von G. Hanani aus der additiven Zahlentheorie', *J. reine angew. Math.* **214/215** (1964) 392–4. Danzer also disposes of a generalized conjecture by W. Narkiewicz, 'Remarks on a conjecture of Hanani in number theory', *Colloquium math.* 7 (1959/60) 161–5; this paper also contains several related results.

II

ADDITION OF SEQUENCES: STUDY OF REPRESENTATION FUNCTIONS BY NUMBER THEORETIC METHODS

1. Introduction

As in Chapter I (see the beginning of § 1 of Chapter I), we use $\mathscr{A}$ to denote a subsequence of the sequence $\mathscr{Z}_0$ of non-negative integers. Although our principal investigations concern infinite sequences $\mathscr{A}$, it will sometimes be appropriate to consider also finite sequences; we shall make it clear in the text when finite sequences are being considered.

Let $h \geqslant 2$ be an integer. In Chapter I we have investigated various relationships between the density properties of a sequence $\mathscr{A}$ and the corresponding sequence $h\mathscr{A}$, where $h\mathscr{A}$ is the h-fold sum-sequence consisting of those integers n for which the number $R_n(h;\mathscr{A})$ of representations of n in the form

$$n = a^{(1)}+\ldots+a^{(h)} \quad (a^{(1)} \in \mathscr{A},\ldots, a^{(h)} \in \mathscr{A}) \tag{1.1}$$

is strictly positive. In this chapter we consider the more general topic of relationships between the rate of growth of a sequence $\mathscr{A}$ (as measured by the rate of growth of the counting function $A(n)$)† and the distribution of values of the representation function $R_n(h;\mathscr{A})$ (considered as a function of n). Most of the work will concern the case $h = 2$. In this case, we write

$$R_n(\mathscr{A}) = R_n(2;\mathscr{A}),$$

and introduce two modified representation functions in the definition below.

Definition 1. We denote by $r_n(\mathscr{A})$ the number of representations of n in the form

$$n = a+a', \quad a < a' \quad (a \in \mathscr{A},\ a' \in \mathscr{A}); \tag{1.2}$$

and by $r'_n(\mathscr{A})$ the number of representations of n in the form

$$n = a+a', \quad a \leqslant a' \quad (a \in \mathscr{A},\ a' \in \mathscr{A}). \tag{1.3}$$

† See § 4 of the 'Notation' section at the beginning of the book.

Remark. We note that

$$(1.4) \qquad r'_n(\mathscr{A}) = \begin{cases} r_n(\mathscr{A})+1 & \text{if } n \text{ is even and } \tfrac{1}{2}n \in \mathscr{A}, \\ r_n(\mathscr{A}) & \text{otherwise,} \end{cases}$$

and

$$(1.5) \qquad R_n(\mathscr{A}) = r_n(\mathscr{A})+r'_n(\mathscr{A}).$$

We shall express each result in terms of whichever representation function (R_n, r_n, or r'_n) is the most appropriate (or the most convenient). Where it might be desirable to obtain a corresponding result in terms of one of the other representation functions, this can usually be done without difficulty.

We have already mentioned the general aim of this chapter. The questions contained in (i), (ii), and (iii) below suggest specific lines of investigation.

(i) What types of relationship between the behaviour of the counting function $A(n)$ of a sequence $\mathscr{A}$ and the behaviour of the corresponding (for given h) representation function R_n are in fact possible?

(ii) There are many well-known theorems concerning the behaviour of the representation functions (for suitable h) arising from certain specific sequences $\mathscr{A}$ (such as, for example, the sequence of integral kth powers, for an integer $k \geqslant 2$). Is it possible to generalize some of these results by proving similar theorems applicable to a reasonably wide class of sequences $\mathscr{A}$ (defined, for example, in terms of the behaviour of $A(n)$ and the distribution of $\mathscr{A}$ among congruence classes)?

(iii) Are there known theorems concerning the representation functions arising from specific sequences $\mathscr{A}$, which were proved by making extensive use of the specific arithmetic character of the sequences in question but which are in fact entirely independent† of some (or all) of these arithmetic properties?

Very little progress has been made with these interesting and important questions. Whilst those results that are known include some of remarkable beauty and power, they concern only some very limited aspects of the above questions.

Question (i) is couched in very general terms, and a large number of more specific questions fall into this category. In particular, various problems of this type arose in connexion with investigations[1] in the

† A similar situation arose in connexion with essential components (cf. Chapter I, § 3).

(1) S. Sidon, 'Ein Satz über trigonometrische Polynome und seine Anwendung in der Theorie der Fourier-Reihen', *Math. Ann.* **106** (1932) 536–9. See also P. Erdös[10], pp. 129, 132.

theory of Fourier series. These problems concern sequences $\mathscr{A}$ for which $r'_n(\mathscr{A})$ is either bounded or majorized by a function which tends to infinity 'slowly' as n tends to infinity. Sidon asks, for example, what rates of growth of $A(n)$ are compatible with $r'_n(\mathscr{A})$ being bounded. Some progress has been made with such questions; quite a lot is known about the possible asymptotic behaviour of $A(n)$ when the sequence $\mathscr{A}$ satisfies

$$r'_n(\mathscr{A}) \leqslant 1 \quad \text{for all } n,$$

although many of the corresponding questions concerning sequences $\mathscr{A}$ satisfying

$$r'_n(\mathscr{A}) \leqslant g \quad \text{for all } n,$$

(when g is a natural number greater than 1) remain unanswered. These and related matters are the subject of § 3.

Another question arising from Sidon's work is whether the relation

$$r'_n(\mathscr{A}) \geqslant 1 \quad \text{for large } n,$$

is compatible with $r'_n(\mathscr{A})$ being majorized by a function which tends to infinity slowly, or is perhaps even compatible with $r'_n(\mathscr{A})$ being bounded. (This is equivalent to asking whether there exists an 'economical' asymptotic basis.) We shall touch on this question in § 4 of this chapter, and return to it in Chapter III (cf. Chapter III, § 2, Theorem 1).

Practically nothing is known in connexion† with (ii). There has been considerable progress, however, with the corresponding question for partition functions (in place of the representation functions R_n, etc.).

As regards (iii), only one result is known to us (apart from certain generalizations and extensions of this result). This concerns the 'error' $\Delta(N)$ in the relation

$$\sum_{0 \leqslant n \leqslant N} R_n(\mathscr{Q}) = \pi N + \Delta(N), \tag{1.6}$$

where $\mathscr{Q}$ denotes, as previously, the sequence consisting of zero and the natural squares. It has long been known that whilst πN is an approximation‡ to the left-hand side of (1.6) when N is large, the accuracy of

† We here exclude the result discussed below in connexion with (iii), although the question (iii) could be regarded as being a special case of (ii).

‡ The left-hand side of (1.6) represents the number of points of the integral lattice within the circle $x^2+y^2 \leqslant N$, and is therefore given approximately by πN when N is large. It is trivial that $\Delta(N) \ll N^{\frac{1}{2}}$, but more precise (and much more profound) estimates are known. The most precise such estimate known at present is $\Delta(N) \ll N^{13/37+\epsilon}$. (Chen, Jing-run, 'The lattice points in a circle', *Sci. Sinica*, **12** (1963) 633–49.)

this approximation is limited† at least to the extent that

$$\limsup_{N\to\infty} \frac{|\Delta(N)|}{(N\log N)^{\frac{1}{4}}} > 0. \tag{1.7}$$

The proof of (1.7) was based on difficult analytic techniques, and depended heavily on the arithmetical character of the sequence of squares. In 1956, however, Erdös and Fuchs(2) showed that there is a very similar limitation to the possible accuracy of any asymptotic formula of type

$$\sum_{0\leqslant n\leqslant N} R_n(\mathscr{A}) = cN + \text{error term},$$

quite irrespective of the nature of the sequence $\mathscr{A}$. The proof of this astonishing and beautiful result (to be stated explicitly in § 4, where it is also proved) is very elegant and surprisingly simple. One of the attractions of the theory of sequences is that there may well be many such unexpected results, of equally remarkable power and generality, still to be discovered. The Erdös–Fuchs theorem will represent the main substance of § 4.

Part of the work of § 3 will involve the application of certain results from the theory of finite fields; we state and prove these results in the following section.

2. Auxiliary results from the theory of finite fields

In § 3 we shall investigate sequences $\mathscr{A}$ which, for a given integer $h \geqslant 2$, have the property that for every integer n, (1.1), subject to the additional condition $a^{(1)} \leqslant a^{(2)} \leqslant \ldots \leqslant a^{(h)}$, has at most one solution. This property may be restated in the form (I) below.

(I) *All the sums* $a^{(1)}+a^{(2)}+\ldots+a^{(h)}$, *where* $a^{(1)} \leqslant a^{(2)} \leqslant \ldots \leqslant a^{(h)}$ ($a^{(1)} \in \mathscr{A}$, $a^{(2)} \in \mathscr{A}, \ldots, a^{(h)} \in \mathscr{A}$) *are distinct.*

For these investigations, it proves very valuable to have at one's disposal a numerous supply of *finite* sequences $\mathscr{A}$, each of which has the property (I) and is 'dense' in a suitable sense. There is a well-known theorem of Singer(3) (Theorem 1 below), establishing the existence of a numerous collection of 'perfect difference sets', which provides

† G. H. Hardy, 'On the expression of a number as the sum of two squares', *Quart. J. Math.* **46** (1915), 263–83, and the remarks on p. 23 of 'On Dirichlet's divisor problem', *Proc. London math. Soc.*, Ser. 2, **15** (1916) 1–25. A. E. Ingham, 'On the classical lattice point problems', *Proc. Cambridge Phil. Soc.* **36** (1940) 131–8.

(2) P. Erdös and W. H. J. Fuchs, 'On a problem of additive number theory', *J. Lond. math. Soc.* **31** (1956) 67–73.

(3) J. Singer, 'A theorem in finite projective geometry and some applications to number theory', *Trans. Am. math. Soc.* **43** (1938) 377–85.

such a supply in the case $h = 2$. Singer's result, which we prove at the end of the section, is the following.

THEOREM 1 (Singer(3)). *If m is a power of a prime, there exist $m+1$ integers $a_1, a_2,..., a_{m+1}$ such that the m^2+m non-zero differences a_i-a_j represent all the non-zero residues modulo m^2+m+1.*

The following theorem is an obvious corollary of Singer's result, for the integers $a_1, a_2,..., a_{m+1}$ of Theorem 1 must also have the property asserted in Theorem 1′.

THEOREM 1′. *If m is a power of a prime, there exist $m+1$ integers a_1, $a_2,..., a_{m+1}$ such that all the sums a_i+a_j, where $1 \leqslant i \leqslant j \leqslant m+1$, are distinct modulo m^2+m+1.*

It is clear that, by a suitable choice of representatives modulo m^2+m+1, the a's can be chosen to satisfy

$$1 \leqslant a_1 < a_2 < ... < a_{m+1} \leqslant m^2+m+1. \tag{2.1}$$

The sequence $\mathscr{A}$ formed by these a's has the property (I) with $h = 2$, and provides a 'dense' sequence of this type.

Singer's result was already available in 1941, when sequences with the property (I) (in the case $h = 2$) were first investigated by Erdös–Turan,(4) but was overlooked at the time. It was not until several years later that Erdös (see ref. (4), Addendum) discovered this oversight and Chowla(5), independently, applied Singer's result (and also Bose's analogue, discussed below) in this connexion; the Erdös–Turan–Singer–Chowla results are included in the following section.

In 1941 Bose(6) proved an analogue of Singer's theorem, which provides an alternative way of constructing 'dense' finite sequences with the property (I). Bose's analogue states that if m is a power of a prime, then there exist integers $a_1, a_2,..., a_m$, such that the $m(m-1)$ non-zero differences a_i-a_j represent all those residues modulo (m^2-1) that are not divisible by $m+1$. (Note that there are exactly

$$(m^2-1)-(m-1) = m(m-1)$$

such residues.) The following theorem (which we prove later, in generalized form), stands in the same relationship to Bose's analogue of Singer's theorem as Theorem 1′ to Theorem 1.

(4) P. Erdös and P. Turan, 'On a problem of Sidon in additive number theory, and some related problems', *J. Lond. math. Soc.* **16** (1941) 212–5. Addendum (by P. Erdös), ibid. **19** (1944) 208.

(5) S. Chowla, 'Solution of a problem of Erdös and Turan in additive number theory', *Proc. natn. Acad. Sci. India* **14** (1944) 1–2.

(6) R. C. Bose, 'An affine analogue of Singer's theorem', *J. Indian math. Soc.* (new series) **6** (1942) 1–15.

THEOREM 2'. *If m is a power of a prime, there exist m integers a_1, $a_2,\ldots, a_m$, such that all the sums a_i+a_j, where $1 \leqslant i \leqslant j \leqslant m$, are distinct modulo m^2-1.*

In 1962 Bose and Chowla[7] obtained generalizations of Theorems 1' and 2' which are applicable to sequences with the property (I) in the case when h is an arbitrary integer satisfying $h \geqslant 2$. We give only the generalization of Theorem 2', as this will suffice for all our applications. In fact *Theorem 3 below is the only theorem in this section required for applications later.*

THEOREM 3 (Bose–Chowla). *Let m be a power of a prime, and let $h \geqslant 2$ be an integer. There exist integers $a_1, a_2,\ldots, a_m$ such that all the sums*

$$(2.2) \qquad a_{j_1}+a_{j_2}+\ldots+a_{j_h}, \quad \text{where} \quad 1 \leqslant j_1 \leqslant j_2 \leqslant \ldots \leqslant j_h \leqslant m,$$

are distinct† modulo m^h-1.

Here again, by a suitable choice of representatives modulo m^h-1, the a's can be chosen to satisfy

$$1 \leqslant a_1 < a_2 < \ldots < a_m \leqslant m^h-1;$$

and this provides a 'dense' set $\mathscr{A}$ with the property (I). It is worth remarking, although we do not require this fact in the sequel, that it is easily verified (cf. Bose–Chowla[7], Theorem 3, proof of part (i)) that the sequence of $m+1$ elements obtained by inserting the additional element m^h into $\mathscr{A}$, also satisfies (I).

The proofs of the above theorems will entail the use of finite commutative (Galois) fields. For those readers who are unfamiliar with the theory of such fields, we include the following brief discussion, concluding with the assertion (E).

We shall assume the basic results (A), (B), (C) below. These results, and their proofs, are all to be found in van der Waerden, *Modern Algebra* (English translation, Ungar, 1949), § 37.

(A) *Corresponding to each prime p and natural number r, there is a unique‡ (commutative) finite field of p^r elements. We denote this field by $GF(p^r)$.*

(B) *If d is a divisor of r, then $GF(p^d)$ is a subfield of $GF(p^r)$.*

(C) *The multiplicative group obtained by omitting the zero element from $GF(p^r)$ is cyclic. We denote this group by $GF^*(p^r)$.*

† Note that, in view of the possibility $j_1 = j_2 = \ldots = j_h$, the integers a are themselves distinct modulo m^h-1.

‡ To within an isomorphism.

(7) R. C. Bose and S. Chowla, 'Theorems in the additive theory of numbers', *Comment. math. helvet.* **37** (1962–3) 141–7.

In addition to these results we use only the following well-known result concerning cyclic groups, which is an immediate consequence of Euclid's algorithm (cf. van der Waerden, loc. cit., p. 21).

(D) *Let $\mathfrak{S}$ be a cyclic group containing more than one element, and let $\mathfrak{S}_1$ be a subgroup of $\mathfrak{S}$. Then $\mathfrak{S}_1$ is also cyclic; and if θ is a generator of $\mathfrak{S}$, the element θ^q of $\mathfrak{S}_1$, where q is the least natural number for which θ^q lies in $\mathfrak{S}_1$, is a generator of $\mathfrak{S}_1$.*

We shall require the following immediate consequences of the above results. Let the element θ of $GF(p^r)$ be a generator of the cyclic group $GF^*(p^r)$; so that the elements of $GF^*(p^r)$ are

$$\theta^a \quad (a = 1, 2, \ldots, p^r-1). \tag{2.3}$$

Now suppose that d is a divisor of r. Since the subgroup $GF^*(p^d)$ of $GF^*(p^r)$ contains p^d-1 elements, it follows from (D) that this subgroup consists of those elements (2.3) for which $a \equiv 0 \pmod q$, where $q = (p^r-1)/(p^d-1)$. Thus two elements θ^a, $\theta^{a'}$ of $GF(p^r)$ are linearly dependent over the subfield $GF(p^d)$ if and only if $a \equiv a' \pmod q$.

Since the unit element appears in (2.3) (actually when $a = p^r-1$), θ is algebraic over any subfield of $GF(p^r)$. Suppose that θ is algebraic of degree h over $GF(p^d)$. Then each of the p^r elements of $GF(p^r)$ appears exactly once among the elements

$$\sum_{j=0}^{h-1} c_j \theta^j$$

where the c_j $(j = 0, 1, \ldots, h-1)$ run independently through the p^d elements of $GF(p^d)$. Thus $h = r/d$.

We have thus deduced (E) below from the results (A)–(D).

(E) *Let d be a divisor of r, so that $GF(p^d)$ is a subfield of $GF(p^r)$; and let θ be a generator of the cyclic group $GF^*(p^r)$. Then the elements θ^a, $\theta^{a'}$ of $GF(p^r)$ are linearly dependent over the subfield $GF(p^d)$ if and only if $a \equiv a' \pmod q$, where $q = (p^r-1)/(p^d-1)$. Furthermore, θ is algebraic of degree r/d over $GF(p^d)$.*

We now turn to the proof of Theorem 3, which is in fact slightly simpler than the proof of Theorem 1.

Proof of Theorem 3. Let $m = p^u$, where p is a prime. The proof will depend on suitable consideration of the subfield $GF(p^u)$ of $GF(p^{hu})$.

Let the element θ of $GF(p^{hu})$ be a generator of the cyclic group

$GF^*(p^{hu})$. Let $\mathscr{A}_h = \mathscr{A}_{h,m} = \{a_1, a_2,..., a_m\}$ denote the set of p^u integers satisfying $1 \leqslant a \leqslant m^h-1$ for which θ^a is representable in the form

$$\theta^a = \theta+c, \quad \text{where} \quad c \in GF(p^u). \tag{2.4}$$

For each $a \in \mathscr{A}_h$, we denote by $L_a(\theta)$ the monic linear polynomial $\theta+c$, considered as a polynomial in θ with coefficients in $GF(p^u)$, associated to a by (2.4).

Now if $j_1, j_2,..., j_h$ and $j_1', j_2',..., j_h'$ are two distinct sets of integers, each satisfying the system of inequalities appearing in the qualifying clause of (2.2), we have (explanation follows)

$$\Big\{\prod_{\nu=1}^{h} L_{a_{j_\nu}}(\theta)\Big\}-\Big\{\prod_{\nu=1}^{h} L_{a_{j_\nu'}}(\theta)\Big\} \neq 0. \tag{2.5}$$

For the left-hand side of (2.5), considered as a polynomial in θ with coefficients in $GF(p^u)$, is of degree at most $h-1$ in θ, and does not vanish identically since there is at most one factorization of a monic polynomial into monic linear factors; whilst θ is of degree h over $GF(p^u)$ (cf. the final assertion of (E) above). It follows from the definition (2.4) of the polynomial $L_a(\theta) = \theta+c$ associated to a that the integers a_1, $a_2,..., a_m$ have the property asserted in Theorem 3.

We have already mentioned that as far as applications in the sequel are concerned, we require only Theorem 3; we can use Theorem 2′, namely the case $h = 2$ of Theorem 3, in place of Theorem 1′. Nevertheless, Singer's result (Theorem 1) is of great interest in itself, and we therefore include a proof of this also.

Proof† *of Theorem* 1. We retain the notation introduced in the opening two paragraphs of the proof of Theorem 3, except that here we take $h = 3$; so that now we consider $GF(p^u)$ as a subfield of $GF(p^{3u})$.

Since θ does not lie in $GF(p^u)$, any two elements $\theta+c$, $\theta+c'$ of $GF(p^{3u})$, where c, c' are distinct elements of $GF(p^u)$, are linearly independent over $GF(p^u)$. It follows (cf. (E) above) that the set of integers $a_1, a_2,..., a_m$ constituting $\mathscr{A}_3$ are incongruent modulo q, where $q = (p^{3u}-1)/(p^u-1) = m^2+m+1$. We define the set

$$\mathscr{A}^* = \{a_1, a_2,..., a_{m+1}\}$$

by inserting into $\mathscr{A}_3$ the additional element $a_{m+1} = 0$. To prove that the elements of $\mathscr{A}^*$ have the property asserted in Theorem 1, namely that the $p^{2u}+p^u = q-1$ differences a_i-a_j $(i \neq j)$ are incongruent

† See also H. Halberstam and R. R. Laxton, 'On perfect difference sets', *Quart. J. Math.* **14** (1963) 86–90.

modulo q, it suffices to show that if each of the integers $a^{(1)}$, $a^{(2)}$, $a^{(3)}$, $a^{(4)}$ lies in $\mathscr{A}^*$, and

$$a^{(1)}-a^{(2)} \equiv a^{(3)}-a^{(4)} \pmod q, \tag{2.6}$$

then either $a^{(1)} = a^{(2)}$ or $a^{(1)} = a^{(3)}$.

We recall that if $a \in \mathscr{A}_3$, $L_a(\theta)$ is defined to be the polynomial $\theta+c$ associated to a by (2.4); and when $a = a_{m+1} = 0$, we define $L_a(\theta)$ to be the unit element. Now (2.6) implies that the two elements

$$L_{a^{(1)}}(\theta)L_{a^{(4)}}(\theta), \qquad L_{a^{(3)}}(\theta)L_{a^{(2)}}(\theta) \tag{2.7}$$

of $GF(p^{3u})$ are linearly dependent over $GF(p^u)$ (cf. (E) above). Each of the two expressions (2.7), considered as a polynomial in θ over $GF(p^u)$, is monic and of degree at most 2. Thus, since θ is algebraic of degree 3 over $GF(p^u)$, it follows that the two polynomials are identical.

Again using the fact that a polynomial has at most one decomposition into monic linear factors over a given field, we see that $L_{a^{(1)}}(\theta)$ must coincide with either $L_{a^{(2)}}(\theta)$ or $L_{a^{(3)}}(\theta)$ (irrespective of whether $L_{a^{(1)}}(\theta)$ is a linear polynomial or the unit element). Thus at least one of $a^{(1)} = a^{(2)}$ or $a^{(1)} = a^{(3)}$ must hold, and the proof of Theorem 1 is complete.

3. Sidon's problems

We have already described in § 1 (in the discussion relating to (i) in § 1) the questions raised by Sidon, concerning the representation function in the addition of sequences. We now consider these and related questions. In particular, we shall investigate the possible asymptotic behaviour of sequences $\mathscr{A}$ satisfying (cf. Definition 1)

$$r'_n(\mathscr{A}) \leqslant 1 \quad \text{for all } n. \tag{3.1}$$

We first introduce some notation and terminology. The term 'B_2-sequences', for sequences satisfying (3.1), and certain related notations have become widely accepted; we therefore use this type of notation here, despite the fact that it does not go too well with the other notations adopted in this book. In this section we shall consider both finite and infinite sequences $\mathscr{A}$; we shall state in the text whether any particular sequence under consideration is finite or infinite.

Definition 2. Let g, h be natural numbers, where $h \geqslant 2$. We denote by $B_h[g]$ the class of all (finite or infinite) sequences $\mathscr{A}$ such that every integer n has at most g representations of the form

$$n = a^{(1)}+\ldots+a^{(h)} \quad (a^{(1)} \in \mathscr{A},\ldots, a^{(h)} \in \mathscr{A}) \tag{3.2}$$

subject to

$$a^{(1)} \leqslant a^{(2)} \leqslant \ldots \leqslant a^{(h)}. \tag{3.3}$$

Definition 3. If $\mathscr{A}$ lies in the class $B_h[g]$, we say that '$\mathscr{A}$ is a $B_h[g]$-sequence'. We shall refer to the class $B_h[1]$ simply as the class B_h, and accordingly, we refer to sequences in the class $B_h[1]$ simply as B_h-sequences.

Remark. We note that, equivalently, a sequence $\mathscr{A}$ is a B_h-sequence if and only if it has the property (I) stated at the beginning of § 2.

Definition 4. We denote by $F_h(n)$ the maximum number of elements that can be selected from the set

$$1, 2, 3, \ldots, n$$

so as to form a B_h-sequence.

Remark. Obviously, if $\mathscr{A}$ is any B_h-sequence, we have

$$A(n) \leqslant F_h(n)$$

for every natural number n.

Our first objective will be to determine the asymptotic behaviour of $F_2(n)$ as $n \to \infty$. We remark that $F_2(n) \ll n^{\frac{1}{2}}$ is trivial, for if $\mathscr{A}$ is a maximal selection of the type described in Definition 4 (with $h = 2$), the $\frac{1}{2}F_2(n)(F_2(n)+1)$ numbers

$$a+a' \quad (a \in \mathscr{A},\ a' \in \mathscr{A}), \quad \text{with } a \leqslant a',$$

are distinct and do not exceed $2n$. In the opposite direction, Sidon himself showed that $F_2(n) \gg n^{\frac{1}{2}}$. We shall see that in fact

$$\lim_{n\to\infty} n^{-\frac{1}{2}}F_2(n) = 1. \tag{3.4}$$

In 1941 Erdös–Turan proved Theorem 4 below, which implies

$$\limsup_{n\to\infty} n^{-\frac{1}{2}}F_2(n) \leqslant 1; \tag{3.5}$$

and they conjectured that (3.4) holds. A few years later Chowla(5) and Erdös (see ref. (4), Addendum), independently of each other, deduced from Singer's theorem (§ 2, Theorem 1) that

$$\liminf_{n\to\infty} n^{-\frac{1}{2}}F_2(n) \geqslant 1, \tag{3.6}$$

thus settling the Erdös–Turan conjecture. In fact, much more precise results are known. Theorem 4 and the case $h = 2$ of Theorem 6 are much more precise than (3.5) and (3.6) respectively, and the combination of these theorems implies

$$F_2(n) = n^{\frac{1}{2}}+O(n^{5/16}). \tag{3.7}$$

As was shown by Bose–Chowla(7) (cf. Theorem 3 and Theorem 5 below), the Singer–Chowla method can be generalized so as to yield

$$\liminf_{n\to\infty} n^{-1/h}F_h(n) \geqslant 1 \quad (h = 2, 3, \ldots). \tag{3.8}$$

There is no corresponding generalization of the Erdös–Turan method of proving (3.5) on the other hand, and it remains an open question whether or not

$$\lim_{n\to\infty} n^{-1/h}F_h(n) = 1$$

when $h > 2$.

It is clear that, in the first instance, the above-mentioned results relate only to finite B_h-sequences. Infinite B_h-sequences will be considered later in this section.

THEOREM 4 (Erdös–Turan(4)). *We have*

$$F_2(n) < n^{\frac{1}{2}}+O(n^{\frac{1}{4}}). \tag{3.9}$$

Proof. We write

$$r = F_2(n), \tag{3.10}$$

and let

$$a_1 < a_2 < \ldots < a_r \tag{3.11}$$

be a maximal selection of the type described in Definition 4; so that the sequence (3.11), which we denote by $\mathscr{A}$, is a B_2-sequence entirely contained in the interval $[1, n]$. Let u be a positive integer less than n, and consider the $n+u$ intervals

$$\mathscr{I}_m = [-u+m, -1+m] \quad (m = 1, 2, \ldots, n+u).$$

Let A_m denote the number of a's in the interval $\mathscr{I}_m$. Since each a_i occurs in exactly u intervals, we have

$$\sum_{m=1}^{n+u} A_m = ru. \tag{3.12}$$

The number of pairs a_i, a_j ($i < j \leqslant r$) lying in $\mathscr{I}_m$ is $\frac{1}{2}A_m(A_m-1)$, and hence the total number of pairs a_i, a_j ($i < j \leqslant r$), with each pair lying in the same $\mathscr{I}_\nu$ ($1 \leqslant \nu \leqslant n+u$), is

$$\tfrac{1}{2}\sum_{m=1}^{n+u} A_m(A_m-1); \tag{3.13}$$

by Cauchy's inequality and (3.12),

$$(ru)^2 = \Big(\sum_{m=1}^{n+u} A_m\Big)^2 \leqslant \Big(\sum_{m=1}^{n+u} 1\Big)\Big(\sum_{m=1}^{n+u} A_m^2\Big) = (n+u)\sum_{m=1}^{n+u} A_m^2,$$

whence

$$\tfrac{1}{2}\sum_{m=1}^{n+u} A_m(A_m-1) \geqslant \tfrac{1}{2}ru\Big(\frac{ru}{n+u}-1\Big).$$

On the other hand, for any pair a_i, a_j counted in (3.13), a_j-a_i is an integer d satisfying $1 \leqslant d \leqslant u-1$, and to each such value of d there corresponds at most one such pair, since the numbers a_j-a_i are all different. The pair which corresponds to a particular d occurs in exactly

$u-d$ of the intervals $\mathscr{I}_\nu$ and hence the total number of pairs counted in the sum (3.13) is at most

$$\sum_{d=1}^{u-1}(u-d) = \tfrac{1}{2}u(u-1).$$

Comparing the upper and lower estimates for the sum (3.13), we have

$$\tfrac{1}{2}ru(ru-n-u) \leqslant \tfrac{1}{2}u(u-1)(u+n),$$

whence (since $1 \leqslant u < n$)

$$r(ru-2n) < u(u+n),$$

so that

$$r < \frac{n}{u}+\left(n+u+\frac{n^2}{u^2}\right)^{\frac{1}{2}}.$$

On choosing $u = [n^{\frac{3}{4}}]$, we obtain

$$r < n^{\frac{1}{2}}+O(n^{\frac{1}{4}}),$$

and this is the desired result (3.9).

As is explained in the discussion following the statement of Theorem 3 in § 2, it is an immediate consequence of that theorem that if h is any integer satisfying $h \geqslant 2$ and $m = p^u$ is a prime power, then there exists a B_h-sequence which consists of m elements and is entirely contained in the interval $[1, m^h-1]$. Theorem 5 below is merely a restatement of this result, whilst the corollary follows at once on observing that if a sequence $\mathscr{A}$ lies in the class B_h, then so does every translate of $\mathscr{A}$.

Theorem 5 (Bose–Chowla[7]). *Let h be an integer satisfying $h \geqslant 2$, and let m be a prime power. Then we have*

$$F_h(m^h-1) \geqslant m. \tag{3.14}$$

Corollary. *Let t be a non-negative integer, and let m, h be defined as in the theorem. Then there exists a B_h-sequence which consists of m elements and is entirely contained in the interval $[t, t+m^h-2]$.*

In fact, as was already mentioned in the discussion following Theorem 3, Bose–Chowla proved the slightly more precise result (retaining the above notation)

$$F_h(m^h) \geqslant m+1;$$

but this improvement is inconsequential for our purposes. It is also worth mentioning that Bose–Chowla deduced an analogous inequality from a generalization of Theorem 1′; if this inequality were used in place of (3.14) for our subsequent applications, it would lead to exactly the same theorems.

It follows from Theorem 5 (on taking $m = p$ and noting that $F_h(n)$ is an increasing function of n) that for each $h \geqslant 2$,

(3.15) $$F_h(p^h) \geqslant p \quad \text{for every prime } p.$$

Thus (3.8) follows immediately in view of the fact† that

(3.16) $$\lim_{p\to\infty} (p'/p) = 1, \quad \textit{where } p, p' \textit{ are consecutive primes.}$$

If in place of (3.16) we use the much more profound result‡

(3.17) $$p'-p = O(p^{5/8}),$$

on the set of all pairs of consecutive primes p, p',

we obtain Theorem 6 below. For, corresponding to any $n \geqslant 2^h$, we can choose a pair of consecutive primes p, p' to satisfy

$$p \leqslant n^{1/h} < p'$$

and thus obtain (on using (3.17))

$$F_h(n) \geqslant F_h(p^h) \geqslant p = n^{1/h} - O(n^{5/(8h)}).$$

THEOREM 6 (Bose–Chowla(7)). *We have, for each integer* $h \geqslant 2$,

(3.18) $$F_h(n) > n^{1/h} - O(n^{5/(8h)}).$$

To sum up the situation in the case $h = 2$, we restate Theorem 4 and the case $h = 2$ of Theorem 6 in the form of a single theorem.

THEOREM 7. *We have*

(3.19) $$n^{\frac{1}{2}} - O(n^{5/16}) < F_2(n) < n^{\frac{1}{2}} + O(n^{\frac{1}{4}}),$$

and in particular, (3.7) *holds*.

We now turn to the investigation of infinite B_2-sequences $\mathscr{A}$. Erdös showed(8) that

(3.20) $$\liminf_{n\to\infty} n^{-\frac{1}{2}} A(n) = 0 \quad \textit{for every (infinite) } B_2\textit{-sequence } \mathscr{A};$$

and, on the other hand, he established(8) by means of an explicit construction, that

(3.21) *there exists an (infinite)* B_2*-sequence* $\mathscr{A}$ *satisfying*

$$\limsup_{n\to\infty} n^{-\frac{1}{2}} A(n) \geqslant \tfrac{1}{2}.$$

† Cf. Appendix, § 2, the corollary to Lemma 5.

‡ A. E. Ingham, 'On the difference between consecutive primes', *Quart. J. Math.* **8** (1937) 255–66. Actually, the exponent 5/8 in (3.17) can be improved slightly, in the light of recent work on the order of magnitude of $\zeta(\frac{1}{2}+it)$. Such an improvement carries over to Theorem 6 (cf. Chapter III, § 14, ref. (7)).

(8) A. Stöhr, 'Gelöste und ungelöste Fragen über Basen der natürlichen Zahlenreihe' II, *J. reine angew. Math.* **194** (1955) 111–40.

Theorems 8 and 9 are due to Erdös, but were communicated by him to Stöhr for inclusion in Stöhr's survey. Accordingly, a number of references in the text are to Erdös.(8)

In the case of each of (3.20) and (3.21), a more precise result is known. Erdös[8] stated that the argument he used to prove (3.20) could be adapted to yield the more precise Theorem 8 below, whilst Krückeberg[9] constructed an infinite B_2-sequence $\mathscr{A}$ satisfying

$$\limsup_{n\to\infty} n^{-\frac{1}{2}}A(n) \geqslant \frac{1}{\sqrt{2}}.$$

Although Krückeberg's result supersedes (3.21), we shall prove these two results separately so as to describe both Erdös's method and Krückeberg's method. Erdös's construction is perhaps still of some interest (despite the fact that it leads to the weaker conclusion) because, unlike Krückeberg's, it does not depend on any of the results proved in § 2 of this chapter; finite fields play a part in Erdös's proof, but only in that quadratic residues modulo p are used in the construction.

THEOREM 8 (Erdös[8]). *The relation*

$$\liminf_{n\to\infty} n^{-\frac{1}{2}}(\log n)^{\frac{1}{2}}A(n) \ll 1 \tag{3.22}$$

holds on the class of all infinite sequences $\mathscr{A}$ in B_2.

Proof. Let N be a large integer. We write

$$\tau_{\mathscr{A}}(N) = \inf_{n\geqslant N} A(n)(\log n/n)^{\frac{1}{2}},$$

and shall show that $\tau_{\mathscr{A}}(N) \ll 1$, where the implied constant is absolute.

Let D_l denote the number of elements of a given B_2-sequence $\mathscr{A}$ lying in the interval $(l-1)N < a \leqslant lN$. The number of solutions of

$$0 < a_i - a_j < N,$$

in elements a_i, a_j of $\mathscr{A}$, is less than N, so that

$$\sum_{l=1}^{\infty} \tfrac{1}{2}D_l(D_l-1) < N. \tag{3.23}$$

But if $D_l > 1$, we have $D_l(D_l-1) \geqslant \frac{1}{2}D_l^2$, and it follows from (3.23) that

$$\sum_{l=1}^{N} D_l^2 = \sum_{\substack{l=1\\ D_l=1}}^{N} 1 + \sum_{\substack{l=1\\ D_l>1}}^{N} D_l^2 \leqslant N + 2\sum_{l=1}^{\infty} D_l(D_l-1) < 5N. \tag{3.24}$$

On the other hand, by Cauchy's inequality,

$$\left(\sum_{l=1}^{N} \frac{1}{l}\right)\left(\sum_{l=1}^{N} D_l^2\right) \geqslant \left(\sum_{l=1}^{N} \frac{D_l}{l^{\frac{1}{2}}}\right)^2. \tag{3.25}$$

(9) F. Krückeberg, 'B_2-Folgen und verwandte Zahlenfolgen', *J. reine angew. Math.* **206** (1961) 53–60.

Furthermore,

$$\sum_{l=1}^{N} \frac{D_l}{l^{\frac{1}{2}}} = \sum_{l=1}^{N} \{A(lN)-A((l-1)N)\}\frac{1}{l^{\frac{1}{2}}}$$

$$= \sum_{l=1}^{N} A(lN)\Big\{\frac{1}{l^{\frac{1}{2}}}-\frac{1}{(l+1)^{\frac{1}{2}}}\Big\}+\frac{A(N^2)}{(N+1)^{\frac{1}{2}}}$$

$$\geqslant \tau_{\mathscr{A}}(N)\sum_{l=1}^{N}\Big(\frac{lN}{\log lN}\Big)^{\frac{1}{2}}\Big\{\frac{1}{l^{\frac{1}{2}}}-\frac{1}{(l+1)^{\frac{1}{2}}}\Big\}$$

$$\gg \tau_{\mathscr{A}}(N)\Big(\frac{N}{\log N}\Big)^{\frac{1}{2}}\sum_{l=1}^{N}\frac{1}{l}.$$

Substituting in (3.25), we obtain

$$\sum_{l=1}^{N} D_l^2 \gg N\tau_{\mathscr{A}}^2(N)$$

and (3.24) now yields the required inequality $\tau_{\mathscr{A}}(N) \ll 1$. This completes the proof of the theorem.

Theorem 9 below has already been stated as (3.21).

THEOREM 9 (Erdös[(8)]). *There exists an (infinite) sequence $\mathscr{A}$ in B_2, for which*

$$\limsup_{n\to\infty} n^{-\frac{1}{2}}A(n) \geqslant \tfrac{1}{2}.$$

Proof. Corresponding to each odd prime p we construct the set $\mathscr{A}_p$ of $p-1$ numbers

(3.26) $$a_k = 2p(k+p)+(k^2)_p \quad (k = 1,2,\ldots,p-1),$$

where $(k^2)_p$ denotes the integer u determined by $u \equiv k^2 \pmod{p}$, $1 \leqslant u \leqslant p-1$. We note that $\mathscr{A}_p$ is a B_2-sequence. For if

$$a_{k_1}+a_{k_2} = a_{k_3}+a_{k_4},$$

we have $$2p(k_1+k_2-k_3-k_4) = (k_3^2)_p+(k_4^2)_p-(k_1^2)_p-(k_2^2)_p,$$

so that $k_1+k_2 = k_3+k_4$ and $k_1^2+k_2^2 \equiv k_3^2+k_4^2 \pmod{p}$; it follows at once that either $k_1 = k_3$ and $k_2 = k_4$ or $k_1 = k_4$ and $k_2 = k_3$.

We note that

(3.27) *$\mathscr{A}_p$ is a subset of the interval $2p^2 < a < 4p^2-p$;*

and if a', a'' are any two distinct elements of $\mathscr{A}_p$, then

(3.28) $$p < |a'-a''| < 2p^2-p.$$

The aim of the subsequent argument is to show that if the sets $\mathscr{A}_p$ corresponding to a sufficiently rapidly increasing sequence $\mathscr{P}$ of primes

p are superimposed, the composite sequence

$$\mathscr{A} = \mathscr{A}(\mathscr{P}) = \bigcup_{p \in \mathscr{P}} \mathscr{A}_p$$

also lies in B_2. This will suffice to prove the theorem; for $\mathscr{A}_p$ contains $p-1$ elements, so that by (3.27),

$$A(4p^2-p-1) \geqslant p-1 \quad \text{for } p \in \mathscr{P}.$$

Suppose then that $p_1 < p_2 < \ldots$ are the primes of $\mathscr{P}$, and that

$$p_{r+1} \geqslant 4p_r^2 \quad (r = 1, 2, \ldots). \tag{3.29}$$

It will suffice to show that if $a^{(1)}$, $a^{(2)}$, $a^{(3)}$, $a^{(4)}$ are elements of $\mathscr{A}(\mathscr{P})$, with $a^{(1)} \neq a^{(2)}$ and $a^{(1)} \neq a^{(3)}$, then

$$a^{(1)}-a^{(2)} = a^{(3)}-a^{(4)} \tag{3.30}$$

cannot hold. For this purpose we may suppose that $a^{(1)}$ is the greatest of the four a's under consideration and, interchanging $a^{(2)}$ and $a^{(3)}$ if necessary, that

$$a^{(1)} > a^{(2)} \geqslant a^{(3)} > a^{(4)}. \tag{3.31}$$

Suppose that $a^{(1)} \in \mathscr{A}_{p_s}$. Since (3.30) cannot be satisfied with the a's all in $\mathscr{A}_{p_s}$, we may assume that $a^{(4)} \notin \mathscr{A}_{p_s}$ (and, in particular, $s > 1$). We consider the two cases arising according as $a^{(3)}$ does or does not lie in $\mathscr{A}_{p_s}$.

Suppose first that $a^{(3)}$ lies in $\mathscr{A}_{p_s}$ (so that, by (3.31), $a^{(2)}$ also lies in $\mathscr{A}_{p_s}$). Then, by (3.28),

$$a^{(1)}-a^{(2)} < 2p_s^2-p_s,$$

whilst by (3.27) and (3.29),

$$a^{(3)}-a^{(4)} > 2p_s^2-4p_{s-1}^2 \geqslant 2p_s^2-p_s;$$

and (3.30) cannot hold. Now suppose that $a^{(3)}$ does not lie in $\mathscr{A}_{p_s}$. Then by (3.27), (3.28), and (3.29),

$$a^{(1)}-a^{(2)} > \min(p_s, 2p_s^2-4p_{s-1}^2) = p_s,$$

whilst by (3.27) and (3.29),

$$a^{(3)}-a^{(4)} < 4p_{s-1}^2 \leqslant p_s;$$

and again, (3.30) cannot hold. This completes the proof of the theorem.

Theorem 10, below, has also been mentioned already.

THEOREM 10 (Krückeberg[9]). *There exists an (infinite) sequence $\mathscr{A}$ in B_2, for which*

$$\limsup_{n\to\infty} n^{-\frac{1}{2}}A(n) \geqslant \frac{1}{\sqrt{2}}.$$

Proof. We begin with an informal discussion of the ideas underlying the proof. The formal proof will commence with the statement of

Lemma 1. We use $|\mathscr{S}|$ to denote the number of elements of a finite sequence $\mathscr{S}$.

Here again, $\mathscr{P}$ will be a sufficiently rapidly increasing sequence of primes

$$p_1, p_2, p_3, \ldots \tag{3.32}$$

(although not the sequence used in the proof of Theorem 9). The sequence $\mathscr{A}$, to be constructed, will again be of the form

$$\mathscr{A} = \mathscr{A}(\mathscr{P}) = \bigcup_{p \in \mathscr{P}} \mathscr{A}_p, \tag{3.33}$$

where each $\mathscr{A}_p$ is a finite B_2-sequence. For each $p \in \mathscr{P}$, $\mathscr{A}_p$ will be defined so as to lie entirely within the interval

$$\mathscr{I}_p = [t(p), t(p)+p^2], \tag{3.34}$$

where $t(n)$ is a suitably chosen positive function of the integer variable n.

The corollary to Theorem 5 (with $m = p$, $h = 2$) tells us that there exists a B_2-sequence $\mathscr{T}_p$, with $|\mathscr{T}_p| = p$, entirely contained in the interval $\mathscr{I}_p$. But the sets $\mathscr{T}_p$ cannot serve as the $\mathscr{A}_p$ above, since their union is not necessarily a B_2-sequence.

In the circumstances specified in the lemma below, however, the union of two finite B_2-sequences $\mathscr{V}$, $\mathscr{W}$ can be modified to become a B_2-sequence by omitting, from $\mathscr{W}$, elements whose total number is bounded in terms of $|\mathscr{V}|$; so that if $|\mathscr{W}|$ is sufficiently large compared to $|\mathscr{V}|$, only a small *proportion* of the elements of $\mathscr{W}$ need be omitted. The sequences $\mathscr{A}_p$, mentioned above, will be chosen by means of successive applications of this lemma; if $\mathscr{A}_{p_1}, \mathscr{A}_{p_2}, \ldots, \mathscr{A}_{p_i}$ are already chosen so that

$$\mathscr{V}_i = \bigcup_{j=1}^{i} \mathscr{A}_{p_j} \tag{3.35}$$

is a B_2-sequence, we can (provided the function $t(n)$ and the prime p_{i+1} are chosen so as to lead to a suitable interval $\mathscr{I}_{p_{i+1}}$) apply the lemma with $\mathscr{V} = \mathscr{V}_i$, $\mathscr{W} = \mathscr{T}_{p_{i+1}}$ in order to find a subsequence $\mathscr{A}_{p_{i+1}}$ of $\mathscr{T}_{p_{i+1}}$, such that

$$\mathscr{V}_{i+1} = \mathscr{V}_i \cup \mathscr{A}_{p_{i+1}}$$

is again a B_2-sequence. The *proportion* of the elements that need be omitted from $\mathscr{T}_{p_j}$ to form $\mathscr{A}_{p_j}$ will tend to zero as $j \to \infty$, provided the sequence (3.32) is sufficiently rapidly increasing; furthermore, this latter requirement will not reduce the effectiveness of the resulting relation

$$\limsup_{j\to\infty} \frac{A(t(p_j)+p_j^2)}{p_j} \geqslant \lim_{j\to\infty} \frac{|\mathscr{A}_{p_j}|}{p_j} = 1 \tag{3.36}$$

(where $A(n)$ is the counting number of the sequence (3.33)). Unfortunately, however, it will be necessary to choose a function $t(n)$ which is sufficiently rapidly increasing, in order to satisfy the hypothesis (3.38) of the lemma, and this is the limiting factor in the effectiveness of (3.36). It will be possible to choose $t(n)$ so that $t(n) \sim n^2$ as $n \to \infty$, and this implies

$$\frac{1}{p_j} \sim \frac{\sqrt{2}}{\sqrt{(t(p_j)+p_j^2)}}, \tag{3.37}$$

which (in conjunction with (3.36)) leads to the result of the theorem. This concludes the informal discussion.

LEMMA 1. *Let $\mathscr{V}$ be a (finite) B_2-sequence entirely contained in the interval $[V_1, V_2]$, and let $\mathscr{W}$ be a B_2-sequence entirely contained in the interval $[W_1, W_2]$; where V_1, V_2, W_1, W_2 are integers satisfying $V_2 \geqslant V_1$ and $W_2 \geqslant W_1$.*

Suppose that

$$W_1 - V_2 > \max(V_2 - V_1, W_2 - W_1). \tag{3.38}$$

Then there exists a subsequence $\mathscr{W}^*$ of $\mathscr{W}$, such that

$$|\mathscr{W}^*| \geqslant |\mathscr{W}| - |\mathscr{V}|(|\mathscr{V}|-1) \tag{3.39}$$

and the sequence $\mathscr{V} \cup \mathscr{W}^*$ is a B_2-sequence.

Proof of the lemma. We consider (cf. (3.30), (3.31) and the appropriate explanation) the equation

$$u^{(1)} - u^{(2)} = u^{(3)} - u^{(4)}, \tag{3.40}$$

where the u's are elements of the sequence $\mathscr{U} = \mathscr{V} \cup \mathscr{W}$ satisfying

$$u^{(1)} > u^{(2)} \geqslant u^{(3)} > u^{(4)}. \tag{3.41}$$

To prove the lemma we must show that it is possible to form a a sequence $\mathscr{W}^*$, by omitting at most

$$|\mathscr{V}|(|\mathscr{V}|-1)$$

elements from $\mathscr{W}$, such that (3.40) subject to (3.41) has no solution with each u lying in the sequence $\mathscr{U}^* = \mathscr{V} \cup \mathscr{W}^*$.

We first show that

(I) *for every solution of* (3.40) *subject to* (3.41), *with the u's lying in $\mathscr{U}$, we have*

$$u^{(1)} \in \mathscr{W}, \quad u^{(2)} \in \mathscr{W}, \quad u^{(3)} \in \mathscr{V}, \quad u^{(4)} \in \mathscr{V}. \tag{3.42}$$

Now $u^{(1)} \in \mathscr{V}$ is impossible, for this (in conjunction with (3.41) and the fact that $V_2 < W_1$) would imply that $u^{(2)}$, $u^{(3)}$, $u^{(4)}$ also lie in $\mathscr{V}$, contradicting the fact that $\mathscr{V}$ is a B_2-sequence. Furthermore, $u^{(4)} \in \mathscr{W}$ would

imply that all the u's lie in $\mathscr{W}$, contradicting the fact that $\mathscr{W}$ is a B_2-sequence. Thus
$$u^{(1)} \in \mathscr{W}, \qquad u^{(4)} \in \mathscr{V}.$$
Now $u^{(2)}$, $u^{(3)}$ cannot both lie in $\mathscr{W}$, for the resulting inequalities
$$u^{(1)}-u^{(2)} \leqslant W_2-W_1, \qquad u^{(3)}-u^{(4)} \geqslant W_1-V_2$$
would contradict (3.40) in view of (3.38); similarly $u^{(2)}$, $u^{(3)}$ cannot both lie in $\mathscr{V}$, as the resulting inequalities
$$u^{(1)}-u^{(2)} \geqslant W_1-V_2, \qquad u^{(3)}-u^{(4)} \leqslant V_2-V_1$$
again contradict (3.40) in view of (3.38). This completes the proof of the statement (I) above.

Now there are exactly $\frac{1}{2}|\mathscr{V}|(|\mathscr{V}|-1)$ distinct pairs of integers $u^{(3)}$, $u^{(4)}$ which satisfy $u^{(3)} > u^{(4)}$ (cf. (3.41)) and lie in $\mathscr{V}$ (cf. (3.42)). To each such pair $u^{(3)}$, $u^{(4)}$, there corresponds at most one pair $u^{(1)}$, $u^{(2)}$ of elements of the B_2-sequence $\mathscr{W}$ (cf. (3.42)) satisfying (3.40); for non-zero differences resulting from distinct pairs of elements of a B_2-sequence are distinct. Thus the union $\mathscr{W}'$ of all such pairs $u^{(1)}$, $u^{(2)}$ contains at most
$$|\mathscr{V}|(|\mathscr{V}|-1)$$
elements, and it is clear that we can form the desired sequence $\mathscr{W}^*$ by omitting the elements of $\mathscr{W}'$ from $\mathscr{W}$. This concludes the proof of the lemma.

We now proceed with the formal proof of Theorem 10. Let

(3.43) $$t(n) = n^2+n \quad (n = 1, 2, 3,...),$$

although we shall use only the fact that $t(n)$ is a positive function satisfying

(3.44) $$\lim_{n\to\infty} (t(n)-n^2) = \infty, \qquad \lim_{n\to\infty} \frac{t(n)}{n^2} = 1.$$

Furthermore, let $\{\epsilon_j\}$ be a sequence of positive numbers which converges to zero, say

(3.45) $$\epsilon_j = j^{-1} \quad (j = 1, 2, 3,...).$$

We shall define (in the order indicated)
$$p_1, \quad \mathscr{A}_{p_1}, \quad p_2, \quad \mathscr{A}_{p_2}, \quad p_3, \quad \mathscr{A}_{p_3},...$$
inductively in such a way that the sequence (3.35) is a B_2-sequence for every i, and

(3.46) $$|\mathscr{A}_{p_j}| > (1-\epsilon_j)p_j \quad (j = 2, 3,...);$$

as was described in the informal discussion, for each $p \in \mathscr{P}$, $\mathscr{A}_p$ will be a subsequence of a sequence $\mathscr{T}_p$ containing p elements in the interval (3.34). This will suffice to prove the theorem, for (3.46) will imply (3.36), whilst the second assertion of (3.44) ensures the truth of (3.37).

We take $p_1 = 2$ and $\mathscr{A}_{p_1}$ to be the empty sequence. Let $i \geqslant 1$, and suppose that

$$p_1, \mathscr{A}_{p_1}, \quad \ldots, \quad p_i, \mathscr{A}_{p_i}$$

have already been chosen (by means of the inductive procedure we are describing). We define p_{i+1} to be the least prime p satisfying (3.47), (3.48) below, where $\mathscr{V}_i$ is defined by (3.35); the existence of such a prime p is ensured by the first assertion of (3.44).

$$t(p)-(t(p_i)+p_i^2) > \max(t(p_i)+p_i^2, p^2), \tag{3.47}$$

$$|\mathscr{V}_i|(|\mathscr{V}_i|-1) < \epsilon_{i+1}p. \tag{3.48}$$

Now $\mathscr{V}_i$ is a B_2-sequence in accordance with the inductive process, and by the corollary to Theorem 5 (with $m = p$, $h = 2$) there exists a B_2-sequence $\mathscr{T}$, such that

$$\mathscr{T} \subset [t(p_{i+1}), t(p_{i+1})+p_{i+1}^2], \qquad |\mathscr{T}| = p_{i+1}.$$

We apply the lemma, with $\mathscr{V} = \mathscr{V}_i$, $\mathscr{W} = \mathscr{T}$; on taking

$$V_1 = 0, \qquad V_2 = t(p_i)+p_i^2, \qquad W_1 = t(p_{i+1}),$$
$$W_2 = t(p_{i+1})+p_{i+1}^2,$$

the hypothesis (3.38) of the lemma, which now takes the form

$$t(p_{i+1})-(t(p_i)+p_i^2) > \max(t(p_i)+p_i^2, p_{i+1}^2)$$

is satisfied since (3.47) holds with $p = p_{i+1}$. We define $\mathscr{A}_{p_{i+1}} = \mathscr{W}^*$, where $\mathscr{W}^*$ is the subsequence of $\mathscr{T}$ arising from this application of the lemma. Then

$$\mathscr{V}_{i+1} = \mathscr{V}_i \cup \mathscr{A}_{p_{i+1}}$$

is a B_2-sequence, and

$$|\mathscr{A}_{p_{i+1}}| \geqslant |\mathscr{T}|-|\mathscr{V}_i|(|\mathscr{V}_i|-1).$$

Since $|\mathscr{T}| = p_{i+1}$, it follows from (3.48), with $p = p_{i+1}$, that (3.46) holds when $j = i+1$. This justifies the inductive procedure, and completes the proof of the theorem.

We conclude the section with a number of interesting open questions which may suggest further lines of investigation. Stöhr has already drawn attention to most of these questions in his survey(8).

(i) For any sequence $\mathscr{A}$, let

$$\lambda(\mathscr{A}) = \liminf_{n\to\infty} n^{-\frac{1}{2}}(\log n)^{\frac{1}{2}}A(n),$$

$$\Lambda(\mathscr{A}) = \limsup_{n\to\infty} n^{-\frac{1}{2}}(\log n)^{\frac{1}{2}}A(n).$$

In view of Theorem 8, the number

$$\mu = \sup_{\mathscr{A}\in B_2} \lambda(\mathscr{A})$$

is finite. It is natural to ask, in the first place, whether or not $\mu = 0$; and an answer to this question would raise others. If $\mu = 0$, there may be results of the type

$$\liminf_{n\to\infty} n^{-\frac{1}{2}}\psi(n)A(n) < \infty \quad \text{for } \mathscr{A} \in B_2, \tag{3.49}$$

where $\psi(n)$ is a positive function which increases more rapidly than $(\log n)^{\frac{1}{2}}$; one would seek a function $\psi(n)$ for which (3.49) is best possible in a suitable sense. If $\mu > 0$, on the other hand, one could ask, for example, whether there exists a B_2-sequence $\mathscr{A}$ such that

$$0 < \lambda(\mathscr{A}), \qquad \Lambda(\mathscr{A}) < \infty.$$

(ii) We have seen that substantial progress has been made with questions concerning $B_2[1]$ sequences. Theorems 4 and 6 (the case $h = 2$ of the latter theorem being the relevant case here) are surprisingly precise, whilst the gap between Theorems 8 and 10 is not unduly large. In comparison, very little is known concerning the corresponding questions relating to $B_2[g]$ sequences. (Here the function $F_2(n)$ is replaced by the function $F_2(g; n)$ which results when the maximal selection in Definition 4 is so as to form a $B_2[g]$ sequence.) Since the class $B_2[1]$ is contained in the class $B_2[g]$ for all $g > 1$, Theorems 6 and 10 remain valid for $B_2[g]$ sequences; but it would be desirable to obtain corresponding results whose precision increases with g. The proofs of Theorems 4 and 8, on the other hand, are applicable only to the case $g = 1$, and the more general question remains open.

(iii) We have already mentioned that it is not yet known whether or not

$$\lim_{n\to\infty} n^{-1/h}F_h(n) = 1$$

when $h > 2$. If the answer is affirmative, it would be natural to seek generalizations, relating to B_h-sequences for $h \geqslant 2$, of Theorems 4 and 8. As Theorem 5 is available for $h \geqslant 2$, it may not be unduly difficult to obtain a generalization of Theorem 10.

(iv) The results (3.50) and (3.51) below are straightforward deductions from Theorems 8 and 9 respectively.

(3.50) *For every B_2-sequence $\mathscr{A} = \{a_j\}$,*

$$\limsup_{j\to\infty} \frac{a_j}{j^2 \log j} > 0.$$

(3.51) *There exists a B_2-sequence $\mathscr{A} = \{a_j\}$ such that*

$$\liminf_{j\to\infty} \frac{a_j}{j^2} < \infty.$$

Erdös[10] has conjectured that corresponding to every $\epsilon > 0$, there exists a B_2-sequence $\mathscr{A}$ for which $a_n \ll n^{2+\epsilon}$ for all n. Two partial results (which we describe below) are known in this connexion, but the conjecture remains open.

It was pointed out by Mian–Chowla[11] that

(3.52) *there exists a B_2-sequence $\mathscr{A} = \{a_n\}$ for which $a_n \ll n^3$.*

Mian–Chowla demonstrated this by means of the following recursive construction. Let $a_1 = 1$, and suppose that $a_1, a_2,..., a_m$ have already been chosen (where $m \geqslant 1$). Take a_{m+1} to be the least natural number differing from all the numbers $a_r+a_s-a_t$ with $1 \leqslant r, s, t \leqslant m$. Since there exist at most m^3 such triads r, s, t, the corresponding numbers $a_r+a_s-a_t$ can account for at most m^3 integers, so that $a_{m+1} < (m+1)^3$. Thus $a_n \leqslant n^3$ for all n, whilst the sequence† $\mathscr{A} = \{a_n\}$ is obviously a B_2-sequence.

The Mian–Chowla result is, of its kind, the most precise known. But Erdös–Rényi proved a more profound result which, whilst rather different in nature, provides a partial answer to Erdös's question. This result, which will be restated (Chapter III, § 2, Theorem 2) and proved in the next chapter, asserts that

(3.53) *corresponding to every $\epsilon > 0$, there exists a number $g = g(\epsilon)$ and a $B_2[g]$-sequence $\mathscr{A} = \{a_n\}$ such that $a_n \ll n^{2+\epsilon}$.*

4. The Erdös–Fuchs theorem

We have already discussed the background of the Erdös–Fuchs theorem (Theorem 11 below) in § 1. The following two comparisons will serve to illustrate the power of this theorem. In the first place, there is the striking contrast between the mild conjecture formulated by Erdös–Turan[4]‡ in 1941, namely that (subject to the same hypotheses as in the theorem)

$$\sum_{n=0}^{N} R_n = cN+O(1)$$

cannot hold (which remained unproved until Erdös–Fuchs obtained their theorem in 1954), and the remarkable precision of the result actually proved by Erdös–Fuchs. Secondly, we recall (cf. § 1) that in

† The first ten terms of the sequence $\mathscr{A}$ constructed in this way are 1, 2, 4, 8, 13, 21, 31, 45, 66, 81.

‡ Loc. cit., p. 214.

(10) P. Erdös, *Colloque sur la Théorie des Nombres* (Bruxelles, 1956), section 3, 127–37.

(11) A. Mian and S. Chowla, 'On the B_2-sequences of Sidon', *Proc. natn. Acad. Sci. India*, Sect. A, **14** (1944) 3–4.

the very special case when $\mathscr{A}$ is the sequence $\mathscr{Q}$ of squares, the corresponding result of Hardy–Landau was that

$$\sum_{n=0}^{N} R_n(\mathscr{Q}) = cN+o\{N^{\frac{1}{4}}(\log N)^{\frac{1}{4}}\}$$

cannot hold; in comparison, the Erdös–Fuchs theorem assumes nothing concerning the nature of $\mathscr{A}$, and yet this complete generality is achieved at the cost of only a very slight loss of precision (only the exponent of $\log N$ differs in the two results—see also reference (14)).

Theorems† of the Erdös–Fuchs type still hold when the principal term cN on the right-hand side of (4.1) is replaced by other suitable functions of N (with appropriate modification of the error term).

THEOREM 11 (Erdös–Fuchs[2]). *Let $\mathscr{A} = \{a_j\}$ be a sequence‡ of non-negative integers, and let c be a positive constant. We denote by $R_n = R_n(\mathscr{A})$ (as in § 1) the number of representations of the integer n in the form*

$$n = a_i + a_j.$$

Then the relation

$$\sum_{n=0}^{N} R_n = cN+o(N^{\frac{1}{4}}\log^{-\frac{1}{2}} N) \tag{4.1}$$

cannot hold.

Proof. We suppose that (4.1) holds, and shall obtain a contradiction.

Since $R_n(\mathscr{A})$ does not depend on the ordering of $\mathscr{A}$, we may suppose that $\mathscr{A} = \{a_j\}$ satisfies

$$0 \leqslant a_1 \leqslant a_2 \leqslant a_3 \leqslant \dots. \tag{4.2}$$

Furthermore, since (4.1) implies that R_n is finite, we may suppose that there cannot be an infinity of successive equalities in (4.2).

Throughout the proof, z denotes a complex variable; we shall be concerned solely with the region $|z| < 1$. We introduce the generating function corresponding to the sequence $\mathscr{A}$, namely

$$g(z) = \sum_{n=1}^{\infty} z^{a_n} = \sum_{m=0}^{\infty} c_m z^m, \tag{4.3}$$

where c_m counts the number of times m appears in $\mathscr{A}$, and is therefore a non-negative integer for each m. Also $c_m \ll m$ by§ (4.1), so that the series (4.3) converges for $|z| < 1$.

† See (12) P. T. Bateman, E. F. Kohlbecker, and J. P. Tull, 'On a theorem of Erdös and Fuchs in additive number theory', *Proc. Am. math. Soc.* **14** (1963) 278–84.

‡ Here we do not assume that the integers in $\mathscr{A}$ are distinct. In fact even the assumption that the elements of $\mathscr{A}$ are integers is superfluous (cf. Erdös–Fuchs[2], loc. cit., p. 68).

§ For $N = a_1+m$, we have $R_N \geqslant c_m$.

The generating function corresponding to the left-hand side of (4.1) is

$$\sum_{N=0}^{\infty}\Big(\sum_{n=0}^{N} R_n\Big)\, z^N = (1-z)^{-1}g^2(z) \quad (|z|<1)$$

whilst the generating function

$$\sum_{N=0}^{\infty} cNz^N = cz(1-z)^{-2} \quad (|z|<1)$$

corresponds to the first term on the right-hand side of (4.1). We may now epitomize (4.1) by the relation

$$(1-z)^{-1}g^2(z) = cz(1-z)^{-2}+\sum_{N=0}^{\infty} v_N z^N \quad (|z|<1),$$

where the v_N are real numbers satisfying

$$v_N = o(N^{\frac{1}{2}}\log^{-\frac{1}{2}}N). \tag{4.4}$$

On multiplying by $(1-z)$, we obtain

$$g^2(z) = P(z)+E(z) \quad (|z|<1), \tag{4.5}$$

where the 'principal' term $P(z)$ is given by

$$P(z) = cz(1-z)^{-1} \tag{4.6}$$

and the 'error' by

$$E(z) = (1-z)\sum_{N=0}^{\infty} v_N z^N. \tag{4.7}$$

It is natural to try to obtain a contradiction from (4.5) by exploiting some difference in the behaviour of the functions $g^2(z)$ and $P(z)$, in the hope that (4.4) can be used to show that $E(z)$ cannot compensate for this distinction.

The fact that $g^2(z)$ is the square of a power series suggests one way of distinguishing between $g^2(z)$ and $P(z)$; using the relation

$$\frac{1}{2\pi}\int_{-\pi}^{\pi} |g(re^{i\alpha})|^2\, d\alpha = \sum_{m=0}^{\infty} |c_m|^2 r^{2m} \quad (0<r<1), \tag{4.8}$$

it will be comparatively easy to show that as $r \to 1-0$, the mean value of $|g^2(z)|$, on the circle $|z| = r$, grows much more rapidly than the corresponding mean value of $|P(z)|$. This fact is not immediately applicable, however, for (4.4) does not enable one to obtain a sufficiently effective estimate for the contribution of $|E(z)|$ on the entire circle $|z| = r$. The relation (4.4) does, however, enable one to show that $|E(z)|$ is small compared to $|P(z)|$ on the part of the circle $|z| = r$ (where $1-r$ is positive and *small*) lying in a neighbourhood of the point $z = 1$.

The crucial step in the proof will be the application of Lemma 2 (below) which provides, for any θ satisfying $0 < \theta \leqslant \pi$, a lower bound for

$$\frac{1}{2\theta}\int_{-\theta}^{\theta} |g(re^{i\alpha})|^2\,d\alpha$$

in terms of the expression (4.8). This makes it possible to obtain the contradiction by comparing the mean values of $|g^2(z)|$ and $|P(z)|$ on a small arc of the circle $|z| = r$, on which (4.4) is effective for estimating the contribution of $|E(z)|$.

LEMMA 2. *Suppose that*

(4.9) $$\phi(z) = \sum_{n=0}^{\infty} b_n z^n,$$

where the coefficients b_n *are non-negative real numbers, and suppose that the series* (4.9) *converges for* $|z| < 1$. *Then, for* $0 < r < 1$, $0 < \theta \leqslant \pi$,

(4.10) $$\int_{-\theta}^{\theta} |\phi(re^{i\alpha})|^2\,d\alpha \gg \theta \int_{-\pi}^{\pi} |\phi(re^{i\alpha})|^2\,d\alpha,$$

where the implicit constant is absolute.

Proof. We define the function $h(\alpha)$ in the interval $[-\pi, \pi]$, by

$$h(\alpha) = h(\theta;\alpha) = \begin{cases} 1-|\alpha/\theta| & \text{for } |\alpha| < \theta \\ 0 & \text{for } \theta \leqslant |\alpha| \leqslant \pi. \end{cases}$$

We shall make use of the following three properties of the function $h(\alpha)$:

(I) $|h(\alpha)| \leqslant 1$ *for* $|\alpha| \leqslant \theta$, *and* $h(\alpha) = 0$ *for* $0 \leqslant |\alpha| \leqslant \pi$.

(II) $\int_{-\pi}^{\pi} |h(\alpha)|^2\,d\alpha \gg \theta \quad (0 < \theta \leqslant \pi)$.

(III) *The function* $h(\alpha)$ *has a Fourier expansion of the form*

(4.11) $$h(\alpha) = \sum_{m=-\infty}^{\infty} d_m e^{im\alpha} \quad (-\pi \leqslant \alpha \leqslant \pi),$$

where

(4.12) $$d_m \geqslant 0 \quad \textit{for every } m.$$

Thus, in particular, $\sum d_m$ converges and the series (4.11) is absolutely convergent for all α in $[-\pi, \pi]$.

Here (I) and (II) are obvious, whilst the d_m are given by

$$d_0 = \theta/2\pi, \qquad d_m = (1-\cos m\theta)/(\pi m^2\theta) \quad \text{if } m \neq 0,$$

so that (III) holds.

By (I), we have

$$(4.13) \qquad \int_{-\theta}^{\theta} |\phi(re^{i\alpha})|^2 \, d\alpha \geqslant \int_{-\pi}^{\pi} |h(\alpha)\phi(re^{i\alpha})|^2 \, d\alpha.$$

Now, in view of the remark following (III) and the fact that $\phi(z)$ has radius of convergence at least 1,

$$h(\alpha)\phi(re^{i\alpha}) = \sum_{l=-\infty}^{\infty} e^{il\alpha}\Big(\sum_{\substack{m=-\infty \\ m+n=l}}^{\infty} \sum_{n=0}^{\infty} d_m b_n r^n\Big)$$

so that using Parseval's formula and the fact that the coefficients d_m, b_n are non-negative, we obtain

$$\begin{aligned}\int_{-\pi}^{\pi} |h(\alpha)\phi(re^{i\alpha})|^2 \, d\alpha &= 2\pi \sum_{l=-\infty}^{\infty} \Big(\sum_{\substack{m,n \\ m+n=l}} d_m b_n r^n\Big)^2 \\ &\geqslant 2\pi \sum_{l=-\infty}^{\infty} \Big(\sum_{\substack{m,n \\ m+n=l}} d_m^2 b_n^2 r^{2n}\Big) \\ &= 2\pi\Big(\sum_{m=-\infty}^{\infty} d_m^2\Big)\Big(\sum_{n=0}^{\infty} b_n^2 r^{2n}\Big) \\ &= \frac{1}{2\pi}\Big(\int_{-\pi}^{\pi} |h(\alpha)|^2 \, d\alpha\Big)\Big(\int_{-\pi}^{\pi} |\phi(re^{i\alpha})|^2 \, d\alpha\Big).\end{aligned}$$

Hence, by (4.13) and (II), we obtain (4.10). (Actually, the value of the integral appearing in (II) is $\frac{4}{3}\theta$, so that the implicit constant on the right-hand side of (4.10) may be taken to be $1/(3\pi)$. But the value of the constant is immaterial for our purposes.)

For reference purposes, we state the following well-known result in the form of a lemma.

LEMMA 3. *Let $d > 0$. The binomial coefficients* $\binom{-d}{n}$ *satisfy*

$$(4.14) \qquad (-1)^n \binom{-d}{n} \sim \frac{1}{\Gamma(d)} n^{d-1} \quad \text{as } n \to \infty.$$

We now proceed with the various stages of the proof of the theorem. Throughout, r will denote a real number satisfying

$$(4.15) \qquad 0 < r < 1.$$

We shall let r tend to $1-0$ in the final stage of the argument, so we may suppose that $1-r$ is small. The number

$$(4.16) \qquad \theta = \theta(r)$$

will be chosen later, subject to the condition

$$(4.17) \qquad 1-r < \theta(r) < \pi;$$

we shall compare the mean values of $|g^2|$ and $|P|$ on the arc

$$\{re^{i\alpha}\colon -\theta \leqslant \alpha \leqslant \theta\}.$$

LEMMA 4. *Let $E(z)$ be defined by (4.7) where the v_N satisfy (4.4). Then, for $|z| < 1$,*

$$|E(z)| \ll |1-z|(1-|z|)^{-5/4}. \tag{4.18}$$

Furthermore, if $\eta = \eta(r)$ is a positive function of r which tends to zero sufficiently slowly as $r \to 1-0$ (the explicit requirement being given by (4.24) in the proof below), we have (subject to (4.17))

$$\int_{-\theta}^{\theta} |E(re^{i\alpha})|\, d\alpha \ll (\theta^{\frac{3}{2}}\eta)(1-r)^{-\frac{3}{4}} \log^{-\frac{1}{2}} \frac{1}{1-r}. \tag{4.19}$$

Proof. We have by (4.4) and Lemma 3,

$$\Big|\sum_{n=0}^{\infty} v_n z^n\Big| \leqslant \sum_{n=0}^{\infty} |v_n|\,|z|^n \ll \sum_{n=0}^{\infty} n^{\frac{1}{4}}|z|^n \ll (1-|z|)^{-5/4},$$

and this proves (4.18). (We note that the full force of (4.4) has not been used, and the inequality (4.18) could be sharpened slightly if anything were to be gained by it.)

By Schwarz's inequality, we have

$$\int_{-\theta}^{\theta} |E(re^{i\alpha})|\, d\alpha \leqslant \left\{\int_{-\theta}^{\theta} |1-re^{i\alpha}|^2\, d\alpha\right\}^{\frac{1}{2}} \left\{\int_{-\theta}^{\theta} \Big|\sum_{n=0}^{\infty} v_n r^n e^{in\alpha}\Big|^2 d\alpha\right\}^{\frac{1}{2}}. \tag{4.20}$$

Now

$$\int_{-\theta}^{\theta} |1-re^{i\alpha}|^2\, d\alpha \ll \theta^3, \tag{4.21}$$

for throughout the range of integration we have, in view of (4.17),

$$|1-re^{i\alpha}| \leqslant |1-r| + r|1-e^{i\alpha}| \ll \theta.$$

On the other hand, the second integral on the right-hand side of (4.20) does not exceed

$$\int_{-\pi}^{\pi} \Big|\sum_{n=0}^{\infty} v_n r^n e^{in\alpha}\Big|^2 d\alpha \leqslant 2\pi \sum_{n=0}^{\infty} |v_n|^2 r^{2n}. \tag{4.22}$$

We split up this last sum into the two sums

$$\Sigma_1 = \sum_{n \leqslant (1-r)^{-\frac{1}{2}}} |v_n|^2 r^{2n}, \qquad \Sigma_2 = \sum_{n > (1-r)^{-\frac{1}{2}}} |v_n|^2 r^{2n}.$$

We now express (4.4) in the form

$$|v_n| \leqslant \epsilon(n) n^{\frac{1}{4}} \log^{-\frac{1}{2}} n \quad (n \geqslant 2), \tag{4.23}$$

where $\epsilon(n)$ is a positive function which converges to zero as $n \to \infty$; and we suppose that the positive function $\eta = \eta(r)$ tends to zero sufficiently slowly, as $r \to 1-0$, to ensure that

$$\sup_{n>(1-r)^{-\frac{1}{4}}} \epsilon(n) < \eta(r) \quad \text{and} \quad (1-r)^{\frac{1}{4}} < \eta(r), \tag{4.24}$$

for all r satisfying (4.15) for which $1-r$ is small. Then

$$\Sigma_1 \ll \sum_{n \leqslant (1-r)^{-\frac{1}{4}}} n^{\frac{1}{2}} r^{2n} < \sum_{n \leqslant (1-r)^{-\frac{1}{4}}} n^{\frac{1}{2}} < (1-r)^{-\frac{3}{8}},$$

and

$$\Sigma_2 \leqslant 2(\eta(r))^2 \left(\log \frac{1}{1-r}\right)^{-1} \sum_{n>(1-r)^{-\frac{1}{4}}} n^{\frac{1}{2}} r^{2n}$$

$$\ll (\eta(r))^2 \left(\log \frac{1}{1-r}\right)^{-1} (1-r^2)^{-\frac{3}{2}}$$

by Lemma 3. Since $1-r^2 \geqslant 1-r$, this completes the proof of the lemma, in view of (4.20), (4.21), and (4.22).

Lemma 5. *Let* $P(z)$ *be defined by* (4.6). *Then*

$$\int_{-\pi}^{\pi} |P(re^{i\alpha})| \, d\alpha \ll \log \frac{1}{1-r}. \tag{4.25}$$

Proof. On noting that the integral on the left-hand side of (4.25) is

$$\int_{-\pi}^{\pi} |cre^{i\alpha}(1-re^{i\alpha})^{-1}| \, d\alpha = cr \int_{-\pi}^{\pi} |(1-re^{i\alpha})^{-\frac{1}{2}}|^2 \, d\alpha$$

$$= 2\pi cr \sum_{n=0}^{\infty} \left\{ \binom{-\frac{1}{2}}{n} r^n \right\}^2,$$

the result follows on applying Lemma 3 (again noting that $1-r^2 \geqslant 1-r$).

Lemma 6. *Let* $g(z)$ *be defined by* (4.3). *Then*

$$\int_{-\theta}^{\theta} |g(re^{i\alpha})|^2 \, d\alpha \gg \theta(1-r)^{-\frac{1}{2}}. \tag{4.26}$$

Proof. Since the coefficients c_m appearing in the definition of g (cf. (4.3)) are non-negative integers, we have $c_m^2 \geqslant c_m$ in (4.8), so that

$$\int_{-\pi}^{\pi} |g(re^{i\alpha})|^2 \, d\alpha \geqslant 2\pi g(r^2).$$

Furthermore, applying (4.5), (4.6), and (4.18), with $z = r^2$, we obtain

$$g(r^2) = \{cr^2(1-r^2)^{-1} + O((1-r^2)^{-\frac{1}{2}})\}^{\frac{1}{2}} \gg (1-r)^{-\frac{1}{2}},$$

so that (4.26) holds when $\theta = \pi$. It now follows from Lemma 2 that (4.26) holds for all θ satisfying $0 < \theta \leqslant \pi$, and the lemma is proved.

To complete the proof of the theorem, we need only apply (4.26), (4.25), and (4.19), to show that the inequality

$$(4.27)\qquad \int_{-\theta}^{\theta} |g^2(re^{i\alpha})|\,d\alpha \leqslant \int_{-\theta}^{\theta} |P(re^{i\alpha})|\,d\alpha + \int_{-\theta}^{\theta} |E(re^{i\alpha})|\,d\alpha,$$

resulting from (4.5), leads to a contradiction as $r \to 1-0$; provided the function $\theta = \theta(r)$ is suitably chosen, subject to (4.17).

Let $\eta = \eta(r)$ be a positive function which converges to zero, but sufficiently slowly to ensure that

$$(4.28)\qquad \{\eta(r)\}^{-1} < \log\frac{1}{1-r}$$

and (4.19) hold for the numbers r under consideration (those satisfying (4.15) for which $1-r$ is small). On applying (4.26), (4.25), and (4.19) to the various terms of (4.27), we obtain

$$(4.29)\qquad \theta(1-r)^{-\frac{1}{2}} \ll \log\frac{1}{1-r} + \eta\theta^{\frac{3}{2}}(1-r)^{-\frac{3}{4}}\left\{\log\frac{1}{1-r}\right\}^{-\frac{1}{2}},$$

subject to (4.17).

We define $\theta = \theta(r)$ so as to equalize the two terms on the right-hand side of (4.29), by writing

$$\theta(r) = \{\eta(r)\}^{-\frac{2}{3}}(1-r)^{\frac{1}{2}}\log\frac{1}{1-r};$$

we note that (in view of (4.28)) this choice is consistent with (4.17), since $1-r$ is small. On substituting in (4.29), we obtain (after dividing by $\log(1-r)^{-1}$)

$$\{\eta(r)\}^{-\frac{2}{3}} \ll 1,$$

contradicting the fact that $\eta(r) \to 0$ as $r \to 1-0$. This completes the proof of the theorem.

The above method is equally applicable when the representation function R_n is replaced by r_n or r'_n (cf. Definition 1); the representation functions corresponding to r_n and r'_n being

$$\tfrac{1}{2}\{g^2(z)-g(z^2)\} \quad\text{and}\quad \tfrac{1}{2}\{g^2(z)+g(z^2)\}$$

respectively. In fact the generating function $\frac{1}{2}\{g^2(z)+g(z^2)\}$ was already used, independently, by Dirac[(13)] and D. Newman in 1950 to show that r'_n cannot be constant from some point onward; $r'_n = c$ for $n > k$ would imply

$$\tfrac{1}{2}\{g^2(z)+g(z^2)\} = P_k(z) + c\frac{z^{k+1}}{1-z},$$

(13) G. A. Dirac, 'Note on a problem in additive number theory', *J. Lond. math. Soc.* **26** (1951) 312–3.

where $P_k(z)$ is a polynomial of degree k, and this immediately leads to a contradiction on letting $z \to -1$, for the left-hand side tends to infinity whereas the right-hand side remains bounded. This simple argument fails when r'_n is replaced by r_n. On the other hand, it is easily seen that R_n assumes an infinity of even values and an infinity of odd values (cf. (1.4) and (1.5)), and this already rules out the possibility of R_n being constant from some point onwards.

Erdös–Fuchs[(2)] obtained much stronger results in this connexion. They showed that if $f(n)$ is any one of the three representation functions R_n, r_n, r'_n, and $c > 0$, then

$$\limsup_{N\to\infty} \frac{1}{N} \sum_{n=0}^{N} (f(n)-c)^2 > 0; \tag{4.30}$$

actually they also showed that (4.30) remains valid, for comparatively trivial reasons, if $c = 0$ and the sequence $\{a_j\}$ satisfies $a_j \ll j^2$.

The proof of this result is not quite as interesting as that of Theorem 11, and we refer the reader to the original paper.† The proof is based on the identity

$$\sum_{n=0}^{\infty} (f(n)-c)^2 r^{2n} = \frac{1}{2\pi} \int_{-\pi}^{\pi} |G(re^{i\alpha})-c(1-re^{i\alpha})^{-1}|^2 \, d\alpha$$

for $0 < r < 1$, where $G(z)$ is the generating function corresponding to $f(n)$.

A related, but probably much more profound, question was raised by the now well-known conjecture of Erdös–Turan which asserts that if $R_n > 0$ for all large n (i.e. if $\mathscr{A}$ is an asymptotic basis of order 2), then

$$\limsup_{n\to\infty} R_n = \infty.$$

An even stronger form of the conjecture is that if $a_j \ll j^2$, then

$$\limsup_{n\to\infty} R_n = \infty;$$

for if $\mathscr{A}$ is an asymptotic basis of order 2, then $a_j \ll j^2$ certainly holds.

It was pointed out by Erdös–Fuchs[(2)] that $a_j \ll j^2$ does *not* imply

$$\limsup_{N\to\infty} \frac{1}{N} \sum_{n=0}^{N} R_n^2 = \infty.$$

This probably means that their method is not applicable to the above-mentioned conjecture, for all known forms of the method yield (implicitly) a lower bound for a variance.

† Erdös–Fuchs[(2)], loc. cit., p. 72.

In conclusion, we mention that Richert[14] has obtained a very interesting multiplicative analogue of the Erdős–Fuchs theorem.

THEOREM (Richert[14]). *If $\{\alpha_n\}$ is a sequence of complex numbers, c is a constant, and*

$$\sum_{n\leqslant N} \alpha_n = N + O(N^{\frac{1}{2}-\delta}) \tag{4.31}$$

for some $\delta > 0$, then

$$\sum_{mn\leqslant N} \alpha_m \alpha_n = N\log N + cN + O(N^{\frac{1}{2}-\epsilon}) \tag{4.32}$$

cannot hold for any $\epsilon > 0$.

It is not known whether or not the hypothesis (4.31) is superfluous, although it can be relaxed to some extent. Nevertheless, the theorem is one of remarkable generality. In the case $\alpha_n = 1$ $(n = 1, 2, 3,...)$ the result corresponds closely to the well-known theorem of Hardy† concerning the divisor problem; the reader will recall that in the special case $\mathscr{A} = \mathscr{Q}$, the Erdős–Fuchs theorem represents a result concerning the circle problem (cf. the footnote relating to (1.6) in § 1, and also the beginning of this section).

† G. H. Hardy, 'On Dirichlet's divisor problem', *Proc. Lond. math. Soc.* (2) **15** (1916) 1–25.

(14) H.-E. Richert, 'Zur multiplikativen Zahlentheorie', *J. reine angew. Math.* **206** (1961) 31–8.

Richert also draws attention to the following very interesting development connected with the Erdős-Fuchs theorem. Richert states on p. 37 of his paper that W. Jurkat has sharpened Theorem 11 by showing that the relation (4.1) cannot hold even with the error term $o(N^{\frac{1}{2}})$.

III

ADDITION OF SEQUENCES: STUDY OF REPRESENTATION FUNCTIONS BY PROBABILITY METHODS

1. Introduction

One often has occasion to ask whether or not there exists an integer sequence possessing certain (e.g. additive) properties; many of the questions considered in other chapters are also of this type. Obviously, the most direct way of obtaining an affirmative answer to such a question is actually to construct a sequence with the required properties. But even when this direct approach proves impracticable, it may still be possible to establish the existence of such a sequence by showing that, in some sense, integer sequences possess the required properties 'on average'. Indeed, in most branches of mathematics, one often finds that it is much easier to prove that an event occurs 'on average' than to give a specific example of such an event.

It is therefore not surprising that probability theory, which consists of averaging arguments of a somewhat specialized kind (cf. § 3), has proved particularly successful in exhibiting possible relationships between the behaviour of the representation function in the addition of sequences and the rates of growth of the sequences in question. Many such existence theorems have been obtained by Erdös(1) and Erdös–Rényi.(2)

The essence of Erdös's idea is to impose a probability measure on the space of integer sequences, such that (in the resulting probability space) almost all integer sequences have some prescribed rate of growth; thus the probable behaviour of the representation function in the addition of sequences of prescribed rates of growth may then be investigated without further reference to these rates of growth. From the point of view of probability theory, the various investigations of this kind are very similar in nature, and a particularly attractive aspect of the work is that only simple basic results from probability theory are required.

(1) P. Erdös, 'Problems and results in additive number theory', *Colloque sur la Théorie des Nombres* (*CBRM*) (Bruxelles, 1956), pp. 127–37.

(2) P. Erdös and A. Rényi, 'Additive properties of random sequences of positive integers', *Acta arith.* **6** (1960) 83–110.

The difficulties of the subject (given Erdös's central idea) arise in another way. The applicability of probability theory depends on the existence of suitable independence or quasi-independence relationships; in many applications to number-theory these independence relationships are approximate only (i.e. of quasi-independence type), and it may not be obvious how to establish them with sufficient accuracy to permit the standard techniques of probability theory to be applied. The nature and extent of these difficulties vary from one application to another.

We include only selected results (stated in § 2), but prove these in complete detail (§§ 10–16). The original proofs in Erdös[1] are in outline form, whilst those of several of the theorems in Erdös–Rényi[2] are incomplete. The one serious gap in Erdös–Rényi arises in connexion with quasi-independence relationships, which are not established. Roughly speaking, those theorems in Erdös–Rényi which involve consideration of quasi-independence fall into two categories; those concerned with the addition of two mutually independent random sequences, and those concerned with the addition of a random sequence to itself. In relation to quasi-independence, the proofs of the former are comparatively simple, whereas the latter lead to a very much more complicated situation where it is no longer practicable to attempt proofs (of quasi-independence relationships) which depend on actually being able to write down explicit expressions for the probabilities under consideration. Although, in Erdös–Rényi, proofs of results in both categories are based on the existence of suitable quasi-independence relationships, only one such relationship is explicitly stated (but the proof† of this relationship, as given in the paper, is not satisfactory); and this is of the simpler type applicable only to theorems in the first category. Moreover, even to establish the corresponding result for the complicated situation arising in connexion with theorems of the second category, would not suffice for all the applications. Of the results based on quasi-independence, all those selected for inclusion in this chapter fall into the second category, so that we shall be concerned solely with the more difficult situation. We shall establish the appropriate quasi-independence relationships in § 15.

† On being unable to understand this proof, appearing on p. 88 of the paper, we requested Professor Rényi to explain the proof in the special case $r = 0$, and are very grateful to him for having sent us a correct proof of the case requested.

The proof of our Lemma 20 (§ 15), which corresponds to this special case in the more difficult situation, is based on a similar approach. Whilst the deduction of the general result presents quite a separate problem (which is the subject of a large part of our § 15), this would be comparatively trivial in the simple situation (relating to theorems of the first category).

As we have already mentioned, we shall not require any very deep results from probability theory. But in view of two circumstances—(*a*) the set of all integer sequences is non-enumerable and (*b*) the method involves the simultaneous consideration of infinities of 'events'—we shall require the full 'axiomatics' of the subject. For those readers who are not familiar with probability theory (or measure theory), we summarize the appropriate parts of the subject in §§ 3–9. In this summary we have endeavoured to take the middle course between two conflicting aims; to view the work within the framework of general probability theory on the one hand, and on the other to make the treatment as economical as the applications permit. To give examples of the concessions made to each point of view, we remark that the interpretation (§ 10) of the space of integer sequences as a product space is uneconomical (a product space is defined (§ 5) in terms of a space of sequences); whilst, on the other hand, the restriction to 'simple' random variables (§ 7) is consistent with all the applications but would be totally unacceptable from the point of view of general probability theory.

The chapter is completely self-contained in all but two respects. We quote the extension theorem (cf. § 4) concerning measures on algebras; and (in the proof of Theorem 3) we apply a deep result, due to Hoheisel, concerning the distribution of primes. Sometimes, however, when a required result represents only a very special case of some theorem in probability theory (or measure theory), we give a proof which takes full advantage of the special circumstances, irrespective of the resulting loss of insight; in such cases, to avoid giving a misleading picture of the subject as a whole, we give outside references to the more general theorems and their proofs (in addition to proving the particular cases we require). These references are confined to a single source (cf. the introduction to § 4).

To sum up, the principal results will be stated in § 2 and proved in §§ 10–16, whilst §§ *3–9 can be ignored by readers with a knowledge of probability theory.*

Notation. Throughout this chapter the integer sequences under consideration are subsequences of the sequence of non-negative integers; and the integer sequences which occur later as elements of probability spaces are, in fact, subsequences of the sequence of natural numbers.

We shall make extensive use of the conventions listed under the heading 'Conventions from set theory' in the 'Notation' section at the beginning of the book.

2. Principal results

We have chosen the four theorems below, from the many results which can be obtained by the methods of this chapter, partly because they are among the more interesting outcomes of the probabilistic approach used, and partly because they fit naturally into the framework of this book. The first three theorems have, indeed, already received brief mention (in Chapters I and II); and we shall see that comparison with a previous investigation (in Chapter I) adds to the interest of the fourth. At the cost of a little repetition, we shall, however, describe the background of each of the four theorems in turn.

Whilst we state our results in the form of existence theorems, without reference to probability theory, it should be borne in mind that a stronger form of these results will be proved; in each case we shall show that, in a suitable probability space, almost all sequences are of the desired type. We also remark that Theorem 4′ (§ 16), the stronger form of Theorem 4, corresponds to a combination of several of the results in Erdös–Rényi.(2)

We recall that if $\mathscr{A}$ is a sequence of natural numbers, then $r_n(\mathscr{A})$ denotes the number of representations of n in the form

$$n = a+a', \qquad a < a' \quad (a \in \mathscr{A},\, a' \in \mathscr{A});$$

and $r'_n(\mathscr{A})$ is the number of representations of n in the form

$$n = a+a', \qquad a \leqslant a' \quad (a \in \mathscr{A},\, a' \in \mathscr{A}).$$

Thus $r'_n(\mathscr{A}) = r_n(\mathscr{A})$, unless n is even and $\frac{1}{2}n \in \mathscr{A}$, when

$$r'_n(\mathscr{A}) = r_n(\mathscr{A})+1.$$

We also recall that $\mathscr{A}$ is said to belong to the class $B_2[g]$, or to be a $B_2[g]$-sequence (where a $B_2[1]$-sequence may be described simply as a B_2-sequence—see Chapter II, § 3) if $r'_n(\mathscr{A}) \leqslant g$ for all n. We see that if $r_n(\mathscr{A}) \leqslant g$ for all n, then $\mathscr{A}$ certainly belongs to the class $B_2[g+1]$. We make this observation because it is more convenient, when using probabilistic methods, to deal with r_n rather than with r'_n, and the remark makes it clear that nothing is lost by doing so in the proof of Theorem 2 below.

We are now in a position to describe our principal results. In the course of some investigations on Fourier Series by S. Sidon,† several questions arose concerning the existence and nature of certain integer

† See Erdös(1), pp. 129, 132; also S. Sidon, 'Ein Satz über trigonometrische Polynome und seine Anwendung in der Theorie der Fourier-Reihen', *Math. Annln.* **106** (1932) 536–9.

sequences $\mathscr{A}$ whose representation functions $\{r_n(\mathscr{A})$ or $r'_n(\mathscr{A})\}$ are bounded or, in some sense, exceptionally small. In particular, he raised the following two questions (i) and (ii), each of which we state and discuss below.

(i) Does there exist an asymptotic basis $\mathscr{A}$ of order 2 (cf. Chapter I, § 6, page 45) which is economical in the sense that, for every $\epsilon > 0$,

$$\lim_{n\to\infty} \frac{r_n(\mathscr{A})}{n^\epsilon} = 0\,?$$

The remarkable Theorem 1 below is much sharper than is required for an answer to this question.

THEOREM 1 (Erdös[1]). *There exists an asymptotic basis of order 2 such that*

$$\log n \ll r_n(\mathscr{A}) \ll \log n \quad \textit{for large } n. \tag{2.1}$$

We note that any sequence $\mathscr{A}$ satisfying (2.1) is in fact automatically an asymptotic basis of order 2, for $r_n(\mathscr{A}) \gg \log n$ implies $r_n(\mathscr{A}) > 0$ for large n.

Another question raised by Sidon was the following:

(ii) To what extent (in terms of the natural number g) must sequences belonging to the class $B_2[g]$ be 'thin' (i.e. have a rapid rate of growth)?

In the case $g = 1$ we know (see Chapter II, the construction in (iv) at the end of § 3) that there exists a B_2-sequence (i.e. $B_2[1]$-sequence) $\mathscr{A}'$ such that $a'_j \ll j^3$ and also (see Chapter II, § 3, (3.50)) that *every* B_2-sequence $\mathscr{A}$ satisfies

$$\limsup_{j\to\infty} \frac{a_j}{j^2 \log j} > 0. \tag{2.2}$$

In the light of these results, Erdös conjectured (see Chapter II, § 3 below (3.51)) that, for every $\epsilon > 0$, there exists a B_2-sequence $\mathscr{A}$ for which $a_j \ll j^{2+\epsilon}$. The conjecture remains unsettled, but Theorem 2, below, provides valuable information in this connexion.

THEOREM 2 (Erdös–Rényi[2]). *Corresponding to every $\epsilon > 0$, there exists a natural number g and a $B_2[g]$-sequence $\mathscr{A}$ such that*

$$a_j \ll j^{2+\epsilon}. \tag{2.3}$$

Corresponding results for h-fold, in place of 2-fold, sums of sequences could be obtained by similar methods.

In connexion with Theorem 3 below, we recall a result, and the appropriate terminology, from Chapter I. For a given sequence $\mathscr{A}$, the sequence $\mathscr{B}$ is said to be 'complementary' to $\mathscr{A}$ if the sequence $\mathscr{A}+\mathscr{B}$

(consisting of all distinct numbers $a+b$, where $a \in \mathscr{A}$, $b \in \mathscr{B}$) contains all natural numbers from some point onward; or, in the notation of Chapter I, if $\mathscr{A}+\mathscr{B} \sim \mathscr{Z}_1$. A theorem of Lorentz (Chapter I, § 3, Theorem 6) shows that to every sequence $\mathscr{A}$ there corresponds a complementary sequence $\mathscr{B}$ which is 'thin' in the sense that its counting number† $B(n)$ satisfies

(2.4) $$B(n) \ll \sum_{\substack{\nu=1 \\ A(\nu)>0}}^{n} \frac{\log A(\nu)}{A(\nu)}.$$

As was remarked in the discussion following the statement of this theorem, in the special case when $\mathscr{A}$ is the sequence of primes, the inequality (2.4) yields $B(n) \ll (\log n)^3$, whereas a special argument due to Erdös[(3)], based on probabilistic considerations of a rather complicated nature, shows that there exists a complementary sequence $\mathscr{B}'$ with $B'(n) \ll (\log n)^2$. This refinement (when $\mathscr{A}$ is the sequence of primes) is the subject of Theorem 3. It was stated by Erdös (cf. Erdös,[(1)] p. 133, the discussion involving the inequality (10)) that the theorem could also be proved in terms of the method described in this chapter. We have carried out this suggestion, and the resulting proof is simple and elegant; although it should be remarked that the proof still uses the deep theorem of Hoheisel mentioned in § 1.

THEOREM 3 (Erdös). *Let‡ $\mathscr{P}'$ denote the sequence of prime numbers. There exists a 'complementary' sequence $\mathscr{B}$ of $\mathscr{P}'$ (in the sense defined above) such that the counting number $B(n)$ of $\mathscr{B}$ satisfies*

(2.5) $$B(n) \ll (\log n)^2.$$

Finally, we discuss the last of the four principal theorems to be proved in this chapter. As in Chapter I, let $\mathscr{Q}$ denote the sequence consisting of 0 and the natural squares, and let $2\mathscr{Q}$ denote the sequence of all distinct integers representable as sums of two elements of $\mathscr{Q}$. A sequence $\mathscr{A}$ is said to be a sequence of *pseudo-squares* if its asymptotic distribution is the same as that of $\mathscr{Q}$ in the sense that

(2.6) $$a_j = \{1+o(1)\}j^2 \quad \text{as } j \to \infty.$$

The considerations which made it convenient§ to include 0 in the

† See § 4 of the 'Notation' section at the beginning of the book.

‡ We use $\mathscr{P}'$ to avoid confusion; in Chapter I, we used $\mathscr{P}$ for the sequence including 0 and 1 as well as the primes.

§ For example, this enabled us to say that $\mathscr{Q}$ is a basis of order 4 (cf. Chapter I, § 1, Definition 2).

(3) P. Erdös, 'Some results on additive number theory', *Proc. Am. math. Soc.* 5 (1954) 847–53.

sequence $\mathscr{Q}$ no longer apply in the present investigation, and we take $\mathscr{A}$ to be a sequence of natural numbers; this will not affect the nature of the results.

It is well known that the number of solutions of

$$a^2+b^2 \leqslant x,$$

in non-negative integers a, b, is asymptotic to $\frac{1}{4}\pi x$, whereas

$$\mathbf{d}(2\mathscr{Q}) = \lim_{x\to\infty} x^{-1} \sum_{\substack{1\leqslant n\leqslant x\\ n\in 2\mathscr{Q}}} 1 = 0. \tag{2.7}$$

The mere knowledge of the rate of growth of $\mathscr{Q}$ would not lead one to expect these contrasting results; for the distribution of the values of $r_n(\mathscr{Q})$ reflects the specific arithmetical character of the natural squares. For example, the factorization of u^2-v^2 (together with the fact that integers of this form have, on average, a disproportionately large number of prime factors) suggests that the diophantine equation

$$(a')^2-(a'')^2 = (b'')^2-(b')^2,$$

subject to suitable restrictions, has a comparatively large number of solutions in a', b', a'', b''. This in turn leads one to expect that

$$\sum_{n\leqslant x} \{r_n(\mathscr{A})\}^2$$

is abnormally large when the sequence $\mathscr{A}$ of pseudo-squares is actually the sequence $\mathscr{Q}$.

J. E. Littlewood was led to suggest, possibly by considerations of this general nature, that the property (2.7) of $\mathscr{Q}$ was untypical of the class of sequences having distributions similar to that of $\mathscr{Q}$, and that a sequence $\mathscr{A}$ of pseudo-squares would, in a suitable sense almost always, have the property $\mathbf{d}(2\mathscr{A}) > 0$. Littlewood's predictions in this respect were confirmed by A. O. L. Atkin(4), who proved also that there exists a sequence $\mathscr{A}'$ of pseudo-squares, lying very 'close' to $\mathscr{Q}$ in the sense that

$$a'_j = j^2+O(\log j), \tag{2.8}$$

such that $\mathbf{d}(2\mathscr{A}') > 0$. Although the precision of this last result (as measured by (2.8)), achieved by a combination of combinatorial and analytic methods, lies beyond the reach of the probabilistic approach developed in this chapter, Theorem 4′ (the stronger form of Theorem 4, stated and proved in § 16) also contains among its assertions one which confirms Littlewood's prediction.

(4) A. O. L. Atkin, Cambridge doctoral thesis, 1952. [*Added in proof*: the relevant material has now appeared as 'On pseudo-squares', *Proc. Lond. math. Soc.* **14A** (1965) 22–7].

THEOREM 4 (Erdös–Rényi). *There exists a sequence* $\mathscr{A}$ *of pseudo-squares (i.e. satisfying* (2.6)) *such that*

(i) $\lambda = \lim_{m\to\infty} \frac{1}{m} \sum_{n=1}^{m} r_n(\mathscr{A}) = \frac{1}{8}\pi,$

(ii) *the density of those integers n for which* $r_n(\mathscr{A}) = d$ *is* $(\lambda^d/d!)e^{-\lambda}$ $(d = 0, 1, 2, \ldots)$,

(iii) $\limsup_{n\to\infty} \frac{r_n(\mathscr{A})\log\log n}{\log n} = 1.$

In particular, the case $d = 0$ *of* (ii) *yields (since* $\mathbf{d}(2\mathscr{A})$ *is the density of those integers n for which* $r'_n(\mathscr{A}) > 0$*, and the integers* $2a$*, with* $a \in \mathscr{A}$*, have zero density)*

$$\mathbf{d}(2\mathscr{A}) = 1-e^{-\lambda}. \tag{2.9}$$

In connexion with Theorem 4, we also refer the reader to the discussion and footnote following Theorem A (Cassels) towards the end of § 5 of Chapter I.

3. Finite probability spaces: informal discussion

Finite probability spaces have some very special features, which render them unrepresentative of the situation in general. In fact, it is only in applications to non-enumerable spaces that the true power of probability theory is revealed. We single out the case of the finite probability space, which will later appear as a special case of the more general theory, merely to enable us to introduce some of the concepts of probability theory in an exceptionally simple setting. In particular, all the definitions in this section will be repeated later (in a wider context).

Let $X = \{x^{(1)}, x^{(2)}, \ldots, x^{(N)}\}$ be a non-empty finite set, consisting of the N elements x. We consider the class S of all subsets of X (including the empty subset). We associate a non-negative number $\mu(E)$ to each subset E of X, in such a manner that

$$\begin{cases} \mu(E_1 \cup E_2) = \mu(E_1)+\mu(E_2) \\ \text{for every pair } E_1,\, E_2 \text{ of disjoint subsets of } X. \end{cases} \tag{3.1}$$

(A set function μ satisfying (3.1) is said to be (finitely) additive.) Clearly, every possible way of doing this is covered by freely choosing the N non-negative numbers $\mu(x^{(r)}) = p_r$ $(r = 1, 2, \ldots, N)$, and then writing

$$\mu(E) = \sum_{x^{(r)}\in E} p_r$$

for every subset E of X.

We may regard the numbers p_r as 'weights' assigned to the elements $x^{(r)}$ of X. The condition (3.1) enables us to apply averaging processes in terms of 'total weights' corresponding to subsets of X. In this connexion the value of $p_1+p_2+...+p_N$ is clearly irrelevant (provided this sum is not zero), and we therefore normalize the system by imposing the further condition

$$\mu(X) = 1. \tag{3.2}$$

A non-negative set function μ satisfying the conditions (3.1) and (3.2) is called a 'probability measure on S'; and S in conjunction with a probability measure on S constitute a 'probability space'.

A subset E of X (which may be empty) is called an 'event', and (in a given probability space) the non-negative number $\mu(E)$ is called the 'probability of the event E'. The elements $x^{(r)}$ of the 'space' X are known as 'elementary events'. In speaking of the 'probability of an event', we may, without ambiguity so far as the entire phrase is concerned, interpret the word 'event' in two different ways. The following example will illustrate both usages of the word.

Consider the space X consisting of the elements 1, 2,..., N (i.e. we take $x^{(r)}$ to be the natural number r). Suppose we construct a 'probability space' by imposing a (specific) 'probability measure' μ on the class S of all subsets of X. Consider the set E consisting of those natural numbers, among 1, 2,..., N, which are congruent to 1 modulo 3: namely,

$$1,\ 4,\ 7,\ldots,\ 3t+1 \tag{3.3}$$

where $t = [(N-1)/3]$. Let $\mu(E)$ be the probability measure of this set E. Now we may refer to $\mu(E)$ either as the probability of the 'event $x \equiv 1 \pmod 3$', or as the probability of the 'event' E consisting of those elements x of X which satisfy the congruence.

Although probability theory may be said to consist of averaging processes applied to measures on abstract sets, these processes are of a rather specialized kind. Roughly speaking, this specialization springs from the fact that probability theory is concerned only with systems in which there are independence (or quasi-independence) relationships of some kind. As an example of such a relationship, a (finite) collection of events E_1, E_2,..., E_k (in a given probability space) is said to be independent if

$$\mu\Big(\bigcap_{j=1}^{n} E_{l_j}\Big) = \prod_{j=1}^{n} \mu(E_{l_j}) \tag{3.4}$$

for every subset $\{l_1,..., l_n\}$ of the set $\{1, 2,..., k\}$. (This is, of course, a much more stringent condition than the mere independence of each pair among $E_1,..., E_k$.) The characteristic feature of probability theory may be described (necessarily vaguely, at this stage) as follows: the presence of suitable independence relationships tends to imply regularity 'in the large' in systems which may be irregular in detail.

We conclude with some remarks concerning the subsequent sections. What we have said concerning finite X is easily generalized so as to apply equally to enumerably† infinite spaces X (where probabilities can be represented by convergent series); but the situation is entirely different in the case of non-enumerable X.

Suppose now that the (non-empty) set X is completely arbitrary. Measures μ satisfying (3.1) and (3.2) remain equally relevant, subject to the reservation that (3.1) should be strengthened‡ in one respect. We must stipulate μ to be 'countably additive', i.e. that

$$\mu\Big(\bigcup_{j=1}^{\infty} E_j\Big) = \sum_{j=1}^{\infty} \mu(E_j) \tag{3.1'}$$

for sequences $\{E_j\}$ of disjoint events. For this is a natural requirement if (as turns out to be the case) it is necessary to consider an enumerably infinite class of events simultaneously; and (3.1′) is not a consequence of (3.1).

But in addition to this fairly natural generalization, we need a subtle modification of our definition of 'probability space' (in view of the possibility of X being non-enumerable). We can no longer insist that every probability measure should be defined for *all* subsets E of X. For this stipulation, in conjunction with (3.1′), would (in the case of non-enumerable X) constitute a most stringent condition, which would admit only measures of a very special type and exclude measures of great significance in probability theory. Instead, we must also consider those measures μ which can be defined only on a suitable subclass S of the class of all subsets of X. (The stipulation that μ should be 'countably additive' is now interpreted to mean that (3.1′) is satisfied for every sequence $\{E_j\}$ of disjoint events for which each E_j, and the union of all the E_j, lies in S.) We must also abandon the process of assigning a 'weight' to each elementary event as a method of generating probability measures. Indeed, for a wide class of such measures we will have $\mu(x) = 0$ for every elementary event x of X.

† This applies to X only. When X is enumerably infinite, the class of all its subsets is non-enumerable.

‡ 'Countably additive' implies 'finitely additive', since all but two of the E_j appearing in (3.1′) may be empty, and (for every bounded countably additive measure) the empty set has zero measure.

As the above remarks indicate, we require a new approach: one which is adequate for the consideration of non-enumerable X, and yet contains (as a special case) what has already been said concerning finite X. To provide such an approach is the objective of the following sections. For a more comprehensive (and more concise) study of the foundations of probability theory, we refer the reader to Kolmogorov's classic monograph.[5]

4. Measure theory: basic definitions

We now give a series of definitions, culminating in the definition of a 'probability space' as a 'measure space' of a special kind. We choose this somewhat indirect approach, which involves a number of those concepts of measure theory not strictly relevant to probability theory, in view of outside references to results and proofs couched in the terminology of measure theory. As already stated in § 1, we shall prove all the results we need to apply (excepting the extension theorem quoted later in this section) and the references are merely intended to identify such results as special cases of much more general theorems, and to indicate the extent to which our proofs take advantage of special circumstances. **All such references (in this and the following sections) will be to Halmos (*Measure Theory*, Van Nostrand, New York, 1950); we refer to this book as H. for short.**

We begin with an informal discussion in which we indicate how the various definitions are motivated. *Most of the definitions occurring in this discussion will be listed at the end of the section; this list of definitions represents the essential content of the section.*

Let X be a non-empty collection of elements x. We shall be concerned solely with subsets of X, and shall refer to X itself as the 'space' under consideration. We impose no restriction whatsoever on X and, in particular, X may have any positive cardinal.

Let S be a class of subsets of X, and let μ be a non-negative set function on S, which assigns a real non-negative number, or† $+\infty$, to each of the sets constituting S. If S satisfies suitable conditions (to be stated later), we shall deem S to constitute a 'measurable space'; and *if μ is countably additive* (cf. the preceding section) and assigns zero to the empty subset of X, we shall say that μ is a 'measure on S'.

† For the development of measure theory, it is desirable to permit μ to assume the 'value' $+\infty$. We need not unduly stress this point, however, as only bounded set functions μ can represent probability measures.

(5) A. N. Kolmogorov, *Grundbegriffe der Wahrscheinlichkeitsrechnung* (Springer, Berlin, 1933). English translation, Chelsea (New York, 1950).

A measurable space S together with a measure μ on S are said to constitute a 'measure space'. We define a probability space to be a 'measure space' for which

$$\mu(X) = 1. \tag{4.1}$$

Note that (4.1) presupposes $X \in S$, so that *only those measurable spaces S which contain X (as one of their constituent sets) can give rise to probability spaces.*

We must still characterize the 'measurable spaces', that is to say those classes S, of subsets of X, on which 'measures' are to be defined.

We have already explained that we must not insist that every measure be defined on the class of *all* subsets of X, for fear of 'losing' important probability measures. For example, Lebesgue measure cannot be defined on the class of all subsets of the 'space' X consisting of the real numbers x with $0 \leqslant x \leqslant 1$.

In fact, various S may admit different types of measures. It is therefore desirable to leave S to a large extent arbitrary; in a particular application to probability theory, S can then be so chosen as to admit the particular probability measure one may wish to impose.

Accordingly, we shall only insist that a 'measurable space' S is to have certain properties which will be useful in the development of the theory of 'measure spaces'. (In effect, a measurable space will be defined as a class S, of subsets of X, having certain useful properties. The term 'measurable' is thus used in a purely formal manner; the statement 'S is measurable' will not imply that S is in any way 'appropriate' to a particular measure.)

We have repeatedly stressed that the aggregate of measurable spaces must be defined in such a way that we do not 'lose' important probability measures. In this connexion, the concept of an 'extended measure' is important. If μ is a measure on S, and $S \subset S_1$, we say that μ can be 'extended onto S_1' if there exists a measure μ_1 on S_1 such that $\mu_1 = \mu$ on S; we call μ_1 an 'extension' of μ.

It is clear that, in such circumstances, we need not regard the measure μ as 'lost' if S is excluded from the aggregate of measurable spaces, as long as S_1 is retained in this aggregate. For S_1 admits the measure μ_1, so that μ is still available on the subclass S of S_1.

(If, for example, we were to make (by definition) the class T of *all* subsets of X constitute the *only* measurable space, we would 'lose' a measure μ, defined on a subclass S of T, only if it cannot be extended onto T; although, as we have pointed out, we would 'lose' many important measures in this way.)

We have already indicated that the conditions to be imposed on S will be dictated by requirements for the development of a theory of 'measure spaces'. These requirements are simple; we must be free to perform, within S, certain set operations (so that S must be closed w.r.t. these operations).

As regards operations involving a finite number of sets, we shall use only combinations of the following two operations:

(i) *Forming the union $E \cup F$ of two subsets E, F of X.*

(ii) *Forming the difference $E - F$ of two subsets E, F, of X.*

(The difference $E-F$ consists of those elements of E which are not in F; in particular, if $F \subset E$, $E-F$ is simply the complement of F w.r.t. E.) It is important to note that forming the intersection $E \cap F$ of two sets E, F (an operation which will be in constant use) may be regarded as a combination of the operations (i) and (ii). For

$$E \cap F = (E \cup F) - \{(E-F) \cup (F-E)\}.$$

But we shall also use an operation involving an enumerably infinite collection of sets, namely:

(iii) *Forming the union $\bigcup_{j=1}^{\infty} E_j$ of a sequence $\{E_j\}$ of subsets of X.*

We note that the intersection $\bigcap_{j=1}^{\infty} E_j$ of such a sequence of sets can be formed by a combination of the operations (ii) and (iii). For

$$\bigcap_{j=1}^{\infty} E_j = E - \bigcup_{j=1}^{\infty} (E - E_j), \quad \text{where} \quad E = \bigcup_{j=1}^{\infty} E_j.$$

Furthermore, the operation (i) corresponds to the special case of the operation (iii) arising when all but two of the sets E_j are empty.

All our requirements will therefore be covered by insisting that S (assumed to be non-empty) should be closed w.r.t. the operations (ii) and (iii). (Note that if S is closed w.r.t. (ii), it must contain the empty set, so that we are justified in treating (i) as a special case of (iii).) Whilst we have now characterized those S that can conveniently be investigated, it is the entire space X that is the real object of such investigations. It is therefore natural to insist that every element of X should appear among the elements of some set in S; in other words that

$$\bigcup_{E \in S} E = X. \tag{4.2}$$

Accordingly, *we define a 'measurable space' to be a class S, of subsets of X, which satisfies* (4.2) *and is closed w.r.t. the operations* (ii) *and* (iii). (Obviously, the condition $X \in S$, which is part of the definition of a probability space, is more stringent than (4.2). The union (4.2) may be non-enumerable.)

A non-empty class S of subsets of X is called a 'ring' if it is closed w.r.t. the operations (i), (ii); and a σ-ring if it is closed w.r.t. the operations (ii), (iii). A ring (or σ-ring) which contains X (as one of its constituent sets) is called an algebra (or σ-algebra). (Obviously, every σ-ring is a ring and every σ-algebra is an algebra.) An important theorem (H., § 13, Theorem A) in measure theory states that every 'σ-finite' measure defined on a ring R can be extended, in exactly one way, onto the minimal† σ-ring containing R. (The definition of 'σ-finite' is given in H., § 7, but for our purposes it suffices to say that every measure satisfying (4.1) is certainly σ-finite.) In relation to probability theory, this theorem tells us that:

Every measure which is defined on an algebra R, and satisfies (4.1), generates (by extension) a unique probability measure on the least σ-algebra containing R.

We therefore cannot 'lose' any probability measures by considering (as we do) only those probability measures defined on σ-algebras, as opposed to all those defined on algebras.

It is worth mentioning that, in view of the identity

$$E - F = \{E^c \cup F\}^c,$$

the above definition of a σ-algebra is equivalent to the following:

A σ-algebra is a non-empty class S, of subsets of X, which is closed w.r.t the operation (iii) *and the operation*

(iv) *forming the complement* (*with respect to X*) E^c *of a subset E of X*.

We also remark that in the case of finite X, considered in the preceding section, *every* probability measure is defined on the σ-algebra consisting of all subsets of X. In this respect, finite (and enumerable) X are exceptional.

Definition 1. A non-empty collection of elements is said to constitute a 'space'.

Definition 2. A 'ring of subsets' of a space X is defined to be a non-empty class R of subsets of X which is closed with respect to the

† This is the intersection of all σ-rings containing R. There is at least one σ-ring containing R, namely the class of all subsets of X.

formation of differences† and FINITE unions. Such a ring R is called an 'algebra' if and only if it contains X as one of its constituent sets.

Definition 3. A 'σ-ring of subsets' of a space X is defined to be a non-empty class S of subsets of X which is closed with respect to the formation of differences and COUNTABLE unions. Such a σ-ring S is called a 'σ-algebra' if and only if it contains X as one of its constituent sets.

Definition 4. A set function μ, defined on a ring R of subsets of a space X, is said to be 'finitely additive' on R if and only if

$$\mu(E \cup F) = \mu(E) + \mu(F)$$

for every pair of disjoint subsets E, F (of X) which both lie in R; and 'countably additive' on R if and only if

$$\mu\Big(\bigcup_{j=1}^{\infty} E_j\Big) = \sum_{j=1}^{\infty} \mu(E_j)$$

for every sequence $\{E_j\}$ of disjoint subsets E_j (of X) for which each E_j lies in R and‡ the union of all the E_j lies in R.

Definition 5. A non-negative set function, which is defined and countably additive on a ring R of subsets of a space X, and§ which assigns zero to the empty set, is called a measure on R.

Definition 6. A 'measurable space' is defined to be a σ-ring S of subsets of X, such that the union of the constituent sets of S is X. We denote such a measurable space by (X, S). For a given S, a subset of X is said to be 'measurable' if and only if it lies in S.

Definition 7. A measurable space (X, S) together with a measure μ on S are said to constitute a 'measure space' (X, S, μ).

Definition 8. A 'measure space' (X, S, μ) is called a 'probability space' if and only if S is a σ-algebra and $\mu(X) = 1$. If (X, S, μ) is a probability space, μ is called a 'probability measure' on S.

To sum up:

A 'probability space' is a non-empty class S of subsets of a non-empty collection X of elements, where S is closed with respect to the formation of complements (with respect to X) and countable unions, together with a non-negative countably additive set function μ, defined on S and satisfying $\mu(X) = 1$.

† Thus R must contain the empty subset of X (although this may be the only element of R).

‡ This final requirement is obviously superfluous in the special case arising when R is a σ-ring.

§ This requirement is obviously superfluous if the set function assumes a finite value, as is the case in all applications to probability theory.

We recall that this definition should be viewed in relation to the extension theorem stated earlier in the section.

5. Measure theory: measures on product spaces

We have characterized those set functions deemed to be 'probability measures' in a wide context, but have as yet given only trivial examples of measures of this kind. Theorem 5 below will enable us to construct a wide class of non-trivial measures. Those readers familiar with the terminology of probability theory will recognize that Theorem 5 asserts that there exists a sequence of independent random variables with prescribed distributions. Both this result and the idea underlying the proof are highly intuitive, but in view of the very technical definition of 'probability space', a completely self-contained proof (not assuming familiarity with integration, product spaces, and sections) is necessarily lengthy. For this reason we only establish the theorem in the (comparatively trivial) special case we require for applications later.

Definition 9. The Cartesian product, denoted by $\bigtimes_{j=1}^{\infty} X_j$, of a sequence $\{X_j\}$ of spaces X_j is defined to be the set of all ordered sequences $(x_1, x_2, \ldots)$, where $x_j \in X_j$ $(j = 1, 2, \ldots)$.

Definition 10. Let $\{X_j\}$ be a sequence of spaces and write

$$X = \bigtimes_{j=1}^{\infty} X_j. \tag{5.1}$$

Let S_j be a σ-algebra of subsets of X_j $(j = 1, 2, \ldots)$. A 'measurable rectangle with respect to the sequence $\{S_j\}$' is defined to be a subset E of X which is representable in the form

$$E = \bigtimes_{j=1}^{\infty} E_j \quad \text{where} \quad E_j \in S_j \ (j = 1, 2, \ldots) \quad \text{and} \quad E_j = X_j \tag{5.2}$$

except for at most a finite set of numbers j.

(We shall only make use of this definition when none of the E_j is empty. But if one (or more) of the E_j is empty, E can be interpreted to be the empty subset of X.)

The representation (5.2) of a (fixed) non-empty measurable rectangle E is unique. (For the special type of product space X to be considered, this will be obvious. In the general case, the uniqueness is only slightly less obvious, being an immediate consequence of the fairly trivial H., § 33, Theorem C.)

The following theorem is equivalent to H., § 38, Theorem B (although, superficially, the latter theorem asserts more than the former).

THEOREM 5. *Let* $\{(X_j, S_j, \mu_j)\}$ *be a sequence of probability spaces, and write*

$$X = \mathop{\times}_{j=1}^{\infty} X_j.$$

Let S be the minimal σ-algebra, of subsets of X, containing every measurable rectangle with respect to the sequence $\{S_j\}$.

Then there exists a unique measure μ *on S with the property that, for every non-empty measurable rectangle E,*

$$\mu(E) = \prod_{j=1}^{\infty} \mu_j(E_j), \tag{5.3}$$

where the E_j *are defined by*

$$E = \mathop{\times}_{j=1}^{\infty} E_j, \qquad E_j \in S_j \quad (j = 1, 2, \ldots). \tag{5.4}$$

We remark that the right-hand side of (5.3) is, in essence, a finite product; for in view of Definition 10, we have

$$\mu_j(E_j) = \mu_j(X_j) = 1$$

except for at most a finite set of numbers j. Furthermore, since

$$\mu(X) = \prod_{j=1}^{\infty} \mu_j(X_j) = 1,$$

the σ-algebra S in conjunction with the measure μ constitute a probability space (X, S, μ).

Proof of a special case of Theorem 5. We shall only need to apply Theorem 5 in the special case arising when each of the spaces X_j consists of exactly two elements. We shall therefore restrict ourselves to this case, which is much easier to establish in view of the fact that all the spaces X_j are finite (we shall make constant use of this circumstance). Accordingly, we suppose that for each j, X_j consists of the two elements x_j', x_j''; so that S_j consists of the four subsets of X_j.

We first introduce some further notation. Suppose that $x_1^*, \ldots, x_k^*$ are elements of $X_1, \ldots, X_k$ respectively, and let V be a subset of X. Then we denote by $V(x_1^*, \ldots, x_k^*)$ the subset of V consisting of those elements $x = \{x_j\}$ of V for which $x_1 = x_1^*, \ldots, x_k = x_k^*$. We note that, in the special case when V is the space X itself, the set $X(x_1^*, \ldots, x_k^*)$ is a rectangle.

For each $n = 1, 2, \ldots$, we denote by $Y_{(n)}$ the class of all those rectangles representable in the form $X(x_1^*, \ldots, x_n^*)$. Thus $Y_{(n)}$ consists of 2^n disjoint rectangles. We denote by $R_{(n)}$ the class consisting of the 2^{2^n} subsets of X obtained by forming all unions of rectangles in $Y_{(n)}$ (including the empty union, which gives rise to the empty subset of X). Thus each non-empty set A in $R_{(n)}$ is representable in the form

$$A = \bigcup_{l=1}^{h} {}_lJ_{(n)}, \tag{5.5}$$

where the ${}_lJ_{(n)}$ are disjoint rectangles in $Y_{(n)}$.

Clearly, A and n *determine the representation* (5.5) *uniquely.*

Whilst retaining the notation $V(x_1^*,..., x_k^*)$ with its former meaning, we shall now also use $V(x_k^*)$ to denote the subset of V consisting of those elements $x = \{x_j\}$ of V for which $x_k = x_k^*$. (The two notations are consistent when $k = 1$.)

Let E be any non-empty rectangle. Choose m such that

(5.6) in the representation (5.2) of E, $E_j = X_j$ for $j > m$.

If $E \notin Y_{(m)}$, we can choose j^* such that $1 \leqslant j^* \leqslant m$ and $E_{j^*} = X_{j^*}$, and 'split' E into two non-empty rectangles by means of the decomposition

$$E = E' \cup E'', \quad \text{where} \quad E' = E(x'_{j^*}), \qquad E'' = E(x''_{j^*}). \tag{5.7}$$

(We recall that x'_j, x''_j are the two elements of X_j.)

Clearly:

(i) If E is any non-empty rectangle satisfying (5.6) with $m = n$, then we can obtain the representation (5.5) with $A = E$, by means of successive 'splittings' of type (5.7).

Furthermore:

(ii) If A is any non-empty set in $R_{(n)}$ and we perform, on each $J_{(n)}$ in the representation (5.5) of A, the 'splitting' (5.7) with $E = J_{(n)}$ and $j^* = n+1$, we obtain the (unique) representation of A as a union of rectangles in $Y_{(n+1)}$.

Thus each rectangle E lies in $R_{(n)}$ for large n; and $R_{(n)} \subset R_{(n+1)}$ for each n. Since $R_{(n)}$ is obviously the minimal algebra containing the rectangles in $Y_{(n)}$, it follows that

$$R = \bigcup_{n=1}^{\infty} R_{(n)} \tag{5.8}$$

is the minimal algebra containing all rectangles E. The σ-algebra S, specified in the theorem, is therefore the minimal σ-algebra containing R.

We shall now complete the proof as follows. We first show that there is a unique set function μ, defined and finitely additive on R, which satisfies (5.3) for all rectangles E. We shall then show that this set function μ is in fact also countably additive on R, and is therefore a measure on R. The proof will then be complete, for in view of the extension theorem (cf. § 4), the measure μ on the algebra R generates (by extension) a unique measure on the minimal σ-algebra S containing R.

Let μ be the set function† defined by (5.3) on the class of all non-empty rectangles E, and which assigns 0 to the empty set. Then, since μ_{j*} is a probability measure on X_{j*}, the following result is an immediate consequence of (5.3).

(iii) For every decomposition of type (5.7), we have

$$\mu(E) = \mu(E')+\mu(E'').$$

We are therefore justified (explanation follows) in extending the set function μ onto R, by associating to each relation (5.5), the relation

$$\mu(A) = \sum_{l=1}^{h} \mu({}_lJ_{(n)}); \tag{5.9}$$

in which the right-hand side constitutes the definition of the left-hand side. For by (ii) and (iii), the right-hand side of (5.9) is independent of n; and by (i) and (iii), the relations (5.9) are consistent with the relations (5.3).

Since for any pair B, C of disjoint sets in R, we can choose n such that both B and C lie in $R_{(n)}$, it is immediately clear from (5.9) that $\mu(B \cup C) = \mu(B)+\mu(C)$.

Thus μ is finitely additive on R. Furthermore, it is clear from the relations (5.9) that μ is the only additive set function on R which satisfies the relations (5.3). It remains to show that μ is countably additive on R. We do this by showing that the union of an infinity of disjoint non-empty sets in R can never‡ lie in R; so that the statement 'μ is countably additive on R' asserts no more than the statement 'μ is finitely additive on R' (cf. Definition 4).

Suppose that $\{A^{(i)}\}$ is a sequence of non-empty disjoint sets in R, and

$$\bigcup_{i=1}^{\infty} A^{(i)} = A, \quad \text{where } A \in R. \tag{5.10}$$

We shall construct, by means of a diagonal argument, an element $x^* = \{x_j^*\}$ of X which lies in A, and yet lies in none of the sets $A^{(i)}$; thus obtaining a contradiction to (5.10).

We write

$$B^{(k)} = A - \bigcup_{i=1}^{k} A^{(i)} \quad (k = 1,2,\ldots). \tag{5.11}$$

† Of course, we cannot assume any additivity property at this stage, even on the class of rectangles.

‡ This is only true because the X_j are finite; the corresponding result does not hold in the general case.

The sequence $\{B^{(k)}\}$ has the property

$$B^{(1)} \supset B^{(2)} \supset B^{(3)} \supset \dots;$$

and we shall say that set sequences having this property are 'monotone decreasing'.

We decompose each of the sets $B^{(i)}$ into the two sets $B^{(i)}(x_1')$, $B^{(i)}(x_1'')$ (cf. the definition given in the second paragraph of the proof), where x_1', x_1'' are (as before) the two elements of X_1. Then each of the sequences $\{B^{(i)}(x_1')\}$ and $\{B^{(i)}(x_1'')\}$ is monotone decreasing, so that at least one of these two sequences consists entirely of non-empty sets. In other words, there exists an element x_1^* of X_1, such that $\{B^{(i)}(x_1^*)\}$ is again a monotone decreasing sequence of non-empty sets. On decomposing this sequence into the two sequences $\{B^{(i)}(x_1^*, x_2')\}$ and $\{B^{(i)}(x_1^*, x_2'')\}$, we see that there exists an element x_2^* of X_2 such that $\{B^{(i)}(x_1^*, x_2^*)\}$ is a monotone decreasing sequence of non-empty sets. Continuing this process, we obtain an element $x^* = \{x_j^*\}$ of X, with the property that for every l, all the sets

$$B^{(1)}(x_1^*,\dots, x_l^*), \qquad B^{(2)}(x_1^*,\dots, x_l^*),\dots$$

are non-empty.

Now for each i, we can choose $n = n(i)$ such that $B^{(i)} \in R_{(n)}$. Then, since $B^{(i)}(x_1^*,\dots, x_n^*)$ is non-empty, we have

$$X(x_1^*,\dots, x_n^*) = B^{(i)}(x_1^*,\dots, x_n^*) \subset B^{(i)}.$$

Thus, since x^* certainly lies in all the rectangles $X(x_1^*,\dots, x_n^*)$, x^* lies in all the sets $B^{(i)}$. It now follows from (5.11) that x^* lies in A but in none of the sets $A^{(i)}$, as required.

This completes the proof of the special case of Theorem 5 under consideration.

6. Measure theory: simple functions

In the Definition 5 (cf. the footnote there appended) we have allowed for the possibility of μ assuming the 'value' $+\infty$ (for some subsets of X), as is appropriate for applications to measure theory. This has fulfilled the purpose of making our definitions correspond to those given in H., a book to which we frequently refer. However, with applications only to probability theory in mind, it is more convenient to restrict our attention to those measures assuming only finite values; every probability measure is of this type. (In fact, probability measures can assume only values lying between 0 and 1, inclusive.)

Henceforth we shall use the term 'measure' to designate 'measure assuming only finite values'. The same assumption is made implicitly with regard to measure spaces. (We avoid the term 'finite measure', which is given a different technical meaning in H.)

Let (X, S) be a measurable space and let f be a real function on X, which assigns a real number $f(x)$ to each element x of X. We say that f is 'simple' w.r.t. the measurable space S if it assumes only a finite number of distinct values and, for each *non-zero* value α assumed by f, the pre-image† $f^{-1}(\alpha)$ (of α in X) lies in S.

Clearly, every simple function has a unique representation of the form

$$f(x) = \begin{cases} \alpha_j & \text{if } x \in E_j \quad (j = 1, 2, \ldots, k) \\ 0 & \text{if } x \in \bigcap_{j=1}^{k} E_j^c \end{cases} \tag{6.1}$$

where $\alpha_1, \ldots, \alpha_k$ are distinct non-zero real numbers and $E_1, \ldots, E_k$ are disjoint subsets of X, each lying in S. (The α_j are the non-zero values assumed by f, and the E_j are the pre-images $f^{-1}(\alpha_j)$.) The relation (6.1) may be written,‡ equivalently, as

$$f(x) = \sum_{j=1}^{k} \alpha_j \chi_{E_j}(x), \tag{6.2}$$

where χ_E denotes the characteristic function of E (i.e. the function which assumes the values 1 or 0 according as x does or does not lie in E).

Now let (X, S, μ) be a measure space, and let f be a simple function on X (strictly speaking, a function on X which is simple with respect to the measurable space S). We define the 'integral' of f by

$$\int_X f\, d\mu = \sum_{j=1}^{k} \alpha_j \mu(E_j), \tag{6.3}$$

where f has the representation (6.2). It is an immediate consequence of this definition that, within the class of simple functions on (X, S), integration is a linear operation and non-negative functions have non-negative integrals.

To obtain a satisfactory theory of integration, it is necessary to generalize the above definition of an integral, so as to apply to a much wider class of functions. However, in the applications we have in mind (namely the proofs of Theorems 1–4), we shall be dealing solely with simple functions. It therefore suffices, for our purpose, to have defined integration within the class of simple functions. (Note that the class of simple functions, on a given measurable space, is closed with respect to the formation of finite linear combinations and finite products.) We

† i.e. the set of those elements x of X for which $f(x) = \alpha$.

‡ As usual, empty sums are interpreted to be zero.

can thus afford to by-pass the theory of abstract Lebesgue integration which, in general, is indispensable for probability theory.

We now repeat the above definitions in summarized form.

Definition 11. Let (X, S, μ) be a measure space, and let $f(x)$ be a real-valued function on X. The function f is said to be 'simple' if it is representable as a finite sum of the form†

$$f(x) = \sum_{j=1}^{k} \alpha_j \chi_{E_j}(x),$$

where the α_j are non-zero real numbers, the E_j are disjoint measurable subsets of X, and χ_E denotes the characteristic function of E. The integral of such a function is defined by

$$\int_X f\, d\mu = \sum_{j=1}^{k} \alpha_j \mu(E_j).$$

Definition 12 below represents the special case of the definition of 'independent measurable functions' [H., § 45] arising when these functions are in fact simple; this being the only case we require. (For those readers who wish to compare the two definitions, we remark that every simple function is measurable and every finite set of real numbers is a Borel set on the real line.)

Definition 12. Let (X, S, μ) be a probability space and let $\mathfrak{C}$ be a collection of simple functions f on X. The functions f of $\mathfrak{C}$ are said to be 'independent' if for each finite subset $\{f_1, f_2, \ldots, f_k\}$ of $\mathfrak{C}$ and every class $\mathfrak{M}_1, \mathfrak{M}_2, \ldots, \mathfrak{M}_k$ of sets $\mathfrak{M}_j$ of real numbers, we have

$$\mu\Big(\bigcap_{j=1}^{k} \{x : f_j(x) \in \mathfrak{M}_j\}\Big) = \prod_{j=1}^{k} \mu(\{x : f_j(x) \in \mathfrak{M}_j\}). \tag{6.4}$$

Theorem 6 below is an immediate consequence of Definitions 11 and 12; here again we take advantage of the fact that we need consider only simple functions. The theorem remains valid in a much wider context [H., § 45, Theorem A], but this is less obvious.

THEOREM 6. *If f_1, f_2 are independent simple functions, on a probability space (X, S, μ), then*

$$\int_X f_1 f_2\, d\mu = \Big(\int_X f_1\, d\mu\Big)\Big(\int_X f_2\, d\mu\Big). \tag{6.5}$$

† In view of the additivity of μ, we need not here insist that the α_j be distinct; the integral will be independent of the particular representation of f. This remark is significant in relation to Theorem 6.

7. Probability theory: basic definitions and terminology

Let (X, S, μ) be a probability space, taken to be fixed throughout. We use x to denote an element of X.

We refer to a subset of X as an 'event'. If an 'event' consists of a single element of X, we call it an 'elementary event'. Those events which lie in S are said to be 'measurable'. We shall be dealing solely with 'measurable events', but will sometimes omit the qualifying adjective. The measure $\mu(E)$, of an event E, is referred to as the 'probability of the event E'.

The following examples should suffice to show how the more descriptive language of probability theory may replace set-theoretic terminology. By the phrase 'with probability 1, the proposition $\mathfrak{P} = \mathfrak{P}(x)$ holds' we mean '$\mu(\{x\colon \mathfrak{P}(x) \text{ is true}\}) = 1$' or, equivalently, $\mu(\{x\colon \mathfrak{P}(x) \text{ is false}\}) = 0$. By the statement 'with probability 1, at most a finite number of the events $E_1, E_2,\dots$ can occur', we mean

$$\mu(\{x\colon x \in E_j \text{ for infinitely many values of } j\}) = 0.$$

A collection $\mathfrak{C}$ of measurable events E is said to be independent (cf. § 3), if for *every* finite subset $\{E_1, E_2,\dots, E_k\}$ of $\mathfrak{C}$, we have

$$\mu\Big(\bigcap_{j=1}^{k} E_j\Big) = \prod_{j=1}^{k} \mu(E_j). \tag{7.1}$$

This may be stated, equivalently, as follows:

Definition 13. A collection of measurable events are said to be 'independent' if their characteristic functions are independent.

A measurable function $f(x)$ (cf. H., § 18), defined on X, is called a 'random variable'.† However, in order to by-pass the theory of abstract Lebesgue integration (cf. § 6), we restrict our attention to those random variables which are simple functions.‡ Whilst all the random variables occurring in the proofs of Theorems 1–4 are of this type (so that we may impose this restriction), we must stress that *the restriction would be quite unacceptable if the general development of probability theory were to be our aim.*

Definition 14. A simple function, defined on X, is called a '(simple) random variable' (this being a random variable of restricted type).

We say that a set of (simple) random variables f are independent, if the functions $f(x)$ are independent.

† This terminology (which is widely accepted) is more appropriate to the experimental sciences than to mathematical theory.

‡ These form a subclass of the class of measurable functions.

The concept of 'moments' plays a crucial part in probability theory. The first moment $\mathbf{M}(f)$ of a (simple) random variable $f(x)$, defined by

$$\mathbf{M}(f) = \int_X f\,d\mu,$$

is referred to as the '*expectation*' or '*mean*' of f.

As regards higher moments, the moments about the mean (central moments),

$$\mathbf{M}(\{f-\mathbf{M}(f)\}^s) = \int_X \{f-\mathbf{M}(f)\}^s\,d\mu \quad (s = 2, 3,\ldots),$$

are of special importance. Of these, we shall be concerned only with the second central moment, known as the '*variance*' of f.

Definition 15. The mean $\mathbf{M}(f)$ and the variance $\mathbf{D}^2(f)$ of a (simple) random variable f, are defined by

$$\mathbf{M}(f) = \int_X f\,d\mu, \qquad \mathbf{D}^2(f) = \mathbf{M}(\{f-\mathbf{M}(f)\}^2). \tag{7.2}$$

We note that

$$\mathbf{D}^2(f) = \mathbf{M}(f^2)-\{\mathbf{M}(f)\}^2. \tag{7.3}$$

It is an immediate consequence of Definition 15, that for every (simple) random variable f and every $\epsilon > 0$,

$$\mu(\{x\colon |f(x)-\mathbf{M}(f)| \geqslant \epsilon\}) \leqslant \frac{1}{\epsilon^2}\mathbf{D}^2(f). \tag{7.4}$$

This fundamental inequality is known as the Tchebycheff (or Bienaymé–Tchebycheff) inequality. Tchebycheff's inequality corresponds to the special case ($k = 1$) of Kolmogorov's inequality (cf. § 9, Theorem 9).

8. Auxiliary lemmas

We now prove a number of lemmas, mostly of a trivial nature, which we require for applications in the following section. Lemmas 1–6 refer to a given probability space (X, S, μ).

LEMMA 1. *If E, F are measurable events, and $E \subset F$, then*

$$\mu(E) \leqslant \mu(F).$$

Proof.

$$\mu(F) = \mu(E)+\mu(F-E).$$

LEMMA 2. *Suppose that $\{E_j\}$ is a sequence of measurable events. Then there exists a sequence $\{F_j\}$ of disjoint measurable events, such that*

$$\bigcup_{j=1}^{\infty} F_j = \bigcup_{j=1}^{\infty} E_j \quad \textit{and} \quad F_j \subset E_j \quad (j = 1, 2,\ldots). \tag{8.1}$$

Proof. The events

$$F_i = E_i - \bigcup_{j=1}^{i-1} E_j \quad (i = 1, 2, \ldots)$$

have the desired property.

LEMMA 3. *Let $\{E_j\}$ be a sequence of measurable events, and suppose that E is a measurable event satisfying $E \subset \bigcup_j E_j$. Then*

$$\mu(E) \leqslant \sum_{j=1}^{\infty} \mu(E_j).$$

Proof. On choosing a sequence $\{F_j\}$ of disjoint events satisfying (8.1), we have (by Lemma 1)

$$\mu(E) \leqslant \mu\Big(\bigcup_{j=1}^{\infty} F_j\Big) = \sum_{j=1}^{\infty} \mu(F_j) \leqslant \sum_{j=1}^{\infty} \mu(E_j).$$

We note that each of Lemmas 2 and 3 includes as a special case (E_j empty for $j > k$) the corresponding result for a terminating sequence $E_1, \ldots, E_k$.

LEMMA 4. *Let $\{f_j\}$ be a sequence of (simple) random variables. Then the condition (C) below is sufficient† to ensure that, with probability 1, the sequence $\{f_j(x)\}$, of real numbers, converges to 0.*

(*C*) *If $E_j(\epsilon)$ is the event $|f_j(x)| \geqslant \epsilon$, then*

$$\mu\Big(\bigcap_{i=1}^{\infty} \bigcup_{j=i}^{\infty} E_j(\epsilon)\Big) = 0 \tag{8.2}$$

for every $\epsilon > 0$.

Proof. Writing
$$E^*(\epsilon) = \bigcap_{i=1}^{\infty} \bigcup_{j=i}^{\infty} E_j(\epsilon),$$
the event N, consisting of those x for which $\{f_j(x)\}$ does not converge to 0, is given by

$$N = \bigcup_{k=1}^{\infty} E^*(1/k).$$

Since each of the sets $E^*(1/k)$ has zero measure, we have, in view of Lemma 3, $\mu(N) = 0$.

LEMMA 5. *Let $\{s_j\}$ be a sequence of (simple) random variables. The condition (C′) below is sufficient† to ensure that, with probability 1, the sequence $\{s_j(x)\}$ of real numbers converges.*

(*C′*) *If $E_i(\epsilon)$ denotes the event*

$$\Lambda_i(x) \geqslant \epsilon, \quad \textit{where} \quad \Lambda_i(x) = \sup_{l > m \geqslant i} |s_l(x) - s_m(x)|,$$

then

$$\lim_{i \to \infty} \mu(E_i(\epsilon)) = 0 \tag{8.3}$$

for every $\epsilon > 0$.

† The condition is, in fact, also necessary.

Proof. It suffices to show that, with probability 1, $\{\Lambda_i(x)\}$ converges to 0. To see that this is in fact the case, we need only note that, in view of $E_{i+1}(\epsilon) \subset E_i(\epsilon)$ $(i = 1, 2, \ldots)$ and Lemma 1, (8.3) implies (8.2).

The following lemma is a special case of H., § 45, Theorem B.

LEMMA 6. *Let $f_1, \ldots, f_k$ be independent (simple) random variables. Let i be a natural number satisfying $1 \leqslant i < k$ and let $\phi(t_1, \ldots, t_i)$, $\psi(t_{i+1}, \ldots, t_k)$ be real-valued functions of the real numbers t_j. Then the (simple) random variables g, h, defined by*

$$g(x) = \phi\{f_1(x), \ldots, f_i(x)\}, \qquad h(x) = \psi\{f_{i+1}(x), \ldots, f_k(x)\}$$

are independent.

Proof. Let $t_1, \ldots, t_k$ be real numbers. We denote by $E = E(t_1, \ldots, t_i)$ the event

$$f_1(x) = t_1, \quad \ldots, \quad f_i(x) = t_i \tag{8.4}$$

and by $F = F(t_{i+1}, \ldots, t_k)$ the event

$$f_{i+1}(x) = t_{i+1}, \quad \ldots, \quad f_k(x) = t_k. \tag{8.5}$$

Then we have

$$\mu(E \cap F) = \mu(E)\mu(F). \tag{8.6}$$

Now let each of $\mathfrak{M}_1$, $\mathfrak{M}_2$ be a set of real numbers, and denote by A, B (respectively) the events $g(x) \in \mathfrak{M}_1$, $h(x) \in \mathfrak{M}_2$. Then A, B are representable in the form

$$A = \bigcup_{r=1}^{l} E_r, \qquad B = \bigcup_{s=1}^{m} F_s$$

where the E_r are disjoint events of type E, and the F_s are disjoint events of type F. Since each pair E_r, F_s satisfies (8.6), it follows that

$$\mu(A \cap B) = \sum_{r=1}^{l} \sum_{s=1}^{m} \mu(E_r \cap F_s) = \mu(A)\mu(B),$$

as required.

We conclude with some lemmas from analysis.

LEMMA 7. *Suppose that the series $\sum_{j=1}^{\infty} a_j$ is convergent, and let $\{m_j\}$ be an unbounded monotone increasing sequence of positive numbers. Then*

$$\lim_{i\to\infty} \frac{1}{m_i} \sum_{j=1}^{i} a_j m_j = 0. \tag{8.7}$$

Proof. Write $r_i = \sum_{j=i}^{\infty} a_j$. Then, writing $m_0 = 0$, we have

$$\sum_{j=1}^{i} a_j m_j = \sum_{j=1}^{i} (r_j - r_{j+1}) m_j = \sum_{j=1}^{i} r_j(m_j - m_{j-1}) - r_{i+1} m_i.$$

Hence, if N is a natural number, we have for $i > N$,

$$\left|\sum_{j=1}^{i} a_j m_j\right| \leqslant \left|\sum_{j=1}^{N} r_j(m_j - m_{j-1})\right| + 2m_i \sup_{j>N} |r_j|.$$

Since $m_i \to \infty$ as $i \to \infty$, this implies

$$\limsup_{i\to\infty} \left|\frac{1}{m_i} \sum_{j=1}^{i} a_j m_j\right| \leqslant 2 \sup_{j>N} |r_j|. \tag{8.8}$$

On letting $N \to \infty$ in (8.8), we obtain (8.7).

In the proof of Lemma 9, we shall make use of Cantor's diagonal argument. The following lemma contains the essence of Cantor's idea.

Lemma 8. *Let each of $\mathscr{G}_1, \mathscr{G}_2, \ldots$ be an increasing sequence of natural numbers, and suppose that $\mathscr{G}_1 \supset \mathscr{G}_2 \supset \mathscr{G}_3 \supset \ldots$. Then there exists an increasing sequence of natural numbers $n_1, n_2, \ldots$ such that*

$$\{n_r : r \geqslant j\} \subset \mathscr{G}_j \quad (j = 1, 2, \ldots).$$

Proof. Suppose that

$$\mathscr{G}_k = \{i_{k1}, i_{k2}, \ldots\} \quad (k = 1, 2, \ldots).$$

Then the numbers $n_k = i_{kk}$ $(k = 1, 2, \ldots)$ have the required property.

Lemma 9. *Suppose that $\{l_j\}$ is a sequence of real numbers satisfying $0 \leqslant l_j \leqslant 1$, and that for each $n = 1, 2, \ldots$, $\{p_j^{(n)}\}$ is a sequence of numbers satisfying $0 \leqslant p_j^{(n)} \leqslant 1$. Write*

$$P^{(n)}(t) = \sum_{j=1}^{\infty} p_j^{(n)} t^j \quad (n = 1, 2, \ldots) \tag{8.9}$$

and

$$P(t) = \sum_{j=1}^{\infty} l_j t^j. \tag{8.10}$$

Then the condition

$$\lim_{n\to\infty} p_j^{(n)} = l_j \quad (j = 1, 2, \ldots) \tag{8.11}$$

is both necessary and sufficient for

$$\lim_{n\to\infty} P^{(n)}(t) = P(t) \tag{8.12}$$

to hold for every number t satisfying $|t| < 1$.

Proof. Let t be a fixed number satisfying $|t| < 1$, and suppose (8.11) holds. Then, for every natural number N,

$$\limsup_{n\to\infty} |P^{(n)}(t) - P(t)| = \limsup_{n\to\infty} \left|\sum_{j=N}^{\infty} (p_j^{(n)} - l_j) t^j\right| \leqslant \sum_{j=N}^{\infty} |t|^j.$$

On letting $N \to \infty$, we obtain (8.12).

We shall prove the necessity of the condition (8.11) without using the theory of functions of a complex variable. (The necessity could be deduced easily from Vitali's theorem,† by expressing the coefficients of $P(t)$ and $P^{(n)}(t)$ as contour integrals on the circle $|t| = \frac{1}{2}$.)

Suppose that (8.12) holds for all t satisfying $|t| < 1$. We suppose that j^* is the least j for which (8.11) is false, and obtain a contradiction. We can choose, successively, sequences $\mathscr{G}_1, \mathscr{G}_2,...$ of natural numbers (for $j < j^*$ we take $\mathscr{G}_j$ to be the sequence of all natural numbers), such that $\mathscr{G}_1 \supset \mathscr{G}_2 \supset \mathscr{G}_3 \supset ...,$

(8.13) $$\theta_j = \lim_{\substack{n\to\infty \\ n \in \mathscr{G}_j}} p_j^{(n)} \quad \text{exists} \quad (j = 1, 2,...)$$

and

(8.14) $$\theta_{j^*} \neq l_{j^*}.$$

Then, by Lemma 8, there exists a sequence $n_1, n_2,...$ of natural numbers, such that

(8.15) $$\lim_{r\to\infty} p_j^{(n_r)} = \theta_j \quad (j = 1, 2,...).$$

It follows from the sufficiency of the condition (8.11), that

(8.16) $$P(t) = \lim_{r\to\infty} P^{(n_r)}(t) = \sum_{j=1}^{\infty} \theta_j t^j \quad \text{for } |t| < 1.$$

In view of (8.14) and the uniqueness of representation of functions as power series, (8.10) and (8.16) are mutually contradictory.

9. Probability theory: some fundamental theorems

We now establish a number of basic results in probability theory, relevant to the proofs of Theorems 1–4. Some of these results could be dispensed with, however, if we were to take a purely objective view of the proofs of Theorems 1–4. For example, Theorem 12, the law of Poisson trials, is not needed at all, in the sense that the particular case of this result required in the sequel (cf. Lemma 16) arises independently as a by-product of the method used (cf. Lemma 14) in the proofs of Theorems 1–4; nevertheless we thought it would be helpful to the reader to view this special case in the light of the more general law. The variant of the strong law of large numbers given in Theorem 11 (and consequently Kolmogorov's inequality) could also be dispensed with. Whilst this theorem is applied in the sequel (cf. Lemma 10), we could, by making full use of special circumstances, have deduced an equally

† E. C. Titchmarsh, *The Theory of Functions*, § 5.21.

effective result from Tchebycheff's inequality. But the technical advantages of such a course would have been offset, to some extent at least, by loss of insight.

We must again stress that, whilst the results of this section are stated in a form appropriate to all the applications we have in mind, the restriction to *simple* random variables (cf. Definition 14) in Theorems 9–12 would be far too severe to provide a satisfactory general theory. In the light of the theory of abstract Lebesgue integration (cf. H., Chapter V), it will be seen that very much less restrictive conditions will suffice to ensure the existence of the various integrals (implicit in the use of $\mathbf{M}$ and $\mathbf{D}^2$) concerned. Both these theorems and their proofs† are therefore valid in a far wider context.

The various theorems of the section refer, of course, to a given probability space (X, S, μ).

The Borel–Cantelli lemma, which plays a crucial role in probability theory, is very well known. We first state the two results, which constitute the Borel–Cantelli lemma, in their standard form as Theorem 7 below. We shall, however, also require a refinement of one of these results. The hypotheses of the second part of Theorem 7 are too restrictive for our purposes. On the only occasion on which we require a result in the converse direction (in the last part of § 16), it will be for an application to events which are not fully independent, but only quasi-independent (in a sense to be made precise later). The appropriate refinement (Theorem 7′ below), due to Erdös–Rényi(6), shows that (9.3) implies (9.4) provided that the events $\{E_j\}$ satisfy merely a certain quasi-independence condition (see (9.5)).

We note that the condition (9.5) is trivially satisfied if the events E_j are independent; indeed, it is trivially satisfied even if the events E_j are only pairwise independent. In particular, it suffices to prove the first assertion of Theorem 7, and Theorem 7′. The second assertion of Theorem 7 may be regarded as a special case of Theorem 7′.

THEOREM 7 (The Borel–Cantelli lemma). *Let $\{E_j\}$ be a sequence of measurable events.*

If

$$\sum_{j=1}^{\infty} \mu(E_j) < \infty, \tag{9.1}$$

† Provided, of course, that Theorem 6 and Lemmas 4, 5, and 6 are replaced by their generalized counterparts.

(6) P. Erdös and A. Rényi, 'On Cantor's series with convergent $\sum 1/q_n$', *Annls. Univ. Scient. Bpest. Rolando Eötvös*, sect. math. 2 (1959) 93–109.

then, with probability 1, *at most a finite number of the events E_j can occur; or, equivalently,*

$$\mu\Big(\bigcap_{i=1}^{\infty}\bigcup_{j=i}^{\infty}E_j\Big)=0. \tag{9.2}$$

In the converse direction, if $\{E_j\}$ is a sequence of INDEPENDENT *events and*

$$Z_n=\sum_{j=1}^{n}\mu(E_j)\to\infty \quad \text{as } n\to\infty, \tag{9.3}$$

then, with probability 1, *infinitely many of the events E_j occur; or, equivalently,*

$$\mu\Big(\bigcap_{i=1}^{\infty}\bigcup_{j=i}^{\infty}E_j\Big)=1. \tag{9.4}$$

THEOREM 7′. *Let $\{E_j\}$ be a sequence of measurable events satisfying* (9.3), *such that*

$$\liminf_{n\to\infty} Z_n^{-2}\sum_{1\leqslant i<j\leqslant n}\{\mu(E_i\cap E_j)-\mu(E_i)\mu(E_j)\}=0. \tag{9.5}$$

Then, with probability 1, *infinitely many of the events E_j occur; or, equivalently,* (9.4) *holds.*

Proof of the first assertion of Theorem 7. By Lemma 3, we have

$$\limsup_{i\to\infty}\mu\Big(\bigcup_{j=i}^{\infty}E_j\Big)\leqslant\limsup_{i\to\infty}\sum_{j=i}^{\infty}\mu(E_j)=0. \tag{9.6}$$

In view of Lemma 1, (9.6) implies (9.2).

Proof of Theorem 7′. On writing

$$f_n(x)=\sum_{i=1}^{n}\chi_{E_i}(x),$$

where χ_{E_i} is the characteristic function of the event E_i, (9.4) may be expressed in the form

$$\mu\Big(\Big\{x\colon \lim_{n\to\infty}f_n(x)=\infty\Big\}\Big)=1. \tag{9.4′}$$

We note that $$\mathbf{M}(f_n)=\sum_{i=1}^{n}\mu(E_i)=Z_n.$$

In view of the quasi-independence of the events E_j (as expressed by (9.5)), it will prove possible to deduce (9.4′) from the fact that, according to (9.3),

$$\lim_{n\to\infty}\mathbf{M}(f_n)=\infty. \tag{9.3′}$$

We will be able to use Tchebycheff's inequality to prove (below) that

$$\liminf_{n\to\infty}\mu(\{x\colon f_n(x)<\tfrac{1}{2}\mathbf{M}(f_n)\})=0. \tag{9.7}$$

Once (9.7) has been established, it requires only an application of the first assertion of the Borel–Cantelli lemma, in conjunction with a well-known device based on the fact that (for each x) $\{f_n(x)\}$ is an increasing sequence, in order to obtain (9.4′). For it follows from (9.7) that

(9.8) *there exists a subsequence* $\{n_j\}$ *of the natural numbers, such that*

$$\sum_{j=1}^{\infty} \mu(\{x: f_{n_j}(x) < \tfrac{1}{2}\mathbf{M}(f_{n_j})\}) < \infty.$$

By the first assertion of Theorem 7, this implies that, with probability 1,

$$f_{n_j}(x) \geqslant \tfrac{1}{2}\mathbf{M}(f_{n_j}) \quad \text{for } j \geqslant j_0(x);$$

so that (9.4′) follows from (9.3′).

It remains only to prove (9.7). We have, by Tchebycheff's inequality (7.4),

$$\mu(\{x: f_n(x) < \tfrac{1}{2}\mathbf{M}(f_n)\}) \leqslant \mu(\{x: |f_n(x) - \mathbf{M}(f_n)| \geqslant \tfrac{1}{2}\mathbf{M}(f_n)\})$$
$$\leqslant 4Z_n^{-2}\mathbf{D}^2(f_n).$$

But

$$\mathbf{D}^2(f_n) = \mathbf{M}(f_n^2) - \{\mathbf{M}(f_n)\}^2$$
$$= \sum_{i=1}^{n}\sum_{j=1}^{n}\{\mu(E_i \cap E_j) - \mu(E_i)\mu(E_j)\}$$
$$\leqslant 2\sum_{1 \leqslant i < j \leqslant n}\{\mu(E_i \cap E_j) - \mu(E_i)\mu(E_j)\} + Z_n,$$

so that, by (9.5) and (9.3),

$$\liminf_{n\to\infty} Z_n^{-2}\mathbf{D}^2(f_n) = 0.$$

This completes the proof of (9.7), and of the theorem.

The following result is merely a combination of the first assertion of Theorem 7 and Lemma 4.

THEOREM 8. *Let* $\{f_j\}$ *be a sequence of (simple) random variables such that, for every* $\epsilon > 0$,

$$\sum_{j=1}^{\infty} \mu(\{x: |f_j(x)| > \epsilon\}) < \infty. \tag{9.9}$$

Then, with probability 1,

$$\lim_{j\to\infty} f_j(x) = 0. \tag{9.10}$$

The following theorem, due to Kolmogorov, represents an important generalization of Tchebycheff's inequality (cf. § 7, formula (7.4)).

THEOREM 9. *Let $f_1, f_2,\ldots, f_k$ be independent (simple) random variables, such that* $\mathbf{M}(f_j) = 0$ *for each* $j = 1, 2,\ldots, k$; *and write*

$$s_i(x) = \sum_{j=1}^{i} f_j(x) \quad (i = 1, 2,\ldots, k). \tag{9.11}$$

Let $\epsilon > 0$, *and denote by* E_i *the event* $|s_i(x)| \geqslant \epsilon$. *Then*

$$\mu\Big(\bigcup_{i=1}^{k} E_i\Big) \leqslant \frac{1}{\epsilon^2} \sum_{i=1}^{k} \mathbf{D}^2(f_i). \tag{9.12}$$

Proof. In the proof of Lemma 2, each event F_i depends only on E_1, $E_2,\ldots, E_i$. Hence, by this lemma (cf. the remark after the proof of Lemma 3), there exist events $F_1,\ldots, F_k$ with the following properties.

(9.13) The F_j are disjoint and

$$\bigcup_{j=1}^{k} F_j = \bigcup_{j=1}^{k} E_j.$$

(9.14) For each $j = 1, 2,\ldots, k$, $F_j \subset E_j$, so that

$$|s_j(x)| \geqslant \epsilon \quad \text{whenever} \quad x \in F_j.$$

(9.15) For each $j = 1, 2,\ldots, k$ the event F_j is defined solely in terms of ϵ and the random variables $f_1, f_2,\ldots, f_j$.

Since the random variables f_j are independent and have zero mean, we have by § 6, Theorem 6,

$$\sum_{j=1}^{k} \mathbf{D}^2(f_j) = \mathbf{M}(s_k^2). \tag{9.16}$$

Furthermore, again using χ_F to denote the characteristic function of F, we have by the first assertion of (9.13),

$$\mathbf{M}(s_k^2) \geqslant \mathbf{M}\Big(s_k^2 \sum_{j=1}^{k} \chi_{F_j}\Big) = \sum_{j=1}^{k} \mathbf{M}(s_k^2\, \chi_{F_j}). \tag{9.17}$$

We shall prove that

$$\mathbf{M}(s_k^2\, \chi_{F_j}) \geqslant \epsilon^2 \mu(F_j) \quad (j = 1, 2,\ldots, k). \tag{9.18}$$

This will suffice to prove the theorem, for (9.16), (9.17), and (9.18) together yield

$$\sum_{j=1}^{k} \mathbf{D}^2(f_j) \geqslant \epsilon^2 \sum_{j=1}^{k} \mu(F_j),$$

and, in view of (9.13), this is equivalent to (9.12).

Now, for $i = 1, 2,\ldots, k$ and every x,

$$s_k^2(x) = \Big(s_i(x) + \sum_{j=i+1}^{k} f_j(x)\Big)^2 \geqslant s_i^2(x) + g_i(x)s_i(x) + h_i(x),$$

where $$g_i = 2 \sum_{j=i+1}^{k} f_j, \qquad h_i = 2 \sum_{i<l<m\leqslant k} f_l f_m;$$

and thus

$$\mathbf{M}(s_k^2\,\chi_{F_i}) \geqslant \mathbf{M}(s_i^2\,\chi_{F_i})+\mathbf{M}(g_i\,s_i\,\chi_{F_i})+\mathbf{M}(h_i\,\chi_{F_i}). \tag{9.19}$$

But the (simple) random variables $s_i\,\chi_{F_i}$ and g_i are independent in the sense that the first is defined in terms of $f_1,\ldots, f_i$ whereas the second is defined in terms of $f_{i+1},\ldots, f_k$. It follows from Lemma 6 that they are independent in the sense of Definition 12. The (simple) random variables χ_{F_i} and h_i are similarly independent. Thus, by Theorem 6,

$$\mathbf{M}(g_i\,s_i\,\chi_{F_i}) = \mathbf{M}(g_i)\mathbf{M}(s_i\,\chi_{F_i})$$

and

$$\mathbf{M}(h_i\,\chi_{F_i}) = \mathbf{M}(h_i)\mathbf{M}(\chi_{F_i}).$$

Since $\mathbf{M}(g_i) = 0$ and (again by Theorem 6) $\mathbf{M}(h_i) = 0$, it follows that the last two terms on the right-hand side of (9.19) vanish.

In view of this and (9.14), we obtain

$$\mathbf{M}(s_k^2\,\chi_{F_i}) \geqslant \mathbf{M}(s_i^2\,\chi_{F_i}) \geqslant \epsilon^2\mathbf{M}(\chi_{F_i}) = \epsilon^2\mu(F_i).$$

This establishes (9.18) and hence the theorem.

There is a group of theorems in probability theory collectively known as the law (subclassified into the weak law and the strong law) of large numbers. A description of the general nature of these theorems is to be found in Kolmogorov's monograph[(5)]. Several variants of the strong law of large numbers can be deduced from Theorem 10 below, but we shall only establish the particular variant (Theorem 11) we require in the sequel.

THEOREM 10. *Let $\{g_j\}$ be a sequence of independent (simple) random variables such that $\mathbf{M}(g_j) = 0$ for each j. Then the condition*

$$\sum_{j=1}^{\infty} \mathbf{D}^2(g_j) < \infty \tag{9.20}$$

is sufficient to ensure that, with probability 1,

$$\sum_{j=1}^{\infty} g_j(x) \text{ converges.} \tag{9.21}$$

Proof. Write $\quad s_l(x) = \sum_{j=1}^{l} g_j(x) \quad (l = 1,2,\ldots).$

Let $\epsilon > 0$. Then for each natural m and all natural numbers k, we have by Theorem 9 (with $f_j(x) = g_{m+j}(x)$),

$$\mu\Big(\Big\{x\colon \sup_{j=1,2,\ldots,k} |s_{m+j}(x)-s_m(x)| \geqslant \epsilon\Big\}\Big) \leqslant \frac{1}{\epsilon^2}\sum_{j=m+1}^{\infty} \mathbf{D}^2(g_j). \tag{9.22}$$

Since the right-hand side of (9.22) tends to zero as m tends to infinity, the assertion of the theorem follows by Lemma 5.

THEOREM 11. *Let $\{f_j\}$ be a sequence of independent (simple) random variables. Suppose that on writing*

$$s_i = \sum_{j=1}^{i} f_j \quad (i = 1, 2, \ldots),$$

we have

$$\mathbf{M}(f_j) > 0 \quad (j = 1, 2, \ldots), \tag{9.23}$$

$$\lim_{i \to \infty} \mathbf{M}(s_i) = \infty, \tag{9.24}$$

and

$$\sum_{i=1}^{\infty} \frac{\mathbf{D}^2(f_i)}{\{\mathbf{M}(s_i)\}^2} < \infty. \tag{9.25}$$

Then, with probability 1,

$$s_i(x) = \{1+o(1)\}\mathbf{M}(s_i) \quad \textit{as } i \to \infty. \tag{9.26}$$

Proof. We write

$$g_i(x) = \frac{f_i(x) - \mathbf{M}(f_i)}{\mathbf{M}(s_i)} \quad (i = 1, 2, \ldots), \tag{9.27}$$

so that

$$\mathbf{M}(g_i) = 0 \quad \text{and} \quad \mathbf{D}^2(g_i) = \frac{\mathbf{D}^2(f_i)}{\{\mathbf{M}(s_i)\}^2}. \tag{9.28}$$

The transformation (9.27), being linear, does not destroy independence, so that in view of (9.28) and (9.25), the sequence $\{g_j\}$ satisfies the premisses of Theorem 10. On applying this theorem we see that, with probability 1,

$$\sum_{i=1}^{\infty} \frac{\{f_i(x) - \mathbf{M}(f_i)\}}{\mathbf{M}(s_i)} \quad \textit{converges.} \tag{9.29}$$

Finally, we note that by (9.23) and (9.24) the sequence $\{m_i\}$, where $m_i = \mathbf{M}(s_i)$, is monotone increasing and unbounded. Thus, on applying Lemma 7, we see that (9.29) implies (9.26).

The following result is known as the law of Poisson trials.

THEOREM 12. *Suppose that for each n = 1, 2,...,*

$$f_1^{(n)}, f_2^{(n)}, \ldots, f_{k_n}^{(n)} \tag{9.30}$$

is a set of independent (simple) random variables, each assuming only the values 0 *and* 1. *Write, for n = 1, 2,...,*

$$\mu(\{x : f_j^{(n)}(x) = 1\}) = p_j^{(n)} \quad (j = 1, 2, \ldots, k_n). \tag{9.31}$$

Suppose that

$$\lim_{n \to \infty} \max_{1 \leqslant j \leqslant k_n} p_j^{(n)} = 0, \tag{9.32}$$

and

$$\lim_{n \to \infty} \sum_{j=1}^{k_n} p_j^{(n)} = \lambda > 0. \tag{9.33}$$

Then the (simple) random variables

$$s^{(n)} = \sum_{j=1}^{k_n} f_j^{(n)} \quad (n = 1, 2, \ldots) \tag{9.34}$$

have in the limit a Poisson distribution with mean value λ, *that is*

$$\lim_{n\to\infty} \mu(\{x\colon s^{(n)}(x) = r\}) = \frac{\lambda^r e^{-\lambda}}{r!} \quad (r = 0, 1, \ldots). \tag{9.35}$$

Remark. It is important to note that (9.32) and (9.33) imply that

$$k_n \to \infty \quad \text{as} \quad n \to \infty. \tag{9.36}$$

Proof. Clearly, the probability of the event $s^{(n)}(x) = r$ is given by the coefficient of t^r in the terminating power series (generating function)

$$P^{(n)}(t) = \prod_{j=1}^{k_n} (1 - p_j^{(n)} + p_j^{(n)} t).$$

We shall show that, for every fixed t,

$$\lim_{n\to\infty} P^{(n)}(t) = e^{-\lambda(1-t)}. \tag{9.37}$$

Since the right-hand side of (9.35) is the coefficient of t^r in the expansion of $e^{-\lambda(1-t)}$ it follows from Lemma 9 that (9.37) implies (9.35).

We have by (9.33)

$$\sum_{j=1}^{k_n} \{p_j^{(n)}\}^2 \leqslant \nu_n, \quad \text{where} \quad \nu_n = \max_{1\leqslant j\leqslant k_n} p_j^{(n)}.$$

Thus, by (9.32), we have for $n > n_0(t)$,

$$\begin{aligned} \log P^{(n)}(t) &= \sum_{j=1}^{k_n} \log\{1 - p_j^{(n)}(1-t)\} \\ &= -(1-t) \sum_{j=1}^{k_n} p_j^{(n)} + O\{\nu_n(1 + |t|^2)\}. \end{aligned}$$

Hence, by (9.32) and (9.33),

$$\lim_{n\to\infty} \log P^{(n)}(t) = -\lambda(1-t).$$

This is equivalent to (9.37), so that the theorem is proved.

10. Probability measures on the space of (positive) integer sequences

We use ω to denote a subsequence of the sequence $\mathscr{Z}_1$ of natural numbers, and Ω to denote the space of all such sequences ω. In each of the proofs of Theorems 1–4, we use a probability space of the type described in Theorem 13 below, with a suitable choice of the sequence (10.1) of real numbers.

THEOREM 13. *Let*

(10.1) $$\alpha_1, \alpha_2, \alpha_3, \ldots$$

be real numbers satisfying

(10.2) $$0 \leqslant \alpha_n \leqslant 1 \quad (n = 1, 2, \ldots).$$

Then there exists a probability space (Ω, S, μ) *with the following two properties.*

(i) *For every natural number n, the event* $B^{(n)} = \{\omega : n \in \omega\}$ *is measurable, and* $\mu(B^{(n)}) = \alpha_n$.

(ii) *The events* $B^{(1)}, B^{(2)}, \ldots$ *are independent.*

We show that this theorem, apart from the different notation and terminology, is merely the special case of Theorem 5 proved in § 5.

Let Y be the space of two elements, $y^{(0)}$ and $y^{(1)}$ say. With each sequence ω we associate the sequence $\{x_j\}$ of elements of Y, where x_j is $y^{(1)}$ or $y^{(0)}$ according as j does or does not lie in ω. The space X consisting of all the sequences $x = \{x_j\}$, is given by

(10.3) $$X = \bigtimes_{j=1}^{\infty} X_j, \quad \text{where} \quad X_j = Y \quad (j = 1, 2, \ldots).$$

Let S_j be the σ-algebra consisting of the four subsets of X_j, and let μ_j be the probability measure on S_j which assigns α_j to the subset of X_j consisting of the single element $y^{(1)}$.

We apply Theorem 5 to the sequence $\{(X_j, S_j, \mu_j)\}$ of probability spaces. In view of the 1–1 correspondence between the elements of X and Ω, we may denote the resulting probability space by (Ω, S, μ). In view of (5.3), this space has the properties (i) and (ii).

We denote by $\rho_n(\omega)$ *the characteristic function of the event* $B^{(n)}$. Then (ii) is equivalent to saying that $\rho_1, \rho_2, \ldots$ are independent (simple) random variables (cf. Definition 13).

All the random variables occurring in the proofs of Theorems 1–4 will be of one of the following types.

(*a*) Random variables constructed from $\rho_1, \rho_2, \ldots$ by the formation of finite sums and products, and finite combinations of these operations. (As was remarked in § 6, the class of (simple) functions is closed with respect to these operations.)

(*b*) The characteristic function of the 'event' of some particular random variable of type (*a*) assuming a specific value. (Characteristic functions of measurable events are simple.)

The construction of random variables will therefore be of a purely combinatorial nature. On the other hand, the method will frequently involve the (implicit) formation of countable unions of 'events'. It is for these reasons that we can dispense with the theory of integration, and yet require the full generality of the definition of 'probability space'.

11. Preparation for the proofs of Theorems 1-4

In this section we introduce some notation and prove a number of miscellaneous combinatorial results which will be in frequent use in subsequent sections.

We shall henceforth be working exclusively with probability spaces (Ω, S, μ) of the type introduced in Theorem 13. *We retain the notation of* § 10, and, in addition, will use the notation introduced in Definitions 16 and 17 below.

Definition 16. Let ω be a constituent sequence of the space Ω, and let n be a natural number. We denote by $s_n^* = s_n^*(\omega)$ the counting number of the sequence ω, so that $s_n^*(\omega)$ is the number of elements of ω which do not exceed n.

Furthermore, we denote (cf. § 2) by $r_n = r_n(\omega)$ the number of representations of n in the form

(11.1) $$n = a'+a'', \qquad a' < a'',$$

where a', a'' are elements of the sequence $\omega = \{a_j\}$.

Remark. Obviously, we have

(11.2) $$s_n^*(\omega) = \sum_{j=1}^{n} \rho_j(\omega),$$

and

(11.3) $$r_n(\omega) = \sum_{1\leqslant j<\frac{1}{2}n} \rho_j(\omega)\rho_{n-j}(\omega);$$

the random variables ρ_j being defined as in § 10.

Definition 17. We define, for every natural number n,

(11.4) $$m_n^* = \mathbf{M}\{s_n^*(\omega)\} = \sum_{j=1}^{n} \alpha_j,$$

(11.5) $$\lambda_n = \mathbf{M}\{r_n(\omega)\} = \sum_{1\leqslant j<\frac{1}{2}n} \alpha_j\,\alpha_{n-j},$$

and

(11.6) $$\lambda_n' = \sum_{1\leqslant j<\frac{1}{2}n} \alpha_j\,\alpha_{n-j}(1-\alpha_j\,\alpha_{n-j})^{-1}.$$

The subsequent work of the section will be subject to the assumptions listed in Hypothesis A below.

Hypothesis A. *The sequence* $\{\alpha_j\}$ *of probabilities (introduced in Theorem* 13) *satisfies the following conditions*

(11.7) $$0 < \alpha_j < 1 \quad \textit{for } j = 1, 2, \ldots,$$

(11.8) *$\{\alpha_j\}$ is monotonic decreasing from some point onward (i.e. for $j \geqslant j_1$),*

(11.9) $$\alpha_j \to 0 \quad as \quad j \to \infty.$$

On applying Theorem 11, with $f_j = \rho_j$, we obtain the following result which will play an important role in the sequel. Note that

$$\mathbf{D}^2(\rho_j) = \alpha_j - \alpha_j^2.$$

LEMMA 10. *If, in addition to Hypothesis A,*

(11.10) $$m_n^* \to \infty \quad as \quad n \to \infty,$$

(11.11) $$\sum_{n=1}^{\infty} \frac{\alpha_n}{(m_n^*)^2} < \infty,$$

then

(11.12) *with probability* 1, $s_n^*(\omega) \sim m_n^*$ *as* $n \to \infty$.

It is clear that if a particular sequence $\omega = \{a_j\}$ (in a given probability space Ω) satisfies $s_n^*(\omega) \sim m_n^*$, then this relation determines the asymptotic behaviour of a_j; for $s_{a_j}^*(\omega) = j$, and in the resulting relation $j \sim m_{a_j}^*$, the right-hand side is a known function of a_j. (This argument will be carried out explicitly, for the type of probability space appropriate to the sequel, in the proof of the last assertion of Lemma 11 below.)

Thus Lemma 10 (above) shows that (subject to suitable provisos) a probability space Ω, of the type described in Theorem 13, may be chosen so as to bring into strong relief those sequences ω having some prescribed rate of growth (by choosing the α_j so as to obtain the desired asymptotic behaviour of m_n^*). Such emphasis (by means of a suitable choice of probability space) of sequences with some specified rate of growth is the essence of Erdös's idea. It provides a starting-point for proofs of the existence of sequences having specified properties, when these properties are, to some extent at least, related to rate of growth.

We shall, in the sequel, make various choices of the sequence $\{\alpha_j\}$ of probabilities, each choice generating a particular probability space Ω of the type described in Theorem 13. In each case it will be necessary (for the appropriate application we have in mind) to determine the asymptotic behaviour of at least one of the parameters m_n^*, λ_n (introduced in Definition 17). Lemmas 11 and 11′ below will be repeatedly applied for this purpose.

LEMMA 11. *Suppose that the sequence $\{\alpha_j\}$, introduced in Theorem 13, is such that*

(11.13) $$\alpha_j = \alpha \frac{(\log j)^{c'}}{j^c} \quad for \quad j \geqslant j_0;$$

where j_0, α, c, c' are constants such that $\alpha > 0$, (11.13) and (11.7) are compatible, and

$$0 < c < 1, \qquad c' \geqslant 0. \tag{11.14}$$

Then, as $n \to \infty$,

$$\lambda_n \sim \tfrac{1}{2}\alpha^2 \frac{\{\Gamma(1-c)\}^2}{\Gamma(2-2c)} (\log n)^{2c'} n^{1-2c}. \tag{11.15}$$

Furthermore, we have

$$m_n^* \sim \frac{\alpha}{1-c} (\log n)^{c'} n^{1-c}. \tag{11.16}$$

Finally, if $c' = 0$, then with probability 1, the elements a_j of the sequence $\omega = \{a_j\}$ satisfy

$$a_j \sim \left(\frac{1-c}{\alpha} j\right)^{1/(1-c)} \quad \text{as } j \to \infty. \tag{11.17}$$

Remark. We note that if (11.13) takes the form $\alpha_j = \frac{1}{2}j^{-\frac{1}{2}}$ (i.e. if $\alpha = \frac{1}{2}$, $c' = 0$, $c = \frac{1}{2}$), then (11.17) takes the form $a_j \sim j^2$ and so expresses the fact that $\{a_j\}$ is a sequence of 'pseudo-squares' (cf. § 2).

Proof. Throughout the proof, n is assumed to be sufficiently large and n is the implicit variable in all errors of type $o(1)$. We write $\delta_n = 1/(\log n)$, but shall make use only of the facts that δ_n is positive and tends to zero sufficiently slowly as $n \to \infty$.

We note that

$$\sum_{j \leqslant n\delta_n} \frac{1}{\{j(n-j)\}^c} \ll \frac{1}{n^c} \sum_{j \leqslant n\delta_n} \frac{1}{j^c} = o(n^{1-2c}); \tag{11.18}$$

so that, since $0 < \alpha_j < 1$ for $1 \leqslant j < j_0$ (and hence

$$\alpha_j \alpha_{n-j} \ll \alpha_{n-j} \ll (\log n)^{c'} n^{-c} = o(n^{1-2c})$$

for $1 \leqslant j < j_0$), we have

$$\lambda_n = \alpha^2 \sum_{n\delta_n < j < \frac{1}{2}n} \frac{\{(\log j)(\log(n-j))\}^{c'}}{(j(n-j))^c} + o\{(\log n)^{2c'} n^{1-2c}\}.$$

Now, in view of the range of summation, we have for each term in the sum on the right,

$$(\log j)(\log(n-j)) = \{1+o(1)\}(\log n)^2.$$

Hence, again using (11.18), we obtain

$$\lambda_n = \alpha^2\{1+o(1)\}(\log n)^{2c'} \sum_{1 \leqslant j < \frac{1}{2}n} \frac{1}{\{j(n-j)\}^c} + o\{(\log n)^{2c'} n^{1-2c}\}. \tag{11.19}$$

But, writing $\phi(t) = \{t(1-t)\}^{-c}$ and $t_j = j/n$, we have

$$n^{2c-1} \sum_{1 \leqslant j < \frac{1}{2}n} \frac{1}{\{j(n-j)\}^c} = \sum_{1 \leqslant j < \frac{1}{2}n} \phi(t_j)(t_j - t_{j-1})$$

$$= \{1+o(1)\} \int_0^{\frac{1}{2}} \phi(t)\, dt$$

$$= \tfrac{1}{2}\{1+o(1)\} \mathrm{B}(1-c, 1-c)$$

$$= \tfrac{1}{2}\{1+o(1)\}\{\Gamma(1-c)\}^2/\Gamma(2-2c).$$

On substituting this relation in (11.19), we obtain (11.15). Furthermore,

$$m_n^* = \sum_{j=1}^{n} \alpha \frac{(\log j)^{c'}}{j^c} + O(1) \tag{11.20}$$

$$= \{1+o(1)\} \frac{\alpha}{1-c} (\log n)^{c'} n^{1-c},$$

and this proves (11.16). We note that (11.16) shows that (11.10) and (11.11) are satisfied.

The final assertion of the lemma follows from (11.16), in view of Lemma 10 and the fact that $s_{a_j}^*(\omega) = j$; for in this way it follows that, if $c' = 0$, we have with probability 1,

$$\frac{\alpha}{1-c} (a_j)^{1-c} \sim j.$$

This completes the proof of the lemma.

It was convenient to exclude the possibility $c = 1$ in the statement of Lemma 11, but we shall require the following trivial supplementary result. The first assertion of Lemma 11′ is an obvious extension of (11.20), and the second assertion follows by Lemma 10. (Again, (11.22) shows that (11.10) and (11.11) are satisfied.)

LEMMA 11′. *Suppose that the sequence* $\{\alpha_j\}$, *introduced in Theorem* 13, *is such that*

$$\alpha_j = \alpha \frac{\log j}{j} \quad \textit{for } j \geqslant j_0, \tag{11.21}$$

where j_0, α *are constants such that* $\alpha > 0$, *and* (11.21), (11.7) *are compatible.*

Then, as $n \to \infty$,

$$m_n^* \sim \tfrac{1}{2}\alpha(\log n)^2. \tag{11.22}$$

Furthermore,

(11.23) *with probability* 1, $s_n^*(\omega) \sim \tfrac{1}{2}\alpha(\log n)^2$ *as* $n \to \infty$.

Our next objective will be to obtain various estimates for the probability of the event $r_n(\omega) = d$. (Our main investigations will hinge on the consideration of such events.) We first give an explicit formula for the probability in question.

LEMMA 12. *For every natural number n and every non-negative integer d,*

(11.24) $$\mu(\{\omega\colon r_n(\omega) = d\}) = \Big(\prod_{1\leqslant k<\frac{1}{2}n}(1-\alpha_k\alpha_{n-k})\Big)\sigma_d^{(n)},$$

where $\sigma_0^{(n)} = 1$ *and, if* $d \geqslant 1$,

(11.25) $$\sigma_d^{(n)} = \sum_{1\leqslant k_1<\ldots<k_d<\frac{1}{2}n}\prod_{j=1}^{d}\frac{\alpha_{k_j}\alpha_{n-k_j}}{1-\alpha_{k_j}\alpha_{n-k_j}}.$$

Proof. The case $d = 0$ is trivial. If $d \geqslant 1$, we have

$$\mu(\{\omega\colon r_n(\omega) = d\}) = \sum_{1\leqslant k_1<\ldots<k_d<\frac{1}{2}n}\mu\{E(k_1,\ldots, k_d)\},$$

where $E(k_1,\ldots, k_d)$ is the event

$$\{\omega\colon \rho_k\rho_{n-k} = 1 \quad \text{if and only if } k \text{ is one of } k_1,\ldots, k_d\}.$$

Clearly

$$\mu\{E(k_1,\ldots, k_d)\} = \Big\{\prod_{j=1}^{d}\alpha_{k_j}\alpha_{n-k_j}\Big\}\Big\{\prod_{\substack{1\leqslant k<\frac{1}{2}n\\ k\neq k_j(j=1,\ldots,d)}}(1-\alpha_k\alpha_{n-k})\Big\}$$

$$= \Big\{\prod_{1\leqslant k<\frac{1}{2}n}(1-\alpha_k\alpha_{n-k})\Big\}\Big\{\prod_{j=1}^{d}\frac{\alpha_{k_j}\alpha_{n-k_j}}{1-\alpha_{k_j}\alpha_{n-k_j}}\Big\}.$$

The required result follows. (If $d \geqslant \frac{1}{2}n$, the sum $\sigma_d^{(n)}$ is empty, and both sides of (11.24) are zero.)

LEMMA 13. *Let* $y_1, y_2,\ldots, y_N$ *be N non-negative numbers. For each natural number d, not exceeding N, we write*

(11.26) $$\sigma_d = \sum_{1\leqslant k_1<\ldots<k_d\leqslant N} y_{k_1}y_{k_2}\cdots y_{k_d},$$

so that σ_d *is the d-th elementary symmetric function of the* y_k.

Then, for each d,

(11.27) $$\frac{1}{d!}\sigma_1^d\Big\{1-\binom{d}{2}\sigma_1^{-2}\sum_{k=1}^{N}y_k^2\Big\} \leqslant \sigma_d \leqslant \frac{1}{d!}\sigma_1^d,$$

where $\binom{d}{2}$ *is interpreted to be zero when* $d = 1$.

Proof. Clearly, for each d, we have

(11.28) $$(d+1)\sigma_{d+1} \leqslant \sigma_1\sigma_d \leqslant (d+1)\sigma_{d+1}+\sigma_{d-1}Q,$$

where $$Q = \sum_{k=1}^{N}y_k^2$$

and σ_0 is interpreted to be 1.

By the left-hand inequality of (11.28), it follows by induction that

(11.29) $$\sigma_d \leqslant \frac{1}{d!}\sigma_1^d.$$

Furthermore, by (11.29) and the right-hand inequality of (11.28) we have

$$(d+1)\sigma_{d+1} \geqslant \sigma_1 \sigma_d - \frac{\sigma_1^{d-1}}{(d-1)!} Q;$$

and it follows by induction that, for $2 \leqslant d \leqslant N$,

$$\sigma_d \geqslant \frac{1}{d!} \sigma_1^d \left\{1 - \binom{d}{2} \sigma_1^{-2} Q\right\}. \tag{11.30}$$

(11.29) and (11.30) are the two assertions of (11.27).

LEMMA 14. *Let n be a natural number and d be a non-negative integer. Then*

$$\mu(\{\omega \colon r_n(\omega) = d\}) \leqslant \frac{(\lambda_n')^d}{d!} e^{-\lambda_n}. \tag{11.31}$$

Furthermore, if $d < \frac{1}{2}n$,

$$\mu(\{\omega \colon r_n(\omega) = d\}) \geqslant \frac{(\lambda_n')^d}{d!} e^{-\lambda_n'} \left\{1 - \binom{d}{2} (\lambda_n')^{-2} Q^*\right\}, \tag{11.32}$$

where

$$Q^* = \sum_{1 \leqslant k < \frac{1}{2}n} \left(\frac{\alpha_k \alpha_{n-k}}{1 - \alpha_k \alpha_{n-k}}\right)^2, \tag{11.33}$$

and $\binom{d}{2}$ is interpreted to be zero if $d < 2$.

Proof. We note that if $d \geqslant \frac{1}{2}n$, the event $r_n(\omega) = d$ is empty and (11.31) is trivial. We may thus suppose that $d < \frac{1}{2}n$. We apply (11.27) (which is trivial† when $d = 0$) with $y_k = \alpha_k \alpha_{n-k}/(1 - \alpha_k \alpha_{n-k})$ to estimate $\sigma_d^{(n)}$ in (11.24); thus obtaining

$$\mu(\{\omega \colon r_n(\omega) = d\}) \leqslant P \frac{(\lambda_n')^d}{d!}$$

and

$$\mu(\{\omega \colon r_n(\omega) = d\}) \geqslant P \frac{(\lambda_n')^d}{d!} \left\{1 - \binom{d}{2} (\lambda_n')^{-2} Q^*\right\},$$

where

$$P = \prod_{1 \leqslant k < \frac{1}{2}n} (1 - \alpha_k \alpha_{n-k}).$$

These inequalities imply (11.31) and (11.32); for on applying

$$e^{-t/(1-t)} < 1 - t < e^{-t} \quad (0 < t < 1)$$

to each factor of P, we obtain

$$e^{-\lambda_n'} < P < e^{-\lambda_n}.$$

LEMMA 15. *If Hypothesis A is satisfied, then*

$$\lambda_n' \sim \lambda_n \quad \textit{as} \quad n \to \infty. \tag{11.34}$$

† Interpreting $\sigma_0 = 1$ and $\binom{0}{2} = 0$.

Proof. If $1 \leqslant k < \frac{1}{2}n$ and n is large, we have by (11.7), (11.8), and (11.9),

$$\alpha_k \alpha_{n-k} \leqslant \alpha_{n-k} \leqslant \alpha_{[\frac{1}{2}n]} = o(1)$$

as $n \to \infty$. This clearly implies the required result.

The following lemma is a special case of the law of Poisson trials (Theorem 12). Although we would obtain Lemma 16 immediately on writing $f_j^{(n)} = \rho_j \rho_{n-j}$ (and $k_n = [\frac{1}{2}n]$) in Theorem 12, we give an entirely different proof based on Lemma 14. Although Lemma 14 was not introduced for this purpose (the lemma will frequently be used in other connexions), it is worth showing that Lemma 14 can also be used to establish the particular case of the law of Poisson trials we require.

Lemma 16. *Suppose Hypothesis A is satisfied and*

$$\lim_{n\to\infty} \lambda_n = \lambda > 0. \tag{11.35}$$

Then, for every non-negative integer d,

$$\lim_{n\to\infty} \mu(\{\omega\colon r_n(\omega) = d\}) = \frac{\lambda^d}{d!} e^{-\lambda}; \tag{11.36}$$

in other words, the simple random variables $r_n(\omega)$ have, in the limit, a Poisson distribution with mean value λ.

Proof. In view of Lemma 15, Lemma 14 implies (11.36) provided

$$(\lambda_n')^{-2} Q^* = o(1) \quad \text{as} \quad n \to \infty. \tag{11.37}$$

But, as in the proof of Lemma 15,

$$Q^* \leqslant \sum_{1\leqslant k<\frac{1}{2}n} (\alpha_k \alpha_{n-k})^2 \leqslant \lambda_n \max_{1\leqslant k<\frac{1}{2}n} (\alpha_k \alpha_{n-k}) = o(\lambda_n);$$

so that (11.35) ensures (11.37).

Several arguments in the sequel will depend on obtaining suitable estimates for probabilities of type

$$\mu(\{\omega\colon r_n(\omega) \geqslant U\}) \quad \text{and} \quad \mu(\{\omega\colon r_n(\omega) \leqslant V\}).$$

In view of the type of estimate available for $\mu(\{\omega\colon r_n(\omega) = d\})$, the appropriate computations involve estimations of partial sums of the exponential series. The following crude estimates will suffice for this purpose.

Lemma 17. *If $0 < \xi \leqslant U$, then*

$$\sum_{d\geqslant U} \frac{\xi^d}{d!} \leqslant \left(\frac{e\xi}{U}\right)^U; \tag{11.38}$$

and if $0 < V \leqslant \xi$, then

$$\sum_{0\leqslant d\leqslant V} \frac{\xi^d}{d!} \leqslant \left(\frac{e\xi}{V}\right)^V. \tag{11.39}$$

Proof. If $0 < \xi \leqslant U$, we have

$$\sum_{d \geqslant U} \frac{\xi^d}{d!} = \sum_{d \geqslant U} \frac{U^d}{d!}\left(\frac{\xi}{U}\right)^d \leqslant \left(\frac{\xi}{U}\right)^U \sum_{d \geqslant U} \frac{U^d}{d!} \leqslant \left(\frac{\xi}{U}\right)^U e^U;$$

and if $0 < V \leqslant \xi$, we have

$$\sum_{0 \leqslant d \leqslant V} \frac{\xi^d}{d!} = \sum_{0 \leqslant d \leqslant V} \frac{V^d}{d!}\left(\frac{\xi}{V}\right)^d \leqslant \left(\frac{\xi}{V}\right)^V \sum_{0 \leqslant d \leqslant V} \frac{V^d}{d!} \leqslant \left(\frac{\xi}{V}\right)^V e^V.$$

We now turn to the proofs of the existence theorems stated in § 2. Each of these theorems will be proved by showing that if the sequence $\{\alpha_j\}$ of probabilities is suitably chosen, almost all sequences ω in the resulting probability space Ω have the desired property.

12. Proof of Theorem 1

We first choose a number α to satisfy $0 < \alpha < 1$ and

(12.1) $$\tfrac{1}{2}\pi\alpha^2 > 1,$$

say $\alpha = \{\frac{1}{2}(1+2/\pi)\}^{\frac{1}{2}}$. We now define the sequence $\{\alpha_j\}$ by

(12.2) $$\alpha_1 = \tfrac{1}{2}, \qquad \alpha_j = \alpha\left(\frac{\log j}{j}\right)^{\frac{1}{2}} \quad \text{for} \quad j \geqslant 2.$$

The precise value of α_j for small j is unimportant, but the above choice of α_1 ensures (11.7), so that Hypothesis A is satisfied. By Lemma 11 (with $c = c' = \frac{1}{2}$), we have

(12.3) $$\lambda_n \sim \tfrac{1}{2}\pi\alpha^2 \log n \quad \text{as} \quad n \to \infty;$$

and in view of (12.1), this ensures the existence of a number $\delta > 0$ such that

(12.4) $$e^{-\lambda_n} \ll n^{-1-\delta}.$$

We establish Theorem 1 by showing that (cf. (2.1)), with probability 1, $\log n \ll r_n(\omega) \ll \log n$ for large n, or equivalently (in view of (12.3) and the fact that $\lambda'_n \sim \lambda_n$ by Lemma 15)

(12.5) $$\lambda'_n \ll r_n(\omega) \ll \lambda'_n \quad \text{for } n > n_0(\omega).$$

We apply the Borel–Cantelli lemma (§ 9, Theorem 7) twice to prove that each of the two assertions of (12.5) holds with probability 1. For this purpose, we must show that if c', c'' are suitably chosen positive constants,

(12.6) $$\sum_{n=1}^{\infty} \mu(\{\omega : r_n(\omega) > c'\lambda'_n\}) < \infty$$

and†

$$\sum_{n=1}^{\infty} \mu(\{\omega: r_n(\omega) < c''\lambda'_n\}) < \infty. \tag{12.7}$$

By (11.31) and (11.38) we obtain the estimate

$$\mu(\{\omega: r_n(\omega) > c'\lambda'_n\}) \leqslant e^{-\lambda_n} \sum_{d > c'\lambda'_n} \frac{(\lambda'_n)^d}{d!}$$

$$\leqslant e^{-\lambda_n}\left(\frac{e}{c'}\right)^{c'\lambda'_n},$$

provided $c' > 1$. Thus, on choosing $c' = e$, we obtain the estimate $e^{-\lambda_n}$ for the summand of (12.6), so that (12.4) ensures (12.6).

On the other hand, by (11.31) and (11.39), we obtain for the summand of (12.7) the estimate

$$e^{-\lambda_n} \sum_{0 \leqslant d \leqslant c''\lambda'_n} \frac{(\lambda'_n)^d}{d!} \leqslant e^{-\lambda_n}\left(\frac{e}{c''}\right)^{c''\lambda'_n},$$

provided $c'' < 1$. Thus it will suffice to show that c'' can be chosen such that, in addition to $0 < c'' < 1$,

$$\left(\frac{e}{c''}\right)^{c''\lambda'_n} \ll n^{\frac{1}{2}\delta};$$

for (12.7) will then follow from (12.4). But $\lambda'_n \leqslant c^* \log n$ (for some constant c^*) so that we need only choose a small positive c'' satisfying

$$(e/c'')^{c''} \leqslant e^{\frac{1}{2}\delta/c^*},$$

as is certainly possible since $(e/t)^t \to 1$ as $t \to +0$.

We have now shown that ω has each of the desired properties with probability 1, and this proves the theorem.

13. Proof of Theorem 2

Let $\epsilon > 0$ be given. We first ensure that, with probability 1, the sequence $\omega = \{a_j\}$ satisfies $a_j \ll j^{2+\epsilon}$, by choosing the α_j so that (11.17) takes the form

$$a_j \sim c^* j^{2+\epsilon} \tag{13.1}$$

where c^* is a positive constant. To achieve this we must have $c = 1-1/(2+\epsilon)$ (and $c' = 0$) in (11.13). The constant α (in (11.13)) remains at our disposal (subject only to $0 < \alpha < 1$), but for the sake

† Needless to say, if each of two (or any finite or countable set of) events occurs with probability 1, then these events occur simultaneously with probability 1; for the complementary events have zero measure (cf. Lemma 3).

of a definite choice we write $\alpha = \frac{1}{2}$. Accordingly, we define the sequence $\{\alpha_j\}$ by

(13.2) $$\alpha_j = \tfrac{1}{2}j^{(2+\epsilon)^{-1}-1} \quad \text{for } j = 1, 2,\ldots.$$

Then Hypothesis A is satisfied and (13.1) holds with probability 1. In addition, by Lemma 11 (with $c = 1-1/(2+\epsilon)$, $c' = 0$) and Lemma 15,

(13.3) $$\lambda'_n \sim \lambda_n \sim c^{**}n^{-\epsilon/(2+\epsilon)};$$

the exponent $-\epsilon/(2+\epsilon)$ being the exponent $1-2c$ in (11.15).

We again appeal to the Borel–Cantelli lemma (§ 9, Theorem 7). It follows from this lemma that if the number K has the property

(13.4) $$\sum_{n=1}^{\infty} \mu(\{\omega\colon r_n(\omega) \geqslant K\}) < \infty,$$

then, with probability 1,

$$r_n(\omega) < K \quad \text{for} \quad n > n_0(\omega).$$

We note that, by (13.3), $\lambda_n \to 0$ and $\lambda'_n \to 0$ as $n \to \infty$. Thus, once more applying (11.31) and (11.38), we obtain for the summand of (13.4) the estimate

$$e^{-\lambda_n}\sum_{d \geqslant K}\frac{(\lambda'_n)^d}{d!} \ll e^{-\lambda_n}\left(\frac{e\lambda'_n}{K}\right)^K \ll (\lambda'_n)^K$$

for large n. It is now clear that (13.3) implies (13.4) provided $K\epsilon/(2+\epsilon) > 1$, or equivalently, provided

$$K > 1+2\epsilon^{-1}.$$

Accordingly we have, with probability 1,

(13.5) $$r_n(\omega) < 2(1+\epsilon^{-1}) \quad \text{for } n > n_1(\epsilon, \omega).$$

This completes the proof of the theorem.

14. Proof of Theorem 3

We define the probabilities α_j by

(14.1) $$\alpha_j = \begin{cases} \frac{1}{2} & \text{for } 1 \leqslant j < j_0 \\ \alpha\dfrac{\log j}{j} & \text{for } j \geqslant j_0, \end{cases}$$

where α is a positive constant to be chosen later, and we choose the minimal $j_0 = j_0(\alpha)$ so that $j_0 \geqslant 3$ and the conditions (14.1), (11.7) are consistent. Then Hypothesis A is satisfied, and (since $j_0 \geqslant 3$) the sequence $\{\alpha_j\}$ decreases for $j \geqslant j_0$. Furthermore, by Lemma 11′, we have

(14.2) $$\sum_{j=1}^{n} \alpha_j \sim \tfrac{1}{2}\alpha(\log n)^2$$

and, with probability 1,

$$s_n^*(\omega) \ll (\log n)^2 \tag{14.3}$$

as required for Theorem 3. (In fact (11.23) is more precise than (14.3).)

The appropriate representation function (in place of $r_n(\omega)$) is now

$$R_n(\omega) = \sum_{p<n} \rho_{n-p}(\omega), \tag{14.4}$$

where the summation is over primes p; and, in analogy to (11.5), we write

$$\Lambda_n = \mathbf{M}\{R_n(\omega)\} = \sum_{p<n} \alpha_{n-p}. \tag{14.5}$$

In place of (11.24) and (11.31), each with $d = 0$, we now have

$$\mu(\{\omega\colon R_n(\omega) = 0\}) = \prod_{p<n} (1-\alpha_{n-p}) \tag{14.6}$$

and

$$\mu(\{\omega\colon R_n(\omega) = 0\}) \leqslant e^{-\Lambda_n}. \tag{14.7}$$

(Here again, (14.6) is trivial and implies (14.7) since $1-t \leqslant e^{-t}$ for real t.)

To establish the theorem, we must prove that (subject to an appropriate choice of α), with probability 1, $R_n(\omega) > 0$ for large n. By the Borel–Cantelli lemma (§ 9, Theorem 7) we need only show that

$$\sum_{n=1}^{\infty} \mu(\{\omega\colon R_n(\omega) = 0\}) < \infty;$$

and in view of (14.7), it suffices to establish the existence of a number $\delta' > 0$ such that

$$e^{-\Lambda_n} \ll n^{-1-\delta'}. \tag{14.8}$$

We shall complete the proof of the theorem by showing that, subject to a suitable choice of α,

$$\Lambda_n > 2\log n \quad \text{for large } n, \tag{14.9}$$

so that (14.8) holds with $\delta' = 1$.

The sum (14.5) is sensitive to local irregularities in the distribution of primes, and in order to obtain (14.9), we must apply the following deep result due to Hoheisel.(7)

(7) G. Hoheisel, 'Primzahlprobleme in der Analysis', *S.B. preuss. Akad. Wiss.* (1930) 580–8.

See also A. E. Ingham, 'On the difference between consecutive primes', *Quart. J. Math.* 8 (1937) 255–66, and W. Haneke, 'Verschärfung der Abschätzung von $\zeta(\frac{1}{2}+it)$', *Acta Arith.* 8 (1962/63) 357–430.

There exists an absolute constant c_0, *satisfying* $0 < c_0 < 1$, *such that*

$$\sum_{\substack{p \\ m<p\leqslant m+m^{1-c_0}}} 1 = \{1+o(1)\}\frac{m^{1-c_0}}{\log m} \quad \textit{as } m \to \infty.$$

In particular, there exists a natural number m_0 *such that*

$$\sum_{m<p\leqslant m+m^{1-c_0}} 1 > \frac{2\,m^{1-c_0}}{3\log m} \quad \textit{if } m \geqslant m_0. \tag{14.10}$$

For the remainder of the proof, we suppose that n is large (the implied constant depending at most on α). We have

$$\Lambda_n \geqslant \sum_{p<n} \alpha_{n-p}\, p^{-1+c_0} \sum_{\substack{m=1 \\ m<p\leqslant m+m^{1-c_0}}}^{\infty} 1$$

since, for each p, there are at most p^{1-c_0} integers m satisfying

$$m < p \leqslant m+m^{1-c_0}$$

(in view of the fact that these inequalities imply

$$p-p^{1-c_0} < p-m^{1-c_0} \leqslant m \leqslant p-1).$$

It follows, on writing

$$l = l(n) = [n^{1-c_0}], \tag{14.11}$$

that
$$\Lambda_n \geqslant \sum_{m=m_0}^{n-2l} \sum_{\substack{p \\ m<p\leqslant m+m^{1-c_0}}} \alpha_{n-p}\, p^{-1+c_0},$$

since the conditions of summation imply $p < n$. Furthermore, since n is large, the conditions of summation also imply $n-p \geqslant j_0$, so that, by (14.10),

$$\begin{aligned}\Lambda_n &\geqslant \sum_{m=m_0}^{n-2l} \alpha_{n-m}(2m)^{-1+c_0} \sum_{\substack{p \\ m<p\leqslant m+m^{1-c_0}}} 1 \\ &\geqslant \sum_{m=m_0}^{n-2l} \alpha_{n-m}(2m)^{-1+c_0}\left(\frac{2\,m^{1-c_0}}{3\log m}\right) \\ &> \frac{1}{3\log n}\sum_{m=m_0}^{n-2l} \alpha_{n-m} \\ &= \frac{1}{3\log n}\sum_{j=2l}^{n-m_0} \alpha_j.\end{aligned}$$

Finally, by (14.2) and (14.11), we have

$$\sum_{j=2l}^{n-m_0} \alpha_j \sim \tfrac{1}{2}\alpha\{1-(1-c_0)^2\}(\log n)^2,$$

so that, since $1-(1-c_0)^2 > c_0$,

$$\Lambda_n > \tfrac{1}{6}\alpha c_0 \log n \quad \text{for large } n. \tag{14.12}$$

To obtain (14.9) we need now only choose $\alpha = 12/c_0$, and this completes the proof of the theorem.

15. Quasi-independence of the variables r_n

We have already remarked, in § 3, that probability theory derives its power from the fact that the presence of suitable dependence (or quasi-independence) relationships tends to imply regularity 'in the large'. The investigation leading to Theorem 4, hinges on the fact that the random variables r_n, r_m are quasi-independent (for the appropriate choice of the probability space Ω) provided only the integers n, m are not too close together. This fact is reflected in each of Theorems 14, 15, and 17, and these are the only quasi-independence results we shall require. (In the applications of these theorems, the choice of Ω will be such that the sequence $\{\alpha_j\}$ tends to zero fairly rapidly, so that α_{m-n} is small provided $m-n$ is large.)

We state Theorems 14 and 15 in the form required for the applications in subsequent sections; but both these theorems are in fact only corollaries of the more powerful Theorem 16, which is the natural outcome of the method of this section (Theorem 15 would be no easier to prove). Theorem 17 is a comparatively trivial result, although appropriate for inclusion in this section.

The probability space Ω will, of course, be of the type introduced in Theorem 13. *All constants (and this includes constants implied by the $\ll$ notation) depend at most on this probability space, which remains fixed throughout the section.* We make the further assumptions stated below.

HYPOTHESIS B. *The sequence $\{\alpha_j\}$ of probabilities (introduced in Theorem 13) satisfies the following conditions:*

(15.1) $$0 < \alpha_1 < 1.$$

(15.2) $\{\alpha_j\}$ *is a monotonic decreasing sequence.*

(15.3) $$\lambda_t = \sum_{1 \leqslant j < \frac{1}{2}t} \alpha_j \alpha_{t-j} \ll 1$$

on the set of all† integers t.

HYPOTHESIS C. *We suppose that m, n are natural numbers satisfying $m > n$.*

THEOREM 14. *Let d denote a non-negative integer, and write*

(15.4) $$F_n(d) = \{\omega: r_n(\omega) = d\}, \qquad F_m(d) = \{\omega: r_m(\omega) = d\}.$$

Then, subject to Hypotheses B, C, the events $F_n(d)$, $F_m(d)$ are quasi-independent in the sense that

(15.5) $$\mu\{F_n(d) \cap F_m(d)\} - \mu\{F_n(d)\}\mu\{F_m(d)\} \ll \alpha_{m-n}.$$

† $\lambda_t = 0$ for all integers $t < 3$, the defining sum being empty.

THEOREM 15. *Let c be a suitably chosen*† *positive constant, and let*

$$\delta(W) = \exp\{-W\log(cW)\} \tag{15.6}$$

for $W \geqslant 0$, $\delta(0)$ *being interpreted as* 1.

Let U, V *be non-negative integers, and write*

$$G_n(U) = \{\omega : r_n(\omega) \geqslant U\}, \qquad G_m(V) = \{\omega : r_m(\omega) \geqslant V\}. \tag{15.7}$$

Then, subject to Hypotheses B, C, the events $G_n(U)$, $G_m(V)$ *are quasi-independent in the sense that*

$$\mu\{G_n(U) \cap G_m(V)\} - \mu\{G_n(U)\}\mu\{G_m(V)\} \ll \alpha_{m-n}\,\delta(U)\delta(V) + \alpha_{m-n}^2\,\delta(V). \tag{15.8}$$

It is convenient to reserve a single symbol for the probability α_{m-n}, which will appear in error terms throughout the section. For this reason, we introduce the relation (15.12) in the statement of Theorem 16.

THEOREM 16. *For any pair of non-negative integers a, b, we write*

$$\phi_n(a) = \mu(\{\omega : r_n(\omega) = a\}), \qquad \phi_m(b) = \mu(\{\omega : r_m(\omega) = b\}), \tag{15.9}$$

$$\phi(a, b) = \phi_{n,m}(a, b) = \mu(\{\omega : r_n(\omega) = a,\ r_m(\omega) = b\}). \tag{15.10}$$

Let U, V *be non-negative integers. Then, subject to Hypotheses B, C,*

$$\sum_{a=U}^{\infty}\sum_{b=V}^{\infty} |\phi(a,b) - \phi_n(a)\phi_m(b)| \ll \beta\delta(U)\delta(V) + \beta^2 \min\{\delta(U), \delta(V)\}; \tag{15.11}$$

where $\delta(W)$ *is defined as in Theorem* 15, *and*

$$\beta = \alpha_{m-n}. \tag{15.12}$$

It is easy to see that Theorems 14 and 15 are in fact corollaries of Theorem 16. On taking $U = V = d$ in (15.11) and omitting all the terms except the term $a = b = d$ from the double sum on the left-hand side, the resulting estimate is in fact more precise than the required estimate (15.5); for the function $\delta(W)$ is bounded. Furthermore, to verify that Theorem 16 implies Theorem 15, we need only note that

$$\mu\{G_n(U)\} = \sum_{a=U}^{\infty} \phi_n(a), \qquad \mu\{G_m(V)\} = \sum_{b=V}^{\infty} \phi_m(b),$$

$$\mu\{G_n(U) \cap G_m(V)\} = \sum_{a=U}^{\infty}\sum_{b=V}^{\infty} \phi(a, b).$$

† The appropriate choice of c will be given in the proof of Lemma 31; this being the first proof in which $\delta(W)$ appears.

Theorem 16 could be improved in one respect, if anything were to be gained by it. The right-hand side of (15.11) could be replaced, for any natural number q, by

$$\beta\delta(U)\delta(V)+\beta^q \min\{\delta(U),\delta(V)\};$$

on the understanding that constants (including the constant c appearing in the definition of the function $\delta(W)$) may now depend on q as well as on the probability space. To establish this, only the proof of Lemma 32 would need to be modified (apart from a trivial corresponding change in Lemma 31). The substitution used at the outset of the proof of Lemma 32 would have to be applied $q-1$ times instead of once; this procedure would lead to the required improvement of the lemma.

Before embarking on the proof of Theorem 16, which will occupy almost the entire section, we state Theorem 17 below. The proof of Theorem 17 is short and simple, but is deferred to the end of the section (so that certain notations introduced in the proof of Theorem 16 will be available).

THEOREM 17. *Subject to Hypotheses B, C,*

$$\mathbf{M}(r_n r_m)-\mathbf{M}(r_n)\mathbf{M}(r_m) \ll \alpha_{m-n}. \tag{15.13}$$

Furthermore, for every natural t,

$$\mathbf{M}(r_t^2)-\{\mathbf{M}(r_t)\}^2 \leqslant \lambda_t. \tag{15.14}$$

We remark that if $n \leqslant 2$, the event $\{\omega\colon r_n(\omega) = 0\}$ constitutes the entire space Ω, and the assertions of the above theorems become trivial (except for (15.14), where the same remark applies to the case $t \leqslant 2$). We may therefore exclude these cases from consideration. In the proof of Theorem 16, we replace Hypothesis C by Hypothesis C′ below (and we assume $t > 2$ in the proof of (15.14)). Actually the proofs, if correctly interpreted, would remain valid even in the excluded cases, but the appropriate interpretations would involve an undue amount of explanation.

HYPOTHESIS C′. *We suppose that m, n are natural numbers satisfying*

$$m > n > 2. \tag{15.15}$$

In order to make the proofs explicit, it was found necessary to introduce quite a number of notations. We introduce these in the form of numbered 'definitions' which will, however, be used in this section only. We have also thought it advisable to insert an unusually large number of back-references to appropriate 'definitions', lemmas, and formulae, but we hope that only a few of these will actually need to be looked up.

Proof of Theorem 16. The relations (15.16) and (15.17) below, which are immediate consequences of Hypothesis B, will be in constant use throughout the proof.

$$1-\alpha_t \gg 1, \qquad 1-\alpha_j\alpha_{t-j} \gg 1 \tag{15.16}$$

on the set of all t (and the set of all pairs t, j) for which α_t (and $\alpha_j\alpha_{t-j}$) is defined.

$$\lambda'_t = \sum_{1\leqslant j<\frac{1}{2}t} \frac{\alpha_j\alpha_{t-j}}{1-\alpha_j\alpha_{t-j}} \ll 1, \tag{15.17}$$

on the set of all integers t.

Definition 18. Every set (or selection) of integers is deemed to be determined by its constituent elements (which will always be distinct), irrespective of order. This applies, in particular, to the pairs and triads of integers introduced in Definitions 21 and 22.

Definition 19. We use h to denote an integer in the range

$$1 \leqslant h < m, \tag{15.18}$$

and we use k, l to denote integers in the ranges

$$1 \leqslant k < \tfrac{1}{2}n, \qquad 1 \leqslant l < \tfrac{1}{2}m \tag{15.19}$$

respectively. In particular, where the symbols h, k, l appear as variables of summation, they will be assumed to range (respectively) over the above intervals (subject to any further conditions explicitly stated).

Definition 20. We use $\mathscr{H}$ to denote a selection (possibly the empty selection) of integers h; in other words, a set of distinct integers lying in the interval $[1, m-1]$. Similarly, $\mathscr{K}$ denotes a selection of integers k, and $\mathscr{L}$ denotes a selection of integers l (where, again, empty selections are not excluded).

We use $|\mathscr{K}|$, $|\mathscr{L}|$ to denote the number of elements of $\mathscr{K}$, $\mathscr{L}$ respectively. $\overline{\mathscr{K}}$ denotes the complement of $\mathscr{K}$ with respect to the range $1 \leqslant k < \frac{1}{2}n$, and $\overline{\mathscr{L}}$ denotes the complement of $\mathscr{L}$ with respect to the range $1 \leqslant l < \frac{1}{2}m$.

The symbols $\mathscr{H}$, $\mathscr{K}$, $\mathscr{L}$ will also appear as variables of summation. $\sum_{\mathscr{H}}$, for example, denotes a summation over all *distinct* selections $\mathscr{H}$ (cf. Definition 18).

LEMMA 18. *We have*

$$\sum_{\substack{h=1\\ h\neq\frac{1}{2}n}}^{n-1} \alpha_h\alpha_{n-h}\alpha_{m-h} \ll \beta. \tag{15.20}$$

In particular,

(15.21)
$$\sum_k \alpha_k \alpha_{n-k} \alpha_{m-k} \ll \beta,$$

(15.22)
$$\sum_k \alpha_{n-k} \alpha_k \alpha_{m-(n-k)} \ll \beta.$$

Proof. By (15.2), (15.3), and (15.12), we have

$$\sum_{\substack{h=1\\h\neq\frac{1}{2}n}}^{n-1} \alpha_h \alpha_{n-h} \alpha_{m-h} \leqslant \alpha_{m-n} \sum_{\substack{h=1\\h\neq\frac{1}{2}n}}^{n-1} \alpha_h \alpha_{n-h} = 2\lambda_n \beta \ll \beta.$$

The following lemma is trivial, by induction on s.

Lemma 19. *If each of* $\theta_1,..., \theta_s$ *lies between* 0 *and* 1,

$$1-\prod_{j=1}^{s} (1-\theta_j) \leqslant \sum_{j=1}^{s} \theta_j.$$

The proof of Lemma 20, which represents the special case $d = 0$ of Theorem 14, is the first step in the proof of Theorem 16.

Lemma 20. *We have*

(15.23)
$$\phi(0,0)-\phi_n(0)\phi_m(0) \ll \beta,$$

where the functions ϕ *are defined by* (15.9) *and* (15.10).

Proof. We write

$$F_n = \{\omega\colon r_n(\omega) = 0\}, \qquad F_m = \{\omega\colon r_m(\omega) = 0\}.$$

Then the event $F_n \cap F_m$ consists of those sequences ω which satisfy simultaneously, the conditions† (cf. Definition 19)

(ii) $\rho_k \rho_{n-k} = 0, \quad \rho_k \rho_{m-k} = 0 \quad$ for every k,

and

(iii) $\rho_l \rho_{m-l} = 0 \quad$ whenever $\frac{1}{2}n \leqslant l < \frac{1}{2}m$.

For each k, the measure of the event

(15.24)
$$\{\omega\colon \rho_k \rho_{n-k} = 0, \quad \rho_k \rho_{m-k} = 0\}$$

is $(1-\alpha_k)+\alpha_k(1-\alpha_{n-k})(1-\alpha_{m-k})$; and for each l, the measure of the event

(15.25)
$$\{\omega\colon \rho_l \rho_{m-l} = 0\}$$

is $1-\alpha_l \alpha_{m-l}$.

The class of all the events (15.24), (15.25) does not constitute a class of independent events. For the integer triplets k, $n-k$, $m-k$ will overlap among themselves (unless m is large compared to n) and the integer

† The conditions (ii), (iii) are numbered to correspond to the conditions (II), (III) appearing later; there is no condition (i).

pairs l, $m-l$ will overlap with the triplets. Nevertheless, in so far as this overlapping is not too extensive, we may expect

$$(15.26)\quad P = \Big\{\prod_k \{(1-\alpha_k)+\alpha_k(1-\alpha_{n-k})(1-\alpha_{m-k})\}\Big\}\prod_{\frac{1}{2}n\leqslant l<\frac{1}{2}m}(1-\alpha_l\,\alpha_{m-l})$$

to be an approximation to $\mu(F_n \cap F_m)$. We shall in fact prove

$$(15.27)\qquad \mu(F_n \cap F_m) = P+O(\beta).$$

The proof of (15.27) will be somewhat lengthy, in view of the complicated structure of overlaps; but, as we now show, it is quite easy to see that (15.27) implies the truth of the lemma.

We have

$$(15.28)\quad \mu(F_n)\mu(F_m) = \Big\{\prod_k \{(1-\alpha_k\,\alpha_{n-k})(1-\alpha_k\,\alpha_{m-k})\}\Big\}\prod_{\frac{1}{2}n\leqslant l<\frac{1}{2}m}(1-\alpha_l\,\alpha_{m-l})$$

and, in view of the identity

$$(1-\theta)+\theta(1-\phi)(1-\psi) = (1-\theta\phi)(1-\theta\psi)+\theta(1-\theta)\phi\psi,$$

$$\prod_k \{(1-\alpha_k)+\alpha_k(1-\alpha_{n-k})(1-\alpha_{m-k})\} = Q\prod_k \{(1-\alpha_k\,\alpha_{n-k})(1-\alpha_k\,\alpha_{m-k})\},$$

where

$$(15.29)\qquad Q = \prod_k\left(1+\frac{\alpha_k(1-\alpha_k)\alpha_{n-k}\,\alpha_{m-k}}{(1-\alpha_k\,\alpha_{n-k})(1-\alpha_k\,\alpha_{m-k})}\right).$$

Thus, recalling the definition (15.26) of P, we see that

$$(15.30)\qquad P = Q\mu(F_n)\mu(F_m).$$

Now

$$Q = \exp\left\{\sum_k O\left(\frac{\alpha_k(1-\alpha_k)\alpha_{n-k}\,\alpha_{m-k}}{(1-\alpha_k\,\alpha_{n-k})(1-\alpha_k\,\alpha_{m-k})}\right)\right\}$$
$$= 1+O\Big(\sum_k \alpha_k\,\alpha_{n-k}\,\alpha_{m-k}\Big),$$

since $(1-\alpha_k\,\alpha_{n-k})(1-\alpha_k\,\alpha_{m-k}) \gg 1$. Thus, by Lemma 18,

$$(15.31)\qquad Q = 1+O(\beta).$$

Since $\mu(F_n)\mu(F_m) \leqslant 1$, the relations (15.27), (15.30), and (15.31) together imply (15.23).

We now turn to the proof of (15.27). We subdivide the entire space Ω into the mutually exclusive events $\Omega_{\mathscr{K}}$ where, for each $\mathscr{K}$, $\Omega_{\mathscr{K}}$ consists of those ω satisfying the condition $\mathrm{I}_{\mathscr{K}}$ below. (In formulating this condition, we deliberately use i for a variable which could more simply be denoted by k; the motivation for this will become apparent only later.)

($\mathrm{I}_{\mathscr{K}}$) $\rho_i = 1$ *whenever* $i \in \mathscr{K}$, $\rho_i = 0$ *whenever* $1 \leqslant i < \frac{1}{2}n$ *and* $i \notin \mathscr{K}$.

A sequence ω which satisfies $\mathrm{I}_{\mathscr{K}}$ will satisfy the condition (ii) (appearing at the beginning of this proof) if and only if:

($\mathrm{II}_{\mathscr{K}}$) $\rho_{n-k} = \rho_{m-k} = 0$ *for every* $k \in \mathscr{K}$.

We next consider which of those sequences ω satisfying $\mathrm{II}_{\mathscr{K}}$ also satisfy (iii). Let $\mathscr{L}_{\mathscr{K}}$ denote the set of those l, satisfying $\frac{1}{2}n \leqslant l < \frac{1}{2}m$, for which the pair l, $m-l$ has a non-empty intersection with at least one of the pairs $n-k$, $m-k$ with $k \in \mathscr{K}$. In view of the respective ranges for l (in $\mathscr{L}_{\mathscr{K}}$) and k, the only intersections that can in fact take place are those arising from the possibilities $l = n-k'$ and $m-l = n-k''$. (For a fixed pair l, $m-l$, either or both of these possibilities may actually occur.) Thus every integer l of $\mathscr{L}_{\mathscr{K}}$ is representable in one of the forms $n-k$ or $m-n+k$ with $k \in \mathscr{K}$. We now restate the above information, concerning the structure of $\mathscr{L}_{\mathscr{K}}$, in the form in which it will be required later. (We remind the reader that, in accordance with Definition 18, any pair of integers is deemed to be determined by its constituent elements, irrespective of the order in which they may appear.)

(15.32) *Every pair* l, $m-l$ *with* $l \in \mathscr{L}_{\mathscr{K}}$ *lies in the set of all pairs* $n-k$, $m-n+k$ *with* $k \in \mathscr{K}$.

Now, a sequence ω which satisfies $\mathrm{II}_{\mathscr{K}}$ automatically satisfies $\rho_l \rho_{m-l} = 0$ for all $l \in \mathscr{L}_{\mathscr{K}}$, and will satisfy (iii) if and only if:

($\mathrm{III}_{\mathscr{K}}$) $\rho_l \rho_{m-l} = 0$ *whenever* $\frac{1}{2}n \leqslant l < \frac{1}{2}m$ *and* $l \notin \mathscr{L}_{\mathscr{K}}$.

Thus, using $E[C_1, C_2,...]$ to denote the event of the conditions $C_1, C_2,...$ being simultaneously satisfied, we have

$$(\Omega_{\mathscr{K}}) \cap (F_n \cap F_m) = E[\mathrm{I}_{\mathscr{K}}, \mathrm{II}_{\mathscr{K}}, \mathrm{III}_{\mathscr{K}}]. \tag{15.33}$$

It is important to note that, for each $\mathscr{K}$, the events $E[\mathrm{I}_{\mathscr{K}}]$, $E[\mathrm{II}_{\mathscr{K}}]$, $E[\mathrm{III}_{\mathscr{K}}]$ are independent. To establish this, we must show that the three sets of suffixes (to ρ) occurring in each of the three conditions $\mathrm{I}_{\mathscr{K}}$, $\mathrm{II}_{\mathscr{K}}$, and $\mathrm{III}_{\mathscr{K}}$, respectively, constitute disjoint sets of integers. But the suffixes appearing in $\mathrm{I}_{\mathscr{K}}$ are all $< \frac{1}{2}n$, whereas all those appearing in $\mathrm{II}_{\mathscr{K}}$ and $\mathrm{III}_{\mathscr{K}}$ are $\geqslant \frac{1}{2}n$; whilst the condition $l \notin \mathscr{L}_{\mathscr{K}}$ certainly removes from $\mathrm{III}_{\mathscr{K}}$ all those suffixes which appear in $\mathrm{II}_{\mathscr{K}}$.

In view of the above remarks, we have

$$\begin{aligned} \mu(F_n \cap F_m) &= \sum_{\mathscr{K}} \mu(\Omega_{\mathscr{K}} \cap F_n \cap F_m) \\ &= \sum_{\mathscr{K}} \mu(E[\mathrm{I}_{\mathscr{K}}, \mathrm{II}_{\mathscr{K}}, \mathrm{III}_{\mathscr{K}}]) \\ &= \sum_{\mathscr{K}} \mu(E[\mathrm{I}_{\mathscr{K}}, \mathrm{II}_{\mathscr{K}}])\mu(E[\mathrm{III}_{\mathscr{K}}]). \end{aligned} \tag{15.34}$$

We give the reference number

$$\sum_{\mathscr{K}} \mu(E[\mathrm{I}_{\mathscr{K}}, \mathrm{II}_{\mathscr{K}}])\mu(E[\mathrm{III}_{\mathscr{K}}]) \tag{15.35}$$

for the final sum appearing in (15.34).

Before proceeding further with the proof of (15.27), we discuss briefly the ideas underlying the remainder of the proof. This discussion does not constitute part of the proof; it is included only to indicate motivation, and will not be referred to in the sequel. Thus the statements made in this discussion do not need to be proved. (We will indicate when the discussion is concluded.)

Suppose that, in the sum (15.35), the condition $\mathrm{III}_{\mathscr{K}}$ is modified to $\mathrm{III}^*_{\mathscr{K}}$ by omitting the condition $l \notin \mathscr{L}_{\mathscr{K}}$ from $\mathrm{III}_{\mathscr{K}}$. (This corresponds to computing $\mu(F_n \cap F_m)$ as though the pairs l, $m-l$ did not intersect any of the triplets k, $n-k$, $m-k$.) Next, suppose that $\mu(E[\mathrm{I}_{\mathscr{K}}, \mathrm{II}_{\mathscr{K}}])$ is replaced by $\mu^*(E[\mathrm{I}_{\mathscr{K}}, \mathrm{II}_{\mathscr{K}}])$, where μ^* is obtained by simply taking the product of the measures of all the constituent events defining $\mathrm{I}_{\mathscr{K}}$, $\mathrm{II}_{\mathscr{K}}$. (This corresponds to treating the constituent events as though they were independent.) We would accordingly expect to obtain

$$P = \sum_{\mathscr{K}} \mu^*(E[\mathrm{I}_{\mathscr{K}}, \mathrm{II}_{\mathscr{K}}])\mu(E[\mathrm{III}^*_{\mathscr{K}}]), \tag{15.36}$$

each side of this relation being the result of computing $\mu(F_n \cap F_m)$ as though the class of events (15.24), (15.25) constituted a class of independent events. To obtain a relation of type (15.27), we need then only estimate first the error introduced by replacing $\mu(E[\mathrm{III}_{\mathscr{K}}])$ by $\mu(E[\mathrm{III}^*_{\mathscr{K}}])$ in (15.35), and estimate secondly the error introduced by replacing $\mu(E[\mathrm{I}_{\mathscr{K}}, \mathrm{II}_{\mathscr{K}}])$ by $\mu^*(E[\mathrm{I}_{\mathscr{K}}, \mathrm{II}_{\mathscr{K}}])$ in the resulting sum.

This concludes the informal discussion.

We first† estimate the error introduced by replacing

$$\mu(E[\mathrm{III}_{\mathscr{K}}]) = \prod_{\substack{\frac{1}{2}n \leqslant l < \frac{1}{2}m \\ l \notin \mathscr{L}_{\mathscr{K}}}} (1-\alpha_l \alpha_{m-l}) \tag{15.37}$$

by

$$P_3 = \prod_{\frac{1}{2}n \leqslant l < \frac{1}{2}m} (1-\alpha_l \alpha_{m-l}) \tag{15.38}$$

in the sum (15.35). By Lemma 19 and (15.32), we have

$$1 - \prod_{l \in \mathscr{L}_{\mathscr{K}}} (1-\alpha_l \alpha_{m-l}) \leqslant \sum_{l \in \mathscr{L}_{\mathscr{K}}} \alpha_l \alpha_{m-l} \leqslant \sum_{k \in \mathscr{K}} \alpha_{n-k} \alpha_{m-n+k},$$

and thus $$0 \leqslant \mu(E[\mathrm{III}_{\mathscr{K}}]) - P_3 \leqslant \sum_{k \in \mathscr{K}} \alpha_{n-k} \alpha_{m-n+k}.$$

† We do not actually carry out the various steps in the exact order in which they appear in the informal discussion above. A relation corresponding to (15.36) will appear in the course of the proof as an automatic by-product of the method.

Hence, recalling that (15.35) is an expression for $\mu(F_n \cap F_m)$, we obtain

(15.39) $$\mu(F_n \cap F_m) = P_3 \sum_{\mathscr{K}} \mu(E[\mathrm{I}_{\mathscr{K}}, \mathrm{II}_{\mathscr{K}}]) + R_1;$$

where

(15.40) $$R_1 \geqslant 0,$$

and

(15.41) $$R_1 \leqslant \sum_{\mathscr{K}} \mu(E[\mathrm{I}_{\mathscr{K}}, \mathrm{II}_{\mathscr{K}}]) \sum_{k \in \mathscr{K}} \alpha_{n-k}\, \alpha_{m-n+k}$$
$$= \sum_{k} \alpha_{n-k}\, \alpha_{m-n+k} \sum_{\substack{\mathscr{K} \\ k \in \mathscr{K}}} \mu(E[\mathrm{I}_{\mathscr{K}}, \mathrm{II}_{\mathscr{K}}]).$$

Now the events $E[\mathrm{I}_{\mathscr{K}}]$, $E[\mathrm{II}_{\mathscr{K}}]$ are independent. Thus, for each fixed k,

(15.42) $$\sum_{\substack{\mathscr{K} \\ k \in \mathscr{K}}} \mu(E[\mathrm{I}_{\mathscr{K}}, \mathrm{II}_{\mathscr{K}}]) = \alpha_k \sum_{\substack{\mathscr{K} \\ k \in \mathscr{K}}} \mu(E[\mathrm{I}_{\mathscr{K}}^{(k)}, \mathrm{II}_{\mathscr{K}}]),$$

where $\mathrm{I}_{\mathscr{K}}^{(k)}$ is derived from $\mathrm{I}_{\mathscr{K}}$ by omitting the condition $\rho_k = 1$. Furthermore, for fixed k, the events $E[\mathrm{I}_{\mathscr{K}}^{(k)}, \mathrm{II}_{\mathscr{K}}]$ appearing in (15.42) are disjoint since the larger events $E[\mathrm{I}_{\mathscr{K}}^{(k)}]$ (corresponding to the various $\mathscr{K}$ containing k) are disjoint. The sum on the right of (15.42) is therefore the sum of the measures of a class of disjoint events, and cannot exceed the measure of the entire space Ω, namely 1.

Hence (15.41), (15.42) yield

$$R_1 \leqslant \sum_{k} \alpha_k\, \alpha_{n-k}\, \alpha_{m-n+k}.$$

In view of Lemma 18 and (15.40), this implies $R_1 = O(\beta)$, and on inserting this estimate in (15.39), we have

(15.43) $$\mu(F_n \cap F_m) = P_3 \sum_{\mathscr{K}} \mu(E[\mathrm{I}_{\mathscr{K}}, \mathrm{II}_{\mathscr{K}}]) + O(\beta).$$

We now estimate the error introduced by replacing

$$\mu(E[\mathrm{I}_{\mathscr{K}}, \mathrm{II}_{\mathscr{K}}]) = \mu(E[\mathrm{I}_{\mathscr{K}}])\mu(E[\mathrm{II}_{\mathscr{K}}])$$

by $$\mu(E[\mathrm{I}_{\mathscr{K}}]) \prod_{k \in \mathscr{K}} \{(1-\alpha_{n-k})(1-\alpha_{m-k})\}$$

in the sum on the right-hand side of (15.43).

For every natural number j, let $\nu(j) = \nu(\mathscr{K}, j)$ be the number of pairs $n-k$, $m-k$ with $k \in \mathscr{K}$, which contain j as one of their elements. Then

$$\mu(E[\mathrm{II}_{\mathscr{K}}]) = \prod_{\substack{j \\ \nu(j) \neq 0}} (1-\alpha_j),$$

whilst $$\prod_{k \in \mathscr{K}} \{(1-\alpha_{n-k})(1-\alpha_{m-k})\} = \prod_{j} (1-\alpha_j)^{\nu(j)}.$$

Clearly $\nu(j)$ assumes only the values 0, 1, and 2. Now the number $n-k$ is representable in the form $m-k'$, with $k' \in \mathscr{K}$, if and only if

$m-n+k \in \mathscr{K}$. Thus $\nu(j) = 2$ if and only if j is representable as $n-k$ with k lying in the set

$$\mathscr{T}_{\mathscr{K}} = \{k: k \in \mathscr{K},\, m-n+k \in \mathscr{K}\}. \tag{15.44}$$

Accordingly, we have

$$\prod_{k\in\mathscr{K}} \{(1-\alpha_{n-k})(1-\alpha_{m-k})\} = \mu(E[\mathrm{II}_{\mathscr{K}}]) \prod_{k\in\mathscr{T}_{\mathscr{K}}} (1-\alpha_{n-k}).$$

On noting that $\mu(E[\mathrm{II}_{\mathscr{K}}]) \leqslant 1$, it follows from Lemma 19 that

$$0 \leqslant \mu(E[\mathrm{II}_{\mathscr{K}}]) - \prod_{k\in\mathscr{K}} \{(1-\alpha_{n-k})(1-\alpha_{m-k})\} \leqslant \sum_{k\in\mathscr{T}_{\mathscr{K}}} \alpha_{n-k}.$$

Thus, on noting that $\mu(E[\mathrm{I}_{\mathscr{K}}]) \leqslant 1$, we have

$$\begin{aligned} S &= \sum_{\mathscr{K}} \mu(E[\mathrm{I}_{\mathscr{K}}, \mathrm{II}_{\mathscr{K}}]) \\ &= \sum_{\mathscr{K}} \mu(E[\mathrm{I}_{\mathscr{K}}]) \prod_{k\in\mathscr{K}} \{(1-\alpha_{n-k})(1-\alpha_{m-k})\} + R_2; \end{aligned} \tag{15.45}$$

where

$$R_2 \geqslant 0, \tag{15.46}$$

and

$$\begin{aligned} R_2 &\leqslant \sum_{\mathscr{K}} \mu(E[\mathrm{I}_{\mathscr{K}}]) \sum_{k\in\mathscr{T}_{\mathscr{K}}} \alpha_{n-k} \\ &= \sum_{k} \alpha_{n-k} \sum_{\substack{\mathscr{K} \\ k\in\mathscr{T}_{\mathscr{K}}}} \mu(E[\mathrm{I}_{\mathscr{K}}]). \end{aligned} \tag{15.47}$$

Now suppose that k is a fixed integer such that the inner sum in the second line of (15.47) is not empty. (If there is no such integer, we have $R_2 = 0$.) Then we have for this inner sum,

$$\sum_{\substack{\mathscr{K} \\ k\in\mathscr{T}_{\mathscr{K}}}} \mu(E[\mathrm{I}_{\mathscr{K}}]) = \alpha_k \alpha_{m-n+k} \sum_{\substack{\mathscr{K} \\ k\in\mathscr{T}_{\mathscr{K}}}} \mu(E[\mathrm{I}_{\mathscr{K}}^{(k,m-n+k)}]),$$

where $\mathrm{I}_{\mathscr{K}}^{(k,m-n+k)}$ is derived from $\mathrm{I}_{\mathscr{K}}$ by omitting the conditions $\rho_k = 1$, $\rho_{m-n+k} = 1$. Here again, the sum on the right is at most 1 since the events $E[\mathrm{I}_{\mathscr{K}}^{(k,m-n+k)}]$ are disjoint. Substituting in (15.47), we obtain

$$R_2 \leqslant \sum_{k} \alpha_k \alpha_{n-k} \alpha_{m-n+k},$$

and hence

$$S = \sum_{\mathscr{K}} \mu(E[\mathrm{I}_{\mathscr{K}}]) \prod_{k\in\mathscr{K}} \{(1-\alpha_{n-k})(1-\alpha_{m-k})\} + O(\beta). \tag{15.48}$$

Now
$$\mu(E[\mathrm{I}_{\mathscr{K}}]) = \Big(\prod_{k\in\mathscr{K}} \alpha_k\Big)\Big(\prod_{k\notin\mathscr{K}} (1-\alpha_k)\Big),$$

so that the sum on the right-hand side of (15.48) is

$$\begin{aligned} &\sum_{\mathscr{K}} \Big[\prod_{k\in\mathscr{K}} \{\alpha_k(1-\alpha_{n-k})(1-\alpha_{m-k})\}\Big] \prod_{k\notin\mathscr{K}} (1-\alpha_k) \\ &\qquad = \prod_{k} \{(1-\alpha_k) + \alpha_k(1-\alpha_{n-k})(1-\alpha_{m-k})\}. \end{aligned}$$

Thus, recalling the definition (15.45) of S, we see that we obtain (15.27) on substituting (15.48) in (15.43). (See also (15.38) and (15.26).) This completes the proof of the lemma.

Definition 21. We denote by $\pi_k^{(n)}$, the pair of integers k, $n-k$; and by $\pi_l^{(m)}$ the pair of integers l, $m-l$ (cf. Definitions 18 and 19).

Furthermore, if $\mathscr{K}$, $\mathscr{L}$ are selections of the type introduced in Definition 20, we write

$$\Pi_{\mathscr{K}}^{(n)} = \bigcup_{k\in\mathscr{K}} \pi_k^{(n)}, \qquad \Pi_{\mathscr{L}}^{(m)} = \bigcup_{l\in\mathscr{L}} \pi_l^{(m)}. \tag{15.49}$$

Thus the set $\Pi_{\mathscr{K}}^{(n)}$ consists of the elements of $\mathscr{K}$ together with their reflections about the point $\frac{1}{2}n$; and the set $\Pi_{\mathscr{L}}^{(m)}$ consists of the elements of $\mathscr{L}$ together with their reflections about the point $\frac{1}{2}m$.

Definition 22. Any triad of integers representable in the form

$$(\pi_k^{(n)}) \cup (\pi_l^{(m)}) \tag{15.50}$$

(where the two pairs π appearing in (15.50) have one integer in common) is said to be a triad τ (cf. Definition 18).

LEMMA 21. *The representation* (15.50) *of a triad* τ *is unique.*

Proof. Suppose that

$$\tau = (\pi_{k'}^{(n)}) \cup (\pi_{l'}^{(m)}) = (\pi_{k''}^{(n)}) \cup (\pi_{l''}^{(m)}).$$

Then the set

$$(\pi_{k'}^{(n)}) \cup (\pi_{k''}^{(n)})$$

contains at most three elements, and hence $k' = k''$. Similarly, we have $l' = l''$.

LEMMA 22. *Corresponding to every triad* τ, *there exists an integer* h, *satisfying* $1 \leqslant h < n$ *and* $h \neq \frac{1}{2}n$, *such that* τ *consists of the integers*

$$h,\ n-h,\ m-h. \tag{15.51}$$

Proof. Suppose τ is the triad (15.50), for given integers k, l. Then, since the pairs appearing in (15.50) intersect, $k = l$ or $n-k = l$ or $n-k = m-l$ (cf. (15.15) and (15.19)). In the first case, τ is the triad (15.51) with $h = k$; whilst in each of the latter two cases, τ is the triad (15.51) with $h = n-k$.

Remark. The converse of Lemma 22 is false, in view of the possibility $h = \frac{1}{2}m$. But, by Lemma 21, each triad τ has exactly one representation (15.51).

Definition 23. Let $\mathscr{K}$, $\mathscr{L}$ be selections of the type introduced in Definition 20. We say that the pair of selections $\mathscr{K}$, $\mathscr{L}$ is *good* if the following three conditions are satisfied.

(i) For every l, $\pi_l^{(m)} \notin \Pi_{\mathscr{K}}^{(n)}$.

(ii) For every k, $\pi_k^{(n)} \notin \Pi_{\mathscr{L}}^{(m)}$.

(iii) The two sets $\Pi_{\mathscr{K}}^{(n)}$, $\Pi_{\mathscr{L}}^{(m)}$ are disjoint.

If the pair of selections $\mathscr{K}$, $\mathscr{L}$ is not good, we say it is *bad*.

Remark. (i) and (ii) are *not* implicit in (iii), of course, in view of the possibilities $l \notin \mathscr{L}$ and $k \notin \mathscr{K}$.

LEMMA 23. *If $\mathscr{K}$, $\mathscr{L}$ is a bad pair of selections, then there exists a triad τ satisfying at least one of the following three conditions.*

(i) *There exist $k_1 \in \mathscr{K}$ and $k_2 \in \mathscr{K}$ such that*

$$\tau \subset (\pi_{k_1}^{(n)}) \cup (\pi_{k_2}^{(n)}).$$

(ii) *There exist $l_1 \in \mathscr{L}$ and $l_2 \in \mathscr{L}$ such that*

$$\tau \subset (\pi_{l_1}^{(m)}) \cup (\pi_{l_2}^{(m)}).$$

(iii) *There exist $k_3 \in \mathscr{K}$ and $l_3 \in \mathscr{L}$ such that*

$$\tau = (\pi_{k_3}^{(n)}) \cup (\pi_{l_3}^{(m)}).$$

Proof. If condition (iii) of Definition 23 is violated, the existence of a triad τ satisfying condition (iii) of the lemma is obvious, in view of Definition 22.

Now suppose that condition (i) of Definition 23 is violated, so that, for some l_0, $\Pi_{\mathscr{K}}^{(n)}$ contains the pair $\pi_{l_0}^{(m)}$. Let k_1, k_2 be defined by

$$l_0 \in \pi_{k_1}^{(n)}, \qquad m-l_0 \in \pi_{k_2}^{(n)}.$$

Then condition (i) of the lemma is clearly satisfied, with these values of k_1, k_2 and

$$\tau = (\pi_{k_1}^{(n)}) \cup (\pi_{l_0}^{(m)}).$$

By symmetry, the violation of condition (ii) of Definition 23 implies the existence of a triad τ satisfying condition (ii) of the lemma.

Definition 24. For any set $\mathscr{S}$ of natural numbers, we write

$$\eta(\mathscr{S}) = \mu(\{\omega\colon \rho_t = 1 \quad \text{for} \quad t \in \mathscr{S}\}) \tag{15.52}$$

and

$$\eta'(\mathscr{S}) = \eta(\mathscr{S})/(1-\eta(\mathscr{S})). \tag{15.53}$$

Definition 25. For any selection $\mathscr{K}$, we write

$$P_{\mathscr{K}} = \prod_{k\in\mathscr{K}} \eta(\pi_k^{(n)}), \qquad P'_{\mathscr{K}} = \prod_{k\in\mathscr{K}} \eta'(\pi_k^{(n)}); \tag{15.54}$$

and for any selection $\mathscr{L}$, we write

(15.55) $$P_{\mathscr{L}} = \prod_{l \in \mathscr{L}} \eta(\pi_l^{(m)}), \qquad P'_{\mathscr{L}} = \prod_{l \in \mathscr{L}} \eta'(\pi_l^{(m)}).$$

Remark. Strictly speaking, the notations $P_{\mathscr{K}}^{(n)}$, $P_{\mathscr{L}}^{(m)}$ would be more appropriate, but there will be no risk of confusion.

Definition 26. For any pair of selections $\mathscr{K}$, $\mathscr{L}$, we define $E_1(\mathscr{K}, \mathscr{L})$ to be the event consisting of those ω for which

(15.56) $$\rho_k \rho_{n-k} = 1 \text{ if and only if } \quad k \in \mathscr{K}$$

and

(15.57) $$\rho_l \rho_{m-l} = 1 \quad \text{if and only if} \quad l \in \mathscr{L};$$

so that $E_1(\mathscr{K}, \mathscr{L})$ consists of those ω for which the pairs $\pi_k^{(n)}$ and $\pi_l^{(m)}$ constituting $\Pi_{\mathscr{K}}^{(n)}$ and $\Pi_{\mathscr{L}}^{(m)}$ are precisely those which contribute 1 to the sums (cf. (11.3)) defining $r_n(\omega)$ and $r_m(\omega)$ respectively.

When the selections $\mathscr{K}$, $\mathscr{L}$ are both empty, we write E_0 for $E_1(\mathscr{K}, \mathscr{L})$; so that E_0 is the event $\{\omega: r_n(\omega) = 0,\ r_m(\omega) = 0\}$.

The following lemma represents a crucial step in our proof of Theorem 16.

LEMMA 24. *For any pair of selections $\mathscr{K}$, $\mathscr{L}$, let $\mathscr{S}_1(\mathscr{K}, \mathscr{L})$ be the set of those integers h satisfying at least one of the following two conditions.*

(i) *There exists $k^* \in \mathscr{K}$ such that† h together with $\pi_{k^*}^{(n)}$ forms a triad τ.*

(ii) *There exists $l^* \in \mathscr{L}$ such that† h together with $\pi_{l^*}^{(m)}$ forms a triad τ.*

Then, for all those pairs $\mathscr{K}$, $\mathscr{L}$ which are good in the sense of Definition 23, we have

(15.58) $$|\mu\{E_1(\mathscr{K}, \mathscr{L})\} - P'_{\mathscr{K}} P'_{\mathscr{L}} \mu(E_0)| \leqslant P_{\mathscr{K}} P_{\mathscr{L}} \sum_{h \in \mathscr{S}_1(\mathscr{K}, \mathscr{L})} \alpha_h.$$

Proof. The following notation will be used in this proof only. For $\theta = 1$ or 0, and for any selections $\mathscr{K}^*$, $\mathscr{L}^*$, we write

$$A_\theta^{(n)}(\mathscr{K}^*) = \{\omega: \rho_k \rho_{n-k} = \theta \quad \text{for} \quad k \in \mathscr{K}^*\},$$

$$A_\theta^{(m)}(\mathscr{L}^*) = \{\omega: \rho_l \rho_{m-l} = \theta \quad \text{for} \quad l \in \mathscr{L}^*\}.$$

($\mathscr{K}^*$ may be empty, in which case $A_\theta^{(n)}(\mathscr{K}^*)$ is the entire space Ω. The analogous interpretation is made when $\mathscr{L}^*$ is empty.)

Let $\mathscr{K}$, $\mathscr{L}$ be a good pair of selections, and write

$$E_1 = E_1(\mathscr{K}, \mathscr{L}), \qquad \mathscr{S}_1 = \mathscr{S}_1(\mathscr{K}, \mathscr{L}).$$

We have, for $\theta = 1$ or 0 (cf. Definition 26),

$$E_\theta = A_\theta^{(n)}(\mathscr{K}) \cap A_\theta^{(m)}(\mathscr{L}) \cap A_0^{(n)}(\overline{\mathscr{K}}) \cap A_0^{(m)}(\overline{\mathscr{L}}).$$

† The h occurring here should not be confused with the (different) h in the representation (15.51) of τ. The representation (15.51) is not relevant here.

We shall prove that

$$\left|\frac{\mu(E_1)}{\mu\{A_1^{(n)}(\mathscr{K})\}\mu\{A_1^{(m)}(\mathscr{L})\}}-\frac{\mu(E_0)}{\mu\{A_0^{(n)}(\mathscr{K})\}\mu\{A_0^{(m)}(\mathscr{L})\}}\right|\leqslant\sum_{h\in\mathscr{S}_1}\alpha_h, \tag{15.59}$$

which, on evaluating the denominators on the left-hand side, is seen to be equivalent to (15.58).

We define the selections $\mathscr{K}'$, $\mathscr{L}'$ as follows (cf. condition (iii) of Definition 23).

(I) Let $\mathscr{K}'$ be the maximal $\mathscr{K}^*$ such that the three sets

$$\Pi_{\mathscr{K}}^{(n)}, \qquad \Pi_{\mathscr{L}}^{(m)}, \qquad \Pi_{\mathscr{K}^*}^{(n)}$$

are disjoint; and let $\mathscr{L}'$ be the maximal $\mathscr{L}^*$ such that the three sets

$$\Pi_{\mathscr{K}}^{(n)}, \qquad \Pi_{\mathscr{L}}^{(m)}, \qquad \Pi_{\mathscr{L}^*}^{(m)}$$

are disjoint.

Clearly

$$\mathscr{K}'\subset\overline{\mathscr{K}}, \qquad \mathscr{L}'\subset\overline{\mathscr{L}}, \tag{15.60}$$

and we write

$$\mathscr{K}''=\overline{\mathscr{K}}-\mathscr{K}', \qquad \mathscr{L}''=\overline{\mathscr{L}}-\mathscr{L}'; \tag{15.61}$$

so that, for example, $\mathscr{K}''$ consists of those k which lie in neither $\mathscr{K}$ nor $\mathscr{K}'$.

In view of these definitions, the following statement holds.

(II) $\pi_k^{(n)}$ intersects $\Pi_{\mathscr{L}}^{(m)}$ if and only if $k\in\mathscr{K}''$,

and $\pi_l^{(m)}$ intersects $\Pi_{\mathscr{K}}^{(n)}$ if and only if $l\in\mathscr{L}''$.

By condition (ii) of Definition 23, for $k\in\mathscr{K}''$, only one of the elements of $\pi_k^{(n)}$ lies in $\Pi_{\mathscr{L}}^{(m)}$; the other element, h say, satisfies the condition (ii) appearing in the definition of $\mathscr{S}_1(\mathscr{K},\mathscr{L})$. Conversely, if an integer h satisfies condition (ii) of the definition of $\mathscr{S}_1(\mathscr{K},\mathscr{L})$, then if

$$\tau=\pi_k^{(n)}\cup\pi_{l^*}^{(m)},$$

we have (by (II) above) $k\in\mathscr{K}''$ and h is the element of $\pi_k^{(n)}$ which does not lie in $\Pi_{\mathscr{L}}^{(m)}$. Furthermore, this element h cannot lie in $\Pi_{\mathscr{K}}^{(n)}$ either, since $\mathscr{K}''\subset\overline{\mathscr{K}}$.

By symmetry, the corresponding results hold in relation to the elements of $\mathscr{S}_1$ satisfying condition (i).

Thus the set $\mathscr{S}_1$ satisfies the following two conditions.

(III) The three sets $\Pi_{\mathscr{K}}^{(n)}$, $\Pi_{\mathscr{L}}^{(m)}$, $\mathscr{S}_1$ are disjoint.

(IV) For each $k\in\mathscr{K}''$, one of the elements of $\pi_k^{(n)}$ lies in $\mathscr{S}_1$; and for each $l\in\mathscr{L}''$ one of the elements of $\pi_l^{(m)}$ lies in $\mathscr{S}_1$.

For $\theta=1$ or 0, we write

$$E'_\theta=A_\theta^{(n)}(\mathscr{K})\cap A_\theta^{(m)}(\mathscr{L})\cap A_0^{(n)}(\mathscr{K}')\cap A_0^{(m)}(\mathscr{L}').$$

Then, by (15.60), we have $E_\theta\subset E'_\theta$. Furthermore, by (15.61), if the sequence ω lies in $E'_\theta - E_\theta$, then either $\rho_k(\omega)\rho_{n-k}(\omega) = 1$ for some $k \in \mathscr{K}''$ or $\rho_l(\omega)\rho_{m-l}(\omega) = 1$ for some $l \in \mathscr{L}''$; so that, in view of (IV), $\rho_h(\omega) = 1$ for some $h \in \mathscr{S}_1$. We write

$$C(\mathscr{S}_1) = \{\omega\colon \rho_h = 1 \quad \text{for at least one } h \in \mathscr{S}_1\}.$$

Then, by (III), we have for $\theta = 1$ or 0,

$$\begin{aligned}(15.62)\qquad 0 &\leqslant \mu(E'_\theta) - \mu(E_\theta)\\ &\leqslant \mu\{A_\theta^{(n)}(\mathscr{K}) \cap A_\theta^{(m)}(\mathscr{L}) \cap C(\mathscr{S}_1)\}\\ &= \mu\{A_\theta^{(n)}(\mathscr{K})\}\mu\{A_\theta^{(m)}(\mathscr{L})\}\mu\{C(\mathscr{S}_1)\}\\ &\leqslant \mu\{A_\theta^{(n)}(\mathscr{K})\}\mu\{A_\theta^{(m)}(\mathscr{L})\}\sum_{h\in\mathscr{S}_1}\alpha_h.\end{aligned}$$

Now, in view of the definition (I) of $\mathscr{K}'$ and $\mathscr{L}'$ the three sets

$$\Pi_{\mathscr{K}}^{(n)}, \qquad \Pi_{\mathscr{L}}^{(m)}, \qquad \Pi_{\mathscr{K}'}^{(n)} \cup \Pi_{\mathscr{L}'}^{(m)}$$

are disjoint. We thus have for $\theta = 1$ or 0,

$$\mu(E'_\theta) = \mu\{A_\theta^{(n)}(\mathscr{K})\}\mu\{A_\theta^{(m)}(\mathscr{L})\}\mu\{A_0^{(n)}(\mathscr{K}') \cap A_0^{(m)}(\mathscr{L}')\},$$

and hence

$$(15.63)\qquad \frac{\mu(E'_1)}{\mu\{A_1^{(n)}(\mathscr{K})\}\mu\{A_1^{(m)}(\mathscr{L})\}} = \frac{\mu(E'_0)}{\mu\{A_0^{(n)}(\mathscr{K})\}\mu\{A_0^{(m)}(\mathscr{L})\}}.$$

Writing
$$\Lambda_\theta = \frac{\mu(E'_\theta) - \mu(E_\theta)}{\mu\{A_\theta^{(n)}(\mathscr{K})\}\mu\{A_\theta^{(m)}(\mathscr{L})\}},$$

we have, by (15.62),
$$|\Lambda_1 - \Lambda_0| \leqslant \sum_{h\in\mathscr{S}_1}\alpha_h.$$

In view of (15.63), this last inequality is equivalent to (15.59) and the proof of the lemma is complete.

In connexion with the application of Lemma 24, we shall require a number of estimates, which are combinatorial in nature.

In view of Lemma 22, the following result is an immediate consequence of (15.20).

Lemma 25. *We have*

$$(15.64)\qquad \sum_\tau \eta(\tau) \ll \beta,$$

where the summation is over all triads τ, *and the function* η *is defined in Definition* 24.

Lemma 26. *With the notation of Definition* 25, *we have for all non-negative integers* a, b,

$$(15.65)\qquad \sum_{\substack{\mathscr{K}\\ |\mathscr{K}|=a}} P'_{\mathscr{K}} \leqslant \frac{(\lambda'_n)^a}{a!},$$

$$(15.66)\qquad \sum_{\substack{\mathscr{L}\\ |\mathscr{L}|=b}} P'_{\mathscr{L}} \leqslant \frac{(\lambda'_m)^b}{b!}.$$

Proof. We have (cf. Definitions 18 and 20 and Lemma 13 (§ 11))

$$\sum_{\substack{\mathscr{K}\\ |\mathscr{K}|=a}} \Big(\prod_{k\in\mathscr{K}} \eta'(\pi_k^{(n)})\Big) \leqslant \frac{1}{a!}\Big(\sum_k \eta'(\pi_k^{(n)})\Big)^a = \frac{(\lambda_n')^a}{a!}.$$

The proof of (15.66) is clearly the same.

Definition 27. For any non-negative integers a, b, we write

$$(15.67)\qquad \delta_n(a) = \sum_{\substack{i=0\\ a-i\geqslant 0}}^{2} \frac{(\lambda_n')^{a-i}}{(a-i)!}, \qquad \delta_m(b) = \sum_{\substack{j=0\\ b-j\geqslant 0}}^{2} \frac{(\lambda_m')^{b-j}}{(b-j)!}.$$

LEMMA 27. *We have for any non-negative integers a, b,*

$$(15.68)\qquad \sum_{\substack{\mathscr{K},\mathscr{L}\text{ bad}\\ |\mathscr{K}|=a,|\mathscr{L}|=b}} P_{\mathscr{K}}' P_{\mathscr{L}}' \ll \beta\delta_n(a)\delta_m(b),$$

where, subject to the additional conditions $|\mathscr{K}| = a$, $|\mathscr{L}| = b$, *the summation is over all those pairs* $\mathscr{K}$, $\mathscr{L}$ *which are bad in the sense of Definition* 23.

Furthermore, if $\mathscr{S}_1(\mathscr{K},\mathscr{L})$ *is the set defined in Lemma* 24,

$$(15.69)\qquad \sum_{\substack{\mathscr{K}\\ |\mathscr{K}|=a}} \sum_{\substack{\mathscr{L}\\ |\mathscr{L}|=b}} P_{\mathscr{K}} P_{\mathscr{L}} \sum_{h\in\mathscr{S}_1(\mathscr{K},\mathscr{L})} \alpha_h \ll \beta\delta_n(a)\delta_m(b).$$

Proof. The sum on the left-hand side of (15.68) does not exceed

$$(15.70)\qquad \sum_\tau \{\sigma^{(1)}(\tau)+\sigma^{(2)}(\tau)+\sigma^{(3)}(\tau)\},$$

where
$$\sigma^{(1)}(\tau) = \sum_{|\mathscr{K}|=a,|\mathscr{L}|=b}^{(1)} P_{\mathscr{K}}' P_{\mathscr{L}}',$$

with the summation $\sum^{(1)}$ extended over all those pairs $\mathscr{K}$, $\mathscr{L}$ for which the condition (i) of Lemma 23 is satisfied for the fixed triad τ in question; and $\sigma^{(2)}(\tau)$, $\sigma^{(3)}(\tau)$ are similarly defined in terms of conditions (ii), (iii) (respectively) of Lemma 23.

Now, if k_1, k_2 are the integers whose existence is asserted in condition (i) of Lemma 23 (these are uniquely determined by τ) we have

$$\sigma^{(1)}(\tau) \leqslant \eta'(\pi_{k_1}^{(n)})\eta'(\pi_{k_2}^{(n)}) \sum_{\substack{\mathscr{K}\\ k_1\in\mathscr{K},k_2\in\mathscr{K}\\ |\mathscr{K}|=a}} \Big\{\prod_{\substack{k\in\mathscr{K}\\ k\neq k_1,k_2}} \eta'(\pi_k^{(n)})\Big\} \sum_{\substack{\mathscr{L}\\ |\mathscr{L}|=b}} P_{\mathscr{L}}'$$

$$\leqslant \eta'(\pi_{k_1}^{(n)})\eta'(\pi_{k_2}^{(n)}) \sum_{\substack{\mathscr{K}\\ |\mathscr{K}|=a-2}} P_{\mathscr{K}}' \sum_{\substack{\mathscr{L}\\ |\mathscr{L}|=b}} P_{\mathscr{L}}',$$

and by (15.16),
$$\eta'(\pi_{k_1}^{(n)})\eta'(\pi_{k_2}^{(n)}) \ll \eta(\tau).$$

Hence, by Lemma 26 ($a \geqslant 2$ if $\sum^{(1)}$ is not empty),

$$\sigma^{(1)}(\tau) \ll \eta(\tau)\frac{(\lambda_n')^{a-2}}{(a-2)!}\frac{(\lambda_m')^b}{b!} \ll \eta(\tau)\delta_n(a)\delta_m(b).$$

By symmetry, we also have

$$\sigma^{(2)}(\tau) \ll \eta(\tau)\delta_n(a)\delta_m(b).$$

Finally, if k_3, l_3 are the integers appearing in condition (iii) of Lemma 23,

$$\sigma^{(3)}(\tau) \leqslant \eta'(\pi_{k_3}^{(n)})\eta'(\pi_{l_3}^{(m)}) \sum_{\substack{\mathscr{K} \\ |\mathscr{K}|=a-1}} P'_{\mathscr{K}} \sum_{\substack{\mathscr{L} \\ |\mathscr{L}|=b-1}} P'_{\mathscr{L}}$$

$$\leqslant \eta(\tau)\frac{(\lambda'_n)^{a-1}}{(a-1)!}\frac{(\lambda'_m)^{b-1}}{(b-1)!}$$

$$\leqslant \eta(\tau)\delta_n(a)\delta_m(b).$$

On substituting these estimates, we see that the sum (15.70) is $\leqslant$

$$\delta_n(a)\delta_m(b)\sum_{\tau}\eta(\tau), \tag{15.71}$$

so that (15.68) follows by Lemma 25.

We now turn to the proof of (15.69). The left-hand side of (15.69) does not exceed

$$\sum_{\tau}\sum_{h\in\tau}\alpha_h\{\sigma^{(\mathrm{i})}(h,\tau)+\sigma^{(\mathrm{ii})}(h,\tau)\}, \tag{15.72}$$

where
$$\sigma^{(\mathrm{i})}(h,\tau) = \sum_{|\mathscr{K}|=a,|\mathscr{L}|=b}^{(\mathrm{i})} P_{\mathscr{K}}P_{\mathscr{L}},$$

with the summation $\sum^{(\mathrm{i})}$ taken over all those pairs $\mathscr{K}$, $\mathscr{L}$ for which condition (i) of Lemma 24 is satisfied for the fixed h, τ in question; and $\sigma^{(\mathrm{ii})}(h,\tau)$ is similarly defined in terms of condition (ii) of Lemma 24.

We have, with the k^* in condition (i) of Lemma 24,

$$\alpha_h\,\sigma^{(\mathrm{i})}(h,\tau) \leqslant \alpha_h\,\eta(\pi_{k^*}^{(n)}) \sum_{\substack{\mathscr{K} \\ |\mathscr{K}|=a-1}} P_{\mathscr{K}} \sum_{\substack{\mathscr{L} \\ |\mathscr{L}|=b}} P_{\mathscr{L}}$$

$$= \eta(\tau) \sum_{\substack{\mathscr{K} \\ |\mathscr{K}|=a-1}} P_{\mathscr{K}} \sum_{\substack{\mathscr{L} \\ |\mathscr{L}|=b}} P_{\mathscr{L}}$$

$$\leqslant \eta(\tau)\delta_n(a)\delta_m(b),$$

since $P_{\mathscr{K}} \leqslant P'_{\mathscr{K}}$, $P_{\mathscr{L}} \leqslant P'_{\mathscr{L}}$ so that Lemma 26 is again applicable. On similarly estimating $\alpha_h\,\sigma^{(\mathrm{ii})}(h,\tau)$, and noting that τ contains only a bounded number of elements (namely three elements), we see that (15.72) is also majorized by (15.71). This completes the proof of (15.69) and of the lemma.

Definition 28. For any non-negative integers a, b, we write

$$\phi^*(a,b) = \sum_{\substack{\mathscr{K},\mathscr{L}\ \mathrm{good} \\ |\mathscr{K}|=a,|\mathscr{L}|=b}} \mu\{E_1(\mathscr{K},\mathscr{L})\}, \tag{15.73}$$

where $E_1(\mathscr{K},\mathscr{L})$ is defined in Definition 26 and the summation, subject to the conditions $|\mathscr{K}| = a$, $|\mathscr{L}| = b$, is over all pairs $\mathscr{K}$, $\mathscr{L}$ which are good in the sense of Definition 23.

Remark. There is an obvious relationship between the function ϕ^* and the function ϕ defined by (15.10). It is clear (from Definition 26) that among the events $E_1(\mathscr{K},\mathscr{L})$ corresponding to distinct pairs $\mathscr{K}$, $\mathscr{L}$, those that are non-empty are mutually exclusive. (The event $E_1(\mathscr{K}, \mathscr{L})$ is certainly empty if there exists $k \notin \mathscr{K}$ such that $\pi_k^{(n)} \subset \Pi_{\mathscr{L}}^{(m)}$, or there exists $l \notin \mathscr{L}$ such that $\pi_l^{(m)} \subset \Pi_{\mathscr{K}}^{(n)}$.) Thus we have

$$\phi(a,b) = \sum_{\substack{\mathscr{K},\mathscr{L} \\ |\mathscr{K}|=a, |\mathscr{L}|=b}} \mu\{E_1(\mathscr{K}, \mathscr{L})\}, \tag{15.74}$$

and

$$0 \leqslant \phi(a,b) - \phi^*(a,b) = \sum_{\substack{\mathscr{K},\mathscr{L} \text{ bad} \\ |\mathscr{K}|=a, |\mathscr{L}|=b}} \mu\{E_1(\mathscr{K}, \mathscr{L})\}. \tag{15.75}$$

LEMMA 28. *We have for any non-negative integers a, b,*

$$\phi_n(a) = \phi_n(0) \sum_{\substack{\mathscr{K} \\ |\mathscr{K}|=a}} P'_{\mathscr{K}}, \tag{15.76}$$

$$\phi_m(b) = \phi_m(0) \sum_{\substack{\mathscr{L} \\ |\mathscr{L}|=b}} P'_{\mathscr{L}}, \tag{15.77}$$

$$\phi^*(a,b) \leqslant \sum_{\substack{\mathscr{K} \\ |\mathscr{K}|=a}} P_{\mathscr{K}} \sum_{\substack{\mathscr{L} \\ |\mathscr{L}|=b}} P_{\mathscr{L}}, \tag{15.78}$$

where the functions ϕ_n, ϕ_m are defined by (15.9).

Proof. We have $\quad \phi_n(a) = \sum_{\substack{\mathscr{K} \\ |\mathscr{K}|=a}} \mu\{E_1^{(n)}(\mathscr{K})\},$

where $E_1^{(n)}(\mathscr{K})$ is the event (15.56). But if, in the definition (15.56) of $E_1^{(n)}(\mathscr{K})$, we replace the condition $\rho_k \rho_{n-k} = 1$ by $\rho_k \rho_{n-k} = 0$ for $k \in \mathscr{K}$, we obtain the event $\{\omega: r_n(\omega) = 0\}$: thus

$$\mu\{E_1^{(n)}(\mathscr{K})\} = P'_{\mathscr{K}}\, \phi_n(0).$$

This proves (15.76), and the proof of (15.77) is the same.

Now write

$$E_2^{(n)}(\mathscr{K}) = \{\omega: \rho_k \rho_{n-k} = 1 \quad \text{for} \quad k \in \mathscr{K}\},$$
$$E_2^{(m)}(\mathscr{L}) = \{\omega: \rho_l \rho_{m-l} = 1 \quad \text{for} \quad l \in \mathscr{L}\}.$$

Clearly (cf. Definition 26),

$$E_1(\mathscr{K}, \mathscr{L}) \subset E_2^{(n)}(\mathscr{K}) \cap E_2^{(m)}(\mathscr{L}).$$

But if $\Pi_{\mathscr{K}}^{(n)}$, $\Pi_{\mathscr{L}}^{(m)}$ are disjoint, and in particular if $\mathscr{K}$, $\mathscr{L}$ is a good pair, the events $E_2^{(n)}(\mathscr{K})$, $E_2^{(m)}(\mathscr{L})$ are independent and hence

$$\mu\{E_1(\mathscr{K}, \mathscr{L})\} \leqslant \mu\{E_2^{(n)}(\mathscr{K})\}\mu\{E_2^{(m)}(\mathscr{L})\} = P_{\mathscr{K}} P_{\mathscr{L}}.$$

In view of the definition of $\phi^*(a,b)$, this proves (15.78).

LEMMA 29. *For any non-negative integers a, b,*

$$\phi^*(a,b) - \phi_n(a)\phi_m(b) \ll \beta \delta_n(a) \delta_m(b). \tag{15.79}$$

Proof. We sum the inequality (15.58) (which was established for all *good* pairs $\mathscr{K}$, $\mathscr{L}$) over all good pairs $\mathscr{K}$, $\mathscr{L}$ satisfying $|\mathscr{K}| = a$, $|\mathscr{L}| = b$.

On using (15.69), and noting that $\mu(E_0) = \phi(0,0)$ (cf. (15.10) and Definition 26), we obtain the estimate

$$\phi^*(a,b)-\phi(0,0) \sum_{\substack{\mathscr{K},\mathscr{L} \text{ good} \\ |\mathscr{K}|=a, |\mathscr{L}|=b}} P'_{\mathscr{K}} P'_{\mathscr{L}} \leqslant \beta\delta_n(a)\delta_m(b).$$

In this relation, we first omit the word 'good' under the summation sign, as (15.68) entitles us to do ($\phi(0,0)$ is a probability and cannot exceed 1); we then replace $\phi(0,0)$ by $\phi_n(0)\phi_m(0)$, as is permissible in view of (15.23) and Lemma 26 (cf. Definition 27). The resulting estimate is (15.79), as can be seen from (15.76) and (15.77).

In the application of Lemma 29 to the proof of (15.11), we need to estimate the error introduced when $\phi^*(a,b)$ is replaced by $\phi(a,b)$. The following lemma will serve this purpose.

LEMMA 30. *For any non-negative integers a, b,*

$$\phi(a,b)-\phi^*(a,b) \leqslant \beta \sum_{\substack{i=1 \\ a-i\geqslant 0}}^{2} \sum_{\substack{j=1 \\ b-j\geqslant 0}}^{2} \phi(a-i, b-j). \tag{15.80}$$

Proof. We refer the reader back to Definitions 20, 21, 22, and 23, which will play a prominent part in this proof.

For any selection $\mathscr{H}$, we denote by $\mathscr{K}(\mathscr{H})$ the maximal $\mathscr{K}$ such that $\Pi^{(n)}_{\mathscr{K}} \subset \mathscr{H}$; and by $\mathscr{L}(\mathscr{H})$ the maximal $\mathscr{L}$ such that $\Pi^{(m)}_{\mathscr{L}} \subset \mathscr{H}$. Thus

$$\mathscr{K}(\mathscr{H}) = \{k: \pi^{(n)}_k \subset \mathscr{H}\}, \qquad \mathscr{L}(\mathscr{H}) = \{l: \pi^{(m)}_l \subset \mathscr{H}\}.$$

We define the event

$$Z(\mathscr{H}) = \{\omega: \rho_h = 1 \quad \text{if and only if } h \in \mathscr{H}\}. \tag{15.81}$$

Obviously, the events $Z(\mathscr{H})$ represent a subdivision of the entire space Ω into disjoint events. We have, for any pair $\mathscr{K}_0$, $\mathscr{L}_0$,

$$\mu\{E_1(\mathscr{K}_0, \mathscr{L}_0)\} = \sum_{\substack{\mathscr{H} \\ \mathscr{K}(\mathscr{H})=\mathscr{K}_0, \mathscr{L}(\mathscr{H})=\mathscr{L}_0}} \mu\{Z(\mathscr{H})\},$$

where $E_1(\mathscr{K}_0, \mathscr{L}_0)$ is defined in accordance with Definition 26.

We say that the selection $\mathscr{H}$ is good or bad according as the pair $\mathscr{K}(\mathscr{H})$, $\mathscr{L}(\mathscr{H})$ is good or bad (in the sense of Definition 23). Then, by (15.75),

$$\phi(a,b)-\phi^*(a,b) = \sum_{\substack{\mathscr{H} \text{ bad} \\ |\mathscr{K}(\mathscr{H})|=a, |\mathscr{L}(\mathscr{H})|=b}} \mu\{Z(\mathscr{H})\}.$$

Now, in each of the cases (i), (ii), and (iii) of Lemma 23, with $\mathscr{K} = \mathscr{K}(\mathscr{H})$, $\mathscr{L} = \mathscr{L}(\mathscr{H})$, we have $\tau \subset \mathscr{H}$. Hence

$$\phi(a,b)-\phi^*(a,b) \leqslant \sum_{\tau} \sigma(\tau), \tag{15.82}$$

where

$$\sigma(\tau) = \sum_{\substack{\mathscr{H} \\ \tau\subset\mathscr{H} \\ |\mathscr{K}(\mathscr{H})|=a,\,|\mathscr{L}(\mathscr{H})|=b}} \mu\{Z(\mathscr{H})\}. \tag{15.83}$$

We now suppose that $\tau \subset \mathscr{H}$, and examine how each of $\mu\{Z(\mathscr{H})\}$, $|\mathscr{K}(\mathscr{H})|$, $|\mathscr{L}(\mathscr{H})|$ is affected when the elements of τ are removed from $\mathscr{H}$.

The effect on the right-hand side of (15.81) is to replace the condition $\rho_h = 1$ by $\rho_h = 0$ when $h \in \tau$; so that

$$\mu\{Z(\mathscr{H})\} = \left(\prod_{h\in\tau} \frac{\alpha_h}{1-\alpha_h}\right)\mu\{Z(\mathscr{H}-\tau)\} \leqslant \eta(\tau)\mu\{Z(\mathscr{H}-\tau)\}. \tag{15.84}$$

Furthermore, if τ has the representation (15.50), τ contains $\pi_k^{(n)}$, $\pi_l^{(m)}$ and intersects at most one further pair $\pi_{k'}^{(n)}$ and at most one further pair $\pi_{l'}^{(m)}$. For example, if h_1 is the element of τ which lies outside $\pi_k^{(n)}$, the integers h_1, $n-h_1$ constitute a pair $\pi_{k'}^{(n)}$ if and only if $h_1 < n$ and $h_1 \neq \frac{1}{2}n$. When the elements of τ are removed from $\mathscr{H}$, the pair $\pi_k^{(n)}$ is certainly removed from $\Pi^{(n)}_{\mathscr{K}(\mathscr{H})}$; but the pair $\pi_{k'}^{(n)}$ (if it exists) will be removed from $\Pi^{(n)}_{\mathscr{K}(\mathscr{H})}$ only if $k' \in \mathscr{K}(\mathscr{H})$. Whatever the case may be, however, at least one and at most two pairs $\pi^{(n)}$ are removed from $\Pi^{(n)}_{\mathscr{K}(\mathscr{H})}$ when the elements of τ are removed from $\mathscr{H}$. Thus

$$|\mathscr{K}(\mathscr{H}-\tau)| = |\mathscr{K}(\mathscr{H})|-1 \quad\text{or}\quad |\mathscr{K}(\mathscr{H})|-2, \tag{15.85}$$

and similarly,

$$|\mathscr{L}(\mathscr{H}-\tau)| = |\mathscr{L}(\mathscr{H})|-1 \quad\text{or}\quad |\mathscr{L}(\mathscr{H})|-2. \tag{15.86}$$

Since the selections $\mathscr{H}^* = \mathscr{H}-\tau$ corresponding to distinct $\mathscr{H}$ (containing the fixed triad τ) are distinct, it follows from (15.84), (15.85), and (15.86) that

$$\begin{aligned}\sigma(\tau) &\leqslant \eta(\tau) \sum_{\substack{\mathscr{H} \\ \tau\subset\mathscr{H} \\ |\mathscr{K}(\mathscr{H})|=a,\,|\mathscr{L}(\mathscr{H})|=b}} \mu\{Z(\mathscr{H}-\tau)\} \\ &\leqslant \eta(\tau) \sum_{\substack{i=1 \\ a-i\geqslant 0}}^{2} \sum_{\substack{j=1 \\ b-j\geqslant 0}}^{2} \sum_{\substack{\mathscr{H} \\ |\mathscr{K}(\mathscr{H})|=a-i,\,|\mathscr{L}(\mathscr{H})|=b-j}} \mu\{Z(\mathscr{H})\} \\ &= \eta(\tau) \sum_{\substack{i=1 \\ a-i\geqslant 0}}^{2} \sum_{\substack{j=1 \\ b-j\geqslant 0}}^{2} \phi(a-i, b-j).\end{aligned}$$

On substituting this estimate in (15.82), we obtain the required result (15.80) in view of Lemma 25.

LEMMA 31. *For all non-negative integers* W,

$$\sum_{\substack{d=0\\ d\geqslant W-4}}^{\infty} \frac{(\lambda_n')^d}{d!} \ll \delta(W), \tag{15.87}$$

$$\sum_{\substack{d=0\\ d\geqslant W-4}}^{\infty} \frac{(\lambda_m')^d}{d!} \ll \delta(W); \tag{15.88}$$

the function $\delta(W)$ *is defined in Theorem* 15.

In particular (*cf. Definition* 27), *for any non-negative integers* U, V,

$$\sum_{a\geqslant U} \delta_n(a) \ll \delta(U), \tag{15.89}$$

$$\sum_{b\geqslant V} \delta_m(b) \ll \delta(V). \tag{15.90}$$

Proof. By (15.17), there exists a constant c', such that

$$\lambda_t' \leqslant c' \quad \text{for all integers } t. \tag{15.91}$$

We define the constant c, appearing in the definition of $\delta(W)$, by

$$c = 1/(4ec'). \tag{15.92}$$

In view of (15.91) and the definition of $\delta(W)$ we may suppose that W is large for the purpose of proving (15.87) and (15.88). Accordingly, we suppose that

$$W > c'+4, \qquad W-4 > \tfrac{1}{2}W, \qquad (ec'W^{-1})^4 > (\tfrac{1}{2})^W.$$

Then, since $W-4 > \lambda_n'$, we have (cf. Lemma 17 (§ 11))

$$\sum_{d\geqslant W-4} \frac{(\lambda_n')^d}{d!} \leqslant \left(\frac{e\lambda_n'}{W-4}\right)^{W-4} \leqslant \left(\frac{ec'}{W-4}\right)^{W-4}$$
$$< \left(\frac{ec'}{W}\right)^{-4}\left(\frac{2ec'}{W}\right)^{W} < \left(\frac{4ec'}{W}\right)^{W} = \delta(W).$$

Since n and m are interchangeable in the above argument, this proves both (15.87) and (15.88). The deduction of (15.89) and (15.90) is trivial.

LEMMA 32. *For any non-negative integers* U, V,

$$\sum_{a\geqslant U}\sum_{b\geqslant V} \{\phi(a,b)-\phi^*(a,b)\} \ll \beta\delta(U)\delta(V)+\beta^2\min\{\delta(U),\delta(V)\}. \tag{15.93}$$

Proof. On writing

$$\phi(u,v) = \phi^*(u,v)+O\Big(\beta \sum_{\substack{i'=1\\ u-i'\geqslant 0}}^{2} \sum_{\substack{j'=1\\ v-j'\geqslant 0}}^{2} \phi(u-i',v-j')\Big)$$

on the right-hand side of (15.80), we obtain

$$\phi(a,b)-\phi^*(a,b) \ll \beta\sigma_1(a,b)+\beta^2\sigma_2(a,b), \tag{15.94}$$

where

$$\sigma_1(a,b) = \sum_{\substack{i=1\\a-i\geqslant 0}}^{2} \sum_{\substack{j=1\\b-j\geqslant 0}}^{2} \phi^*(a-i,b-j),$$

$$\sigma_2(a,b) = \sum_{\substack{i=2\\a-i\geqslant 0}}^{4} \sum_{\substack{j=2\\b-j\geqslant 0}}^{4} \phi(a-i,b-j).$$

Now by (15.78) and Lemma 26 (since $P_{\mathscr{X}} \leqslant P'_{\mathscr{X}}$, $P_{\mathscr{Y}} \leqslant P'_{\mathscr{Y}}$ always), we have

$$\phi^*(u,v) \leqslant \frac{(\lambda'_n)^u}{u!}\frac{(\lambda'_m)^v}{v!}$$

for all non-negative integers u, v. Hence by Definition 27, (15.89), and (15.90),

$$\sum_{a\geqslant U}\sum_{b\geqslant V}\sigma_1(a,b) \leqslant 4 \sum_{\substack{u=0\\u\geqslant U-2}}^{\infty} \sum_{\substack{v=0\\v\geqslant V-2}}^{\infty} \phi^*(u,v) \tag{15.95}$$

$$\leqslant 4 \sum_{u\geqslant U}\sum_{v\geqslant V} \delta_n(u)\delta_m(v) \ll \delta(U)\delta(V).$$

On the other hand, since (cf. (15.9) and (15.10)),

$$\phi_n(u) = \sum_{v=0}^{\infty} \phi(u,v), \qquad \phi_m(v) = \sum_{u=0}^{\infty} \phi(u,v),$$

we have

$$\sum_{a\geqslant U}\sum_{b\geqslant V}\sigma_2(a,b) \leqslant 9 \sum_{\substack{u=0\\u\geqslant U-4}}^{\infty} \sum_{\substack{v=0\\v\geqslant V-4}}^{\infty} \phi(u,v)$$

$$\leqslant 9 \min\Bigl(\sum_{\substack{u=0\\u\geqslant U-4}}^{\infty} \phi_n(u), \ \sum_{\substack{v=0\\v\geqslant V-4}}^{\infty} \phi_m(v)\Bigr).$$

It now follows from Lemmas 28, 26, and 31, that

$$\sum_{a\geqslant U}\sum_{b\geqslant V}\sigma_2(a,b) \ll \min\{\delta(U),\delta(V)\}. \tag{15.96}$$

Clearly (15.94), (15.95), and (15.96) imply (15.93).

In view of (15.89) and (15.90), the following lemma is an immediate consequence of Lemma 29.

Lemma 33. *For any non-negative integers U, V,*

$$\sum_{a\geqslant U}\sum_{b\geqslant V} |\phi^*(a,b)-\phi_n(a)\phi_m(b)| \ll \beta\delta(U)\delta(V). \tag{15.97}$$

Obviously, the combination of Lemmas 32 and 33 yields Theorem 16. (As was pointed out in the remark following Definition 28, the terms in the double sum on the left-hand side of (15.93) are non-negative.)

Proof of Theorem 17. We have

$$\mathbf{M}(r_n r_m) - \mathbf{M}(r_n)\mathbf{M}(r_m) = \sum_k \sum_l \gamma(k,l) = \sum_{k,l}{}^* \gamma(k,l), \tag{15.98}$$

where $\gamma(k,l) = \mathbf{M}(\rho_k\rho_{n-k}\rho_l\rho_{m-l}) - \mathbf{M}(\rho_k\rho_{n-k})\mathbf{M}(\rho_l\rho_{m-l})$, and the sum $\sum^*$ is extended over the set $\mathscr{J}$ consisting of all pairs k, l for which $\pi_k^{(n)}$, $\pi_l^{(m)}$ intersect.

Now, by Definition 22 and Lemma 21, the relation

$$\tau = (\pi_k^{(n)}) \cup (\pi_l^{(m)}) \tag{15.99}$$

induces a 1–1 correspondence between the set of all triads τ and the set $\mathscr{J}$. Furthermore, (15.99) implies

$$\mathbf{M}(\rho_k \rho_{n-k} \rho_l \rho_{m-l}) = \eta(\tau),$$

and hence $0 < \gamma(k,l) < \eta(\tau)$. Thus

$$0 \leqslant \mathbf{M}(r_n r_m) - \mathbf{M}(r_n)\mathbf{M}(r_m) \leqslant \sum_{\tau} \eta(\tau),$$

and in view of Lemma 25, this proves (15.13).

To prove (15.14), we need only note that (using n in place of t) on writing

$$\gamma_1(k',k'') = \mathbf{M}(\rho_{k'} \rho_{n-k'} \rho_{k''} \rho_{n-k''}) - \mathbf{M}(\rho_{k'} \rho_{n-k'})\mathbf{M}(\rho_{k''} \rho_{n-k''}),$$

we have

$$\gamma_1(k',k'') = 0 \quad \text{if} \quad k' \neq k'', \quad \text{and} \quad \gamma_1(k,k) < \eta(\pi_k^{(n)}).$$

For it follows immediately that

$$\mathbf{M}(r_n^2) - \{\mathbf{M}(r_n)\}^2 = \sum_{k'} \sum_{k''} \gamma_1(k',k'') < \sum_{k} \eta(\pi_k^{(n)}) = \lambda_n.$$

16. Proof of Theorem 4—sequences of pseudo-squares

Whilst the results of this section are capable of various obvious generalizations, we shall restrict our attention to sequences of pseudo-squares. Accordingly (cf. the remark following Lemma 11), we define the sequence $\{\alpha_j\}$, of probabilities, by

$$\alpha_j = \tfrac{1}{2} j^{-\frac{1}{2}} \quad (j = 1, 2, \ldots). \tag{16.1}$$

We note that this choice is consistent with Hypothesis A, so that the results of § 11 are at our disposal. In particular, applying Lemma 11 with $\alpha = \frac{1}{2}$, $c = \frac{1}{2}$, $c' = 0$, we obtain $\lambda_n \sim \frac{1}{8}\pi$ as $n \to \infty$; in other words,

$$\lim_{n\to\infty} \lambda_n = \lim_{n\to\infty} \mathbf{M}(r_n) = \lambda, \tag{16.2}$$

where

$$\lambda = \tfrac{1}{8}\pi. \tag{16.3}$$

It follows that Hypothesis B of § 15 is satisfied, so that the results of that section are also at our disposal.

We prove Theorem 4 by establishing Theorem 4′ below.

THEOREM 4′. *Suppose that* Ω *is the probability space generated, in accordance with Theorem* 13, *by the choice* (16.1) *of the probabilities* α_j. *Let* λ *be the value of the limit* (16.2), *so that* λ *is given by* (16.3). *Then, with probability* 1, *the sequence* $\omega = \{a_j\}$ *has each of the following four properties.*†

(i) ω *is a sequence of pseudo-squares in the sense that*

$$a_j \sim j^2 \quad \text{as } j \to \infty. \tag{16.4}$$

(ii) *For every non-negative integer* d, *the sequence* $\mathscr{S}_d$, *consisting of those positive integers* n *for which* $r_n(\omega) = d$, *has asymptotic density* $\lambda^d e^{-\lambda}/d!$

(iii) *The asymptotic mean of the representation function* $r_n(\omega)$ *is given by*

$$\lim_{N\to\infty} \frac{1}{N} \sum_{n=1}^{N} r_n(\omega) = \lambda. \tag{16.5}$$

(iv) *The representation function* $r_n(\omega)$ *satisfies a relation of the iterated logarithm type, namely*

$$\limsup_{n\to\infty} \frac{r_n(\omega)\log\log n}{\log n} = 1. \tag{16.6}$$

Theorem 4′ is really a combination‡ of results, namely that each of (i)–(iv) holds with probability 1. We shall prove each of these results separately. In fact (as has already been remarked) the first of these results is an immediate consequence of Lemma 11 since, when $\alpha = \frac{1}{2}$, $c = \frac{1}{2}$, $c' = 0$, the assertion (11.17) coincides with (16.4). We need only prove, therefore, that each of (ii), (iii), and (iv) holds with probability 1.

For every natural number n and every non-negative integer d, we denote by $\epsilon_n(d;\omega)$ the characteristic function of the event $\{\omega: r_n(\omega) = d\}$ (so that, for given n and d, the random variable $\epsilon_n(d;\omega)$ assumes the values 1 or 0 according as the relation $r_n(\omega) = d$ does or does not hold). We write

$$S_N(d;\omega) = \sum_{n=1}^{N} \epsilon_n(d;\omega); \tag{16.7}$$

so that, for given d and ω, S_N counts the number of positive integers $n \leqslant N$ for which $r_n(\omega) = d$. Using this notation, the assertion (ii) of Theorem 4′ may be written in the form

$$\lim_{N\to\infty} N^{-1} S_N(d;\omega) = \frac{\lambda^d}{d!} e^{-\lambda} \quad (d = 0, 1, 2, \ldots). \tag{16.8}$$

We also write

$$T_N(\omega) = \sum_{n=1}^{N} r_n(\omega), \tag{16.9}$$

† See Definition 16 for the definition of $r_n(\omega)$.

‡ Cf. footnote appended to (12.6), (12.7).

so that the assertion (iii) of Theorem 4′ takes the form

$$\lim_{N\to\infty} N^{-1}T_N(\omega) = \lambda. \tag{16.10}$$

We remark that by Lemma 11 (again with $\alpha = \frac{1}{2}$, $c = \frac{1}{2}$, $c' = 0$),

$$m_n^* = \sum_{j=1}^{n} \alpha_j \sim n^{\frac{1}{2}} \quad \text{as } n \to \infty, \tag{16.11}$$

and by Lemma 16,

$$\lim_{n\to\infty} \mathbf{M}\{\epsilon_n(d;\omega)\} = \frac{\lambda^d}{d!}e^{-\lambda} \quad (d = 0, 1, 2,\ldots). \tag{16.12}$$

Obviously, (16.12) and (16.2) imply

$$\lim_{N\to\infty} \mathbf{M}\{N^{-1}S_N(d;\omega)\} = \frac{\lambda^d}{d!}e^{-\lambda} \tag{16.13}$$

and

$$\lim_{N\to\infty} \mathbf{M}\{N^{-1}T_N(d;\omega)\} = \lambda. \tag{16.14}$$

In view of (16.13) and (16.14), the law of large numbers suggests that each of (16.8) and (16.10) holds with probability 1, as required. We are not able to apply Theorem 11, however, since neither the set of random variables

$$\epsilon_1(d;\omega),\ \epsilon_2(d;\omega),\ldots \tag{16.15}$$

(for given d) nor the set of random variables

$$r_1(\omega),\ r_2(\omega),\ldots \tag{16.16}$$

is a set of independent random variables in the strict sense of Definition 12. Kolmogorov's inequality (Theorem 9), which played a central part in the proof of Theorem 11 (cf. the estimates (9.12), (9.22)) is not available in the absence of strict independence. Instead, we must go back to Tchebycheff's inequality (7.4), making use of the quasi-independence of the sets (16.15), (16.16) to estimate $\mathbf{D}^2(f)$ when f is a sum of random variables chosen from one of these sets. The fact that, for given d and ω, the functions S_N and T_N are monotonic increasing functions of N is an additional useful circumstance, and enables us to use the following lemma in place of Theorem 11.

Lemma 34. *Suppose that the sequence $\{s_n(\omega)\}$ of random variables satisfies the following three conditions.*

(I) *For each ω, $s_n(\omega)$ is a monotonic increasing function of n.*

(II) $$\lim_{n\to\infty} \mathbf{M}(n^{-1}s_n) \quad \textit{exists.}$$

(III) *There exists $\delta > 0$ such that*

$$\mathbf{D}^2(s_n) \ll n^{2-\delta}. \tag{16.17}$$

Then, with probability 1,

(16.18) $$\lim_{n\to\infty} n^{-1}s_n(\omega) = \lim_{n\to\infty} \mathbf{M}(n^{-1}s_n).$$

Proof. In view of (I), it suffices to prove that for some sequence

(16.19) $$n_1 < n_2 < \dots$$

satisfying

(16.20) $$\lim_{j\to\infty} n_{j+1}/n_j = 1,$$

we have

(16.21) with probability 1, $$\lim_{j\to\infty} n_j^{-1}s_{n_j}(\omega) = \lim_{n\to\infty} \mathbf{M}(n^{-1}s_n);$$

for $n_j \leqslant n < n_{j+1}$ implies

$$(n_j/n)n_j^{-1}s_{n_j}(\omega) \leqslant n^{-1}s_n(\omega) \leqslant (n_{j+1}/n)n_{j+1}^{-1}s_{n_{j+1}}(\omega).$$

Furthermore, by Theorem 8, in order to establish (16.21) we need only show that, for every $\epsilon > 0$,

(16.22) $$\sum_{j=1}^{\infty} \mu(\{\omega\colon |n_j^{-1}s_{n_j}(\omega) - \mathbf{M}(n_j^{-1}s_{n_j})| > \epsilon\}) < \infty.$$

But (III) enables us to use Tchebycheff's inequality (7.4) to estimate the summand of (16.22). We obtain in this way the upper estimate

$$\frac{1}{\epsilon^2}\mathbf{D}^2(n_j^{-1}s_{n_j}) \ll n_j^{-\delta}$$

for the summand on the left-hand side of (16.22). Thus we can ensure the truth of (16.22) by choosing $n_j = j^{[2/\delta]+1}$ $(j = 1, 2,\dots)$. This choice of the sequence $\{n_j\}$ satisfies (16.19) and (16.20) so that the proof of the lemma is complete.

Proof that, with probability 1, *the assertion* (ii) *of Theorem* 4′ *holds.*

We must show that for each $d = 0, 1,\dots$, (16.8) holds with probability 1. In view of (16.13), it suffices to show that Lemma 34 is applicable, for each d, with $s_n(\omega) = S_n(d;\omega)$. For this purpose, we need only verify that the hypothesis (III) of that lemma is satisfied, and we shall do this by proving that, for each $d = 0, 1,\dots$,

(16.23) $$\mathbf{D}^2\{S_N(d;\omega)\} \ll N^{\frac{3}{2}}.$$

Since $\mathbf{D}^2(f) = \mathbf{M}(f^2) - \{\mathbf{M}(f)\}^2$, we have

(16.24) $$\mathbf{D}^2\{S_N(d;\omega)\} = \sum_{n=1}^{N} \mathbf{D}^2\{\epsilon_n(d;\omega)\} + \Delta_N(d),$$

where (with a self-explanatory simplification of notation)

$$\Delta_N = 2 \sum_{1 \leqslant n < m \leqslant N} \{\mathbf{M}(\epsilon_n \epsilon_m) - \mathbf{M}(\epsilon_n)\mathbf{M}(\epsilon_m)\}.$$

The first sum on the right-hand side of (16.24) is easily estimated. Obviously, since ϵ_n is a characteristic function,

$$\sum_{n=1}^{N} \mathbf{D}^2\{\epsilon_n(d;\omega)\} \leqslant \sum_{n=1}^{N} \mathbf{M}\{\epsilon_n(d;\omega)\} \leqslant N \tag{16.25}$$

by (16.12). As regards the term $\Delta_N(d)$ on the right-hand side of (16.24), this is an error term arising only because the set of random variables (16.15) are not fully independent; these random variables are, however, quasi-independent in the sense of Theorem 14 (§ 15), and we can use this theorem to estimate Δ_N. The assertion (15.5) of Theorem 14 states, in terms of our present notation,

$$\mathbf{M}(\epsilon_n \epsilon_m) - \mathbf{M}(\epsilon_n)\mathbf{M}(\epsilon_m) \ll \alpha_{m-n},$$

so that

$$\Delta_N \ll \sum_{1 \leqslant n < m \leqslant N} \alpha_{m-n} \leqslant N \sum_{j=1}^{N-1} \alpha_j \ll N^{\frac{3}{2}}, \tag{16.26}$$

by (16.11). This estimate, in conjunction with (16.25), establishes (16.23). This completes the proof.

Proof that, with probability 1, *the assertion* (iii) *of Theorem* 4′ *holds.*

The proof of this result, namely that (16.10) holds with probability 1, will run parallel to the proof we have just completed, but with Theorem 17 (§ 15) replacing Theorem 14 (§ 15) for the purpose of estimating the error arising from the fact that the random variables under consideration are not strictly independent. Here it suffices (cf. (16.14)) to verify that the hypothesis (III) of Lemma 34 is satisfied when $s_n(\omega) = T_n(\omega)$, and we shall do this by proving that

$$\mathbf{D}^2\{T_N(\omega)\} \ll N^{\frac{3}{2}}. \tag{16.27}$$

In analogy to (16.24), we have

$$\mathbf{D}^2\{T_N(\omega)\} = \sum_{n=1}^{N} \mathbf{D}^2\{r_n(\omega)\} + \Delta'_N, \tag{16.28}$$

where
$$\Delta'_N = 2 \sum_{1 \leqslant n < m \leqslant N} \{\mathbf{M}(r_n r_m) - \mathbf{M}(r_n)\mathbf{M}(r_m)\}.$$

Each of the sums on the right-hand side of (16.28) is easily estimated with the aid of Theorem 17 (§ 15). The assertions (15.14) and (15.13) of this theorem, namely

$$\mathbf{D}^2(r_n) \ll \lambda_n, \qquad \mathbf{M}(r_n r_m) - \mathbf{M}(r_n)\mathbf{M}(r_m) \ll \alpha_{m-n},$$

imply (in view of (16.2))

$$\sum_{n=1}^{N} \mathbf{D}^2(r_n) \ll \sum_{n=1}^{N} \lambda_n \ll N$$

and (cf. (16.26)) $$\Delta'_N \ll \sum_{1\leqslant n<m\leqslant N} \alpha_{m-n} \ll N^{\frac{3}{2}}.$$

On combining these two estimates we obtain (16.27), and the proof is complete.

Proof that, with probability 1, *the assertion* (iv) *of Theorem* 4′ *holds.*

We shall now again make use of the function λ'_n defined by (11.6), and note that by Lemma 15 (in view of (16.2) and (16.3)),

(16.29) $$\lim_{n\to\infty} \lambda'_n = \tfrac{1}{8}\pi.$$

We shall also be using the function

$$\delta(W) = \exp\{-W\log(cW)\}$$

introduced in Theorem 15 (§ 15) (the constant c is absolute, now that an explicit choice for the α_j has been made); and we recall that, by Lemma 31 (§ 15),

(16.30) $$\sum_{d\geqslant W} \frac{(\lambda'_n)^d}{d!} \ll \delta(W)$$

for all natural numbers W.

We first prove the partial result that,

(16.31) *with probability* 1, $$\limsup_{n\to\infty} \frac{r_n(\omega)\log\log n}{\log n} \leqslant 1.$$

By the Borel–Cantelli lemma (§ 9, Theorem 7), it suffices to prove that for every $\epsilon > 0$,

(16.32) $$\sum_{n=3}^{\infty} \mu\left[\left\{\omega\colon r_n(\omega) > (1+\epsilon)\frac{\log n}{\log\log n}\right\}\right] < \infty.$$

We denote by W_n the *least* integer satisfying

$$W_n > (1+\epsilon)\frac{\log n}{\log\log n},$$

so that the summand of (16.32) may be written in the form

$$\mu(\{\omega\colon r_n(\omega) \geqslant W_n\}).$$

We estimate this summand by applying (11.31), and obtain in this way

(16.33) $$\mu(\{\omega\colon r_n(\omega) \geqslant W_n\}) \leqslant e^{-\lambda_n} \sum_{d\geqslant W_n} \frac{(\lambda'_n)^d}{d!} \ll \delta(W_n),$$

in view of (16.2) and (16.30). Now it is clear from the definition of W_n that $W_n \log(cW_n) > (1+\frac{1}{2}\epsilon)\log n$ for large n, and hence

$$\delta(W_n) \ll n^{-1-\frac{1}{4}\epsilon}.$$

Thus (16.33) implies (16.32) and the proof of (16.31) is complete.

The proof that

(16.34) *with probability* 1, $$\limsup_{n\to\infty} \frac{r_n(\omega)\log\log n}{\log n} \geqslant 1$$

is more difficult. We must show that for every small positive ϵ,

(16.35) *with probability* 1, $$r_n(\omega) > (1-\epsilon)\frac{\log n}{\log\log n}$$

for infinitely many values of n.

We shall first prove that

(16.36) $$\sum_{n=3}^{\infty} \mu\left[\left\{\omega: r_n(\omega) > (1-\epsilon)\frac{\log n}{\log\log n}\right\}\right] = \infty,$$

and this will be straightforward. We cannot, however, use Theorem 7 (the usual form of the Borel–Cantelli lemma) to deduce (16.35) from (16.36), since the random variables r_n are not strictly independent. Instead, we shall have to show that the r_n satisfy the quasi-independence condition (9.5) of the refined form, Theorem 7′, of the Borel–Cantelli lemma; and this will entail the use of the substantial Theorem 15 (§ 15).

Throughout the proof, ϵ denotes a small positive number, and implicit constants may depend on ϵ. Here again, we introduce a simplified notation in connexion with the summand of (16.36), denoting by U_n the least natural number satisfying

$$U_n > (1-\epsilon)\frac{\log n}{\log\log n}.$$

We record for future use that, for all n from some point onward,

(16.37) $$U_n \log U_n + U_n < (1-\tfrac{3}{4}\epsilon)\log n$$

and

(16.38) $$U_n \log(cU_n) > (1-\tfrac{5}{4}\epsilon)\log n,$$

c being the absolute constant appearing in the definition of $\delta(W)$. In conformity with the notation of Theorem 15 (§ 15), we write $G_n = G_n(U_n)$ for the event

$$\{\omega: r_n(\omega) \geqslant U_n\}.$$

Furthermore, we write, for any natural number N,

$$Z_N = \sum_{n=1}^{N} \mu(G_n),$$

and for any pair n, m of natural numbers,

$$\eta_{n,m} = \mu(G_n \cap G_m) - \mu(G_n)\mu(G_m).$$

We shall establish (16.34) by showing that

$$Z_N \gg N^{\frac{3}{4}\epsilon}, \tag{16.39}$$

and

$$\sum_{1 \leqslant n < m \leqslant N} \eta_{n,m} = o(Z_N^2) \quad \text{as } N \to \infty; \tag{16.40}$$

the first of these relations implies $Z_N \to \infty$ as $N \to \infty$ and thus confirms (16.36), whilst the second verifies that the condition (9.5) of Theorem 7′ is satisfied, so that (16.36) implies (16.35).

We obtain the lower bound for $\mu(G_n)$ needed to establish (16.39) by using the crude estimate

$$\mu(G_n) \geqslant \mu(\{\omega\colon r_n(\omega) = U_n\})$$

in conjunction with the assertion (11.32) of Lemma 14. We first show that, if $d = U_n$ and n is large, the factor

$$1 - \binom{d}{2}(\lambda_n')^{-2} Q^*$$

appearing on the right-hand side of (11.32) is $\gg 1$. In fact we have

$$Q^* \ll \sum_{1 \leqslant k < \frac{1}{2}n} (\alpha_k \alpha_{n-k})^2 = \frac{1}{4} \sum_{1 \leqslant k < \frac{1}{2}n} \frac{1}{k(n-k)} \ll \frac{\log n}{n},$$

whilst (when $d = U_n$)

$$\binom{d}{2} \ll (\log n)^2 \quad \text{and} \quad (\lambda_n')^{-2} \ll 1.$$

Thus the application of (11.32) yields

$$\mu(G_n) \gg \frac{(\lambda_n')^{U_n}}{(U_n)!} \gg \left(\frac{1}{e}\right)^{U_n} U_n^{-U_n} \quad (n \geqslant 3), \tag{16.41}$$

since $\lambda_n' \sim \frac{1}{8}\pi$ and hence $\lambda_n' > e^{-1}$ for large n. On writing the second lower estimate in (16.41) in the form $\exp(-U_n \log U_n - U_n)$, we see that, by (16.37),

$$\mu(G_n) \gg n^{-1+\frac{3}{4}\epsilon}, \tag{16.42}$$

which in turn implies (16.39).

It remains only to prove (16.40). On applying Theorem 15 (§ 15) with $U = U_n$, $V = U_m$, we obtain

$$\eta_{n,m} \ll \alpha_{m-n}\,\delta(U_n)\delta(U_m) + \alpha_{m-n}^2\,\delta(U_m).$$

Since $\alpha_n = \frac{1}{2}n^{-\frac{1}{2}}$ and, by (16.38),

$$\delta(U_n) = \exp\{-U_n \log(cU_n)\} \ll n^{-1+\frac{5}{8}\epsilon},$$

it follows that in order to prove (16.40) it suffices to show that each of the sums

$$\textstyle\sum_1 = \sum_{1\leqslant n<m\leqslant N} (m-n)^{-\frac{1}{2}} n^{-1+\frac{4}{3}\epsilon} m^{-1+\frac{4}{3}\epsilon}$$

and

$$\textstyle\sum_2 = \sum_{1\leqslant n<m\leqslant N} (m-n)^{-1} m^{-1+\frac{4}{3}\epsilon}$$

is $o(Z_N^2)$. Replacing the summation variables n, m by $t = m-n$, m, we may write these sums in the form

$$\begin{aligned}\textstyle\sum_1 &= \sum_{m=2}^{N} m^{-1+\frac{4}{3}\epsilon} \sum_{1\leqslant t<m} t^{-\frac{1}{2}}(m-t)^{-1+\frac{4}{3}\epsilon}\\ &= \sum_{m=2}^{N} m^{-\frac{3}{2}+\frac{8}{3}\epsilon} \sum_{1\leqslant t<m} \left(\frac{t}{m}\right)^{-\frac{1}{2}}\left(1-\frac{t}{m}\right)^{-1+\frac{4}{3}\epsilon}\frac{1}{m},\end{aligned}$$

and

$$\textstyle\sum_2 = \sum_{m=2}^{N} m^{-1+\frac{4}{3}\epsilon} \sum_{1\leqslant t<m} t^{-1}.$$

Now the inner sum of the last expression for $\sum_1$ converges to the Beta integral $\mathrm{B}(\frac{1}{2}, \frac{4}{3}\epsilon)$ as $m \to \infty$, so that (since ϵ is small) $\sum_1 \ll 1$; whilst the inner sum in the expression for $\sum_2$ is $\ll \log N$, so that $\sum_2 \ll N^{\frac{4}{3}\epsilon} \log N$. Thus, by (16.39),

$$\textstyle\sum_1 + \sum_2 = o(Z_N^2)$$

as required, and the proof of Theorem 4′ is complete.

IV

SIEVE METHODS

1. Introduction

In this chapter we discuss the sieve methods of Viggo Brun and A. Selberg, and the 'large sieves' of Linnik and Rényi. Of these, the theorems of Linnik and Rényi in § 10 fall most naturally within the scope of this book; they are theorems of surprising generality, and are of intrinsic interest quite apart from their applications to sieve problems. The sieves of Brun and Selberg, on the other hand, are effective only when applied to sequences of rather a special kind. Nevertheless, the method is one of some generality and great beauty, and we therefore give an account of its mechanism. We do not include any applications to specific sieve problems, although we prove two general theorems of Selberg (Theorems 3 and 4) which are applicable to a wide variety of such problems. For applications in other chapters, we require only Theorem 1. Whilst this result is usually ascribed to the Brun-type sieve (and indeed can be derived in this way), we shall see that it can be proved by simpler and more direct means.

Every theorem in this chapter really relates to a finite *set* of integers even though, for technical convenience, we may order the elements of such a set to obtain a 'finite *sequence*'. In applications, these finite sets usually arise as the first n terms of some *infinite sequence*. The integer sequences considered in this chapter are not all monotone but, in view of the above remark, this is inconsequential.

The historical origin of sieve methods is the well-known 'sieve of Eratosthenes'. Eratosthenes noted that the prime numbers between $n^{\frac{1}{2}}$ and n can be isolated by removing from the sequence 2, 3,..., n every number which is a multiple of a prime not exceeding $n^{\frac{1}{2}}$; or, in more descriptive language, by 'sifting the sequence 2, 3,..., n by the primes not exceeding $n^{\frac{1}{2}}$'. This remark leads immediately to an expression for the number $\pi(n)-\pi(n^{\frac{1}{2}})$ of primes p satisfying $n^{\frac{1}{2}} < p \leqslant n$, which we obtain below (using modern notation).

Eratosthenes's principle may be expressed as

$$1+\pi(n)-\pi(n^{\frac{1}{2}}) = \sum_{a\leqslant n} s^{(0)}(a), \tag{1.1}$$

where $s^{(0)}(a)$ is the 'sifting function' defined by

$$s^{(0)}(a) = \begin{cases} 1 & \text{if } a \text{ is not divisible by any prime } p \leqslant n^{\frac{1}{2}} \\ 0 & \text{if } a \text{ is divisible by some prime } p \leqslant n^{\frac{1}{2}}. \end{cases} \tag{1.2}$$

On writing

$$\Pi = \prod_{p \leqslant n^{\frac{1}{2}}} p, \tag{1.3}$$

we have

$$s^{(0)}(a) = \sum_{d|(a,\Pi)} \mu(d). \tag{1.4}$$

We thus obtain the relation

$$\begin{aligned} 1+\pi(n)-\pi(n^{\frac{1}{2}}) &= \sum_{a \leqslant n} \sum_{d|(a,\Pi)} \mu(d) \\ &= \sum_{d|\Pi} \mu(d) \sum_{\substack{a \leqslant n \\ d|a}} 1 = \sum_{d|\Pi} \mu(d) \left[\frac{n}{d}\right]. \end{aligned} \tag{1.5}$$

This formula is of no practical value, however. At first sight, we might hope to estimate $\pi(n)-\pi(n^{\frac{1}{2}})$ by replacing $[n/d]$ by n/d in the last sum of (1.5). This would lead to the relation

$$1+\pi(n)-\pi(n^{\frac{1}{2}}) = n \sum_{d|\Pi} \frac{\mu(d)}{d} + E = n \prod_{p|\Pi} \left(1-\frac{1}{p}\right) + E \tag{1.6}$$

where

$$E = \sum_{d|\Pi} \mu(d) \left\{\left[\frac{n}{d}\right] - \frac{n}{d}\right\}. \tag{1.7}$$

But although each term on the right-hand side of (1.7) is $O(1)$, the number $2^{\pi(n^{\frac{1}{2}})}$ of terms in the sum is very large compared to n. Furthermore, it seems hopeless to try to make use of the cancellation effect of $\mu(d)$ in an attempt to obtain an effective estimate for E. In fact, if we estimate E indirectly, by *assuming* the prime number theorem, we see that the cancellation effect is somewhat limited in extent. For by a well-known result[1] of Mertens (cf. the proof of Lemma 4 in the Appendix),

$$n \prod_{p|\Pi} \left(1-\frac{1}{p}\right) \sim e^{-\gamma} \frac{n}{\log n^{\frac{1}{2}}},$$

so that by (1.6) and the prime number theorem (cf. Appendix, Lemma 5), E is also of the order of magnitude $n/(\log n)$.

Before discussing the innovations of Brun and Selberg, we formulate the notion of a 'sieve' in a wider context. Let $h(\nu)$ be a polynomial in ν, which assumes only integral values for integral ν (e.g. a polynomial

(1) F. Mertens, 'Ein Beitrag zur analytischen Zahlentheorie', *J. reine angew. Math.* **78** (1874) 46–62. Or see Hardy and Wright, *An Introduction to the Theory of Numbers*, 4th ed. (Clarendon Press, Oxford, 1960), Theorem 429.

with integral coefficients). Let $\mathscr{A} = \{a_\nu\}$ be the sequence defined by $a_\nu = h(\nu)$ ($\nu = 1, 2,...$). Let $\mathscr{P}$ denote a set of r primes

$$p_1 < p_2 < ... < p_r.$$

We consider the 'sieve' in which all multiples of the primes p_i of $\mathscr{P}$ are deleted from the finite sequence

$$a_1, a_2, ..., a_n.$$

Various important problems in number theory require effective estimates for the number $S(\mathscr{A}, \mathscr{P}; n)$ of unsifted elements in such a 'sieve'. For example, Brun proved that the number of prime twins p, $p+2$ (where each of p, $p+2$ is a prime) not exceeding n is $O(n/\log^2 n)$ by obtaining this upper estimate for $S(\mathscr{A}, \mathscr{P}; n)$ when $h(\nu) = \nu(\nu-2)$ and $\mathscr{P}$ consists of all primes $p \leqslant n^{\frac{1}{2}}$.

Let $N(d)$ denote the number of solutions of

$$h(\nu) \equiv 0 \pmod{d}, \qquad 1 \leqslant \nu \leqslant d,$$

so that $N(d)$ is a multiplicative function of d. We may assume that $N(p) < p$ for each $p \in \mathscr{P}$, for otherwise $S(\mathscr{A}, \mathscr{P}; n) = 0$. Furthermore, the primes p for which $N(p) = 0$ are not relevant to the sieve and may be discarded from $\mathscr{P}$. For the remaining primes of $\mathscr{P}$, we have $1 \leqslant N(p) < p$. It is convenient to write

$$N(d) = d/f(d),$$

so that $f(d)$ is a multiplicative function satisfying

$$1 < f(p) \leqslant p \quad \text{for} \quad p \in \mathscr{P}. \tag{1.8}$$

Now among the a_ν with ν not exceeding $[n/d]d$, exactly $[n/d]N(d)$ are divisible by d; and among the a_ν with $[n/d]d < \nu \leqslant n$, at most $N(d)$ are divisible by d. Hence

$$\sum_{\substack{\nu \leqslant n \\ d | a_\nu}} 1 = \frac{n}{f(d)} + R_d, \tag{1.9}$$

where the error term R_d satisfies

$$|R_d| \leqslant d/f(d). \tag{1.10}$$

The only knowledge of the sequence $\mathscr{A}$ that we shall require is that it satisfies (1.9) with $f(d)$ multiplicative and R_d suitably restricted in magnitude. Our discussion would apply equally well to any finite sequence

$$a_1, a_2, ..., a_n$$

satisfying a suitable relation of type (1.9). Unfortunately (1.9), in conjunction with the type of conditions on R_d and $f(d)$ that are necessary (for the success of the methods to be described), is highly restrictive.

The case $a_\nu = h(\nu)$ ($\nu = 1, 2, 3, \ldots$), which has just been discussed, is the only one of importance for which the above requirements are satisfied and an effective estimate (cf. (1.10)) for $|R_d|$ is obtainable by purely elementary means. It is, however, practicable to take $\mathscr{A}$ to be a suitable subsequence of the sequence $\{h(\nu)\}$, such as the sequence $\{h(q_j)\}$, where q_j is the jth prime; although the corresponding estimates for $|R_d|$ are relatively poor and can be established only by the use of difficult analytic techniques. We shall say more about this later.

Let
$$\Pi(\mathscr{P}) = \prod_{p \in \mathscr{P}} p$$
and let $s^{(0)}(a)$ be the 'sifting function' defined by

$$s^{(0)}(a) = \sum_{d \mid (a, \Pi(\mathscr{P}))} \mu(d). \tag{1.11}$$

The sieve of Eratosthenes, in this more general context, yields (cf. (1.5) and (1.9))

$$\begin{aligned} S(\mathscr{A}, \mathscr{P}; n) &= \sum_{\nu \leqslant n} s^{(0)}(a_\nu) \\ &= n \sum_{d \mid \Pi(\mathscr{P})} \frac{\mu(d)}{f(d)} + \sum_{d \mid \Pi(\mathscr{P})} \mu(d) R_d. \end{aligned} \tag{1.12}$$

Here again, the sum representing the error is over the divisors of $\Pi(\mathscr{P})$, and the large number of such divisors will render the formula useless unless r is exceedingly small compared with n.

The methods of Brun and Selberg lead (under suitable conditions) to both upper and lower estimates for $S(\mathscr{A}, \mathscr{P}; n)$ even when r is far too large for (1.12) to be applicable. For the purpose of this discussion, we restrict ourselves (temporarily) to the problem of estimating $S(\mathscr{A}, \mathscr{P}; n)$ from above.

The Brun–Selberg methods stem from a reappraisal of the structure of the relation (1.12). In the first place, if $s(a)$ is *any* function of type

$$s(a) = \sum_{d \mid (a, \Pi(\mathscr{P}))} \lambda(d) \tag{1.13}$$

(where the function $\lambda(d)$ is, for the moment, quite arbitrary), we still have

$$\sum_{\nu \leqslant n} s(a_\nu) = n \sum_{d \mid \Pi(\mathscr{P})} \frac{\lambda(d)}{f(d)} + \sum_{d \mid \Pi(\mathscr{P})} \lambda(d) R_d. \tag{1.14}$$

Furthermore, for the purpose of estimating $S(\mathscr{A}, \mathscr{P}; n)$ from above, we may use any such function $s^{(+)}(a)$ (to be referred to as a sifting function), with the property†

$$s^{(+)}(a) \geqslant s^{(0)}(a) \quad \text{for all } a. \tag{1.15}$$

† In practice, it is convenient to insist on equality in (1.15) when $(a, \Pi(\mathscr{P})) = 1$.

For any sifting function $s^{(+)}$, we have

(1.16) $$S(\mathscr{A},\mathscr{P};n) \leqslant \sum_{\nu\leqslant n} s^{(+)}(a_\nu).$$

One is therefore free to use any sifting function $s^{(+)}$ of type (1.13) which satisfies (1.15); subject to the reservation that $s^{(+)}(a)$ should be a sufficiently good approximation to $s^{(0)}(a)$ to ensure that the resulting estimate (1.16) is not too crude.

This freedom in the choice of the function $s^{(+)}$ (and, implicitly, of $\lambda(d)$), may be used to reduce the magnitude of the error

(1.17) $$\sum_{d\mid\Pi(\mathscr{P})} |\lambda(d)R_d|.$$

Both Brun and Selberg ensure that the expression (1.17) is not too large by defining $\lambda(d) = 0$ when d is outside a subset $\mathscr{D}^*$, containing somewhat fewer than n elements, of the divisors of $\Pi(\mathscr{P})$.

Brun defines $\lambda(d) = \mu(d)$ within $\mathscr{D}^*$, and ensures (1.15) by means of an appropriate choice of $\mathscr{D}^*$. Selberg, on the other hand, chooses $\mathscr{D}^*$ in the simplest possible manner, and selects a function λ, consistently with this choice of $\mathscr{D}^*$, from a whole class of functions so chosen that the corresponding sifting functions $s(a)$ satisfy (1.15) for trivial reasons. In a successful choice of λ, $\lambda(d)$ differs from $\mu(d)$, although within $\mathscr{D}^*$ it remains, in a sense, an approximation to $\mu(d)$.

It is clear that the appropriate choice of $\mathscr{D}^*$ in Brun's method, and of $\lambda(d)$ in Selberg's method, is far from obvious. The choice must be such that, *in addition* to (1.15):

(i) $\mathscr{D}^*$ does not contain too many elements,

(ii) $\sum_{d\in\mathscr{D}^*} \lambda(d)/f(d)$ is not too large.

The methods of estimating $S(\mathscr{A},\mathscr{P};n)$ from below are analogous, although the analogy is really close only in the case of Brun's method.

Brun(2) and Selberg(3) discovered their methods in 1920 and 1947 respectively. Selberg's upper-bound method is superior to that of Brun in every way; it is surprisingly simple and leads to more precise results than the much more complicated method of Brun. Indeed, Selberg's upper-bound method has a certain air of finality, and it would require a completely new idea to improve it further. Selberg's lower-bound method, on the other hand, although again more powerful than that of Brun, still leaves much to be desired. It is by no means certain that

(2) V. Brun, 'Le crible d'Eratosthène et le théorème de Goldbach', *Skr. Vidensk. Selsk. Christ.* 1, No. 3 (1920).

(3) A. Selberg, 'On an elementary method in the theory of primes', *K. norske Vidensk. Selsk. Forh. Trondhjem*, 19 (1947) 64–7.

the most effective sifting function has yet been found. Furthermore, in nearly all applications, deep analytic methods are necessary to obtain the required lower bound for

$$\sum_{d \in \mathscr{D}^*} \frac{\lambda(d)}{f(d)},$$

whereas in the Brun method this is achieved by purely elementary means.†

We add some further remarks regarding the historical background of the Brun–Selberg methods. The sieve of Eratosthenes, with its large error term, is almost useless; and at one time seemed likely to remain so. It was a great achievement when Brun, in 1920, devised his method and applied it successfully to several difficult and important problems. The method (as improved by Selberg) still represents an indispensable tool in number theory.

Two famous conjectures in number theory are:

(*a*) *there exists an infinity of prime twins* p, $p+2$;

(*b*) *every large‡ even integer* n *is the sum of two primes.*

Although both these problems are still quite intractable, weaker results in this direction can be obtained by sieve methods. We have already seen that the sequence $a_\nu = \nu(\nu-2)$ is relevant to the first problem, and similarly the sequence $a_\nu = \nu(n-\nu)$ is relevant to the second. In each case we sift the sequence

$$a_1, a_2, \ldots, a_n$$

by the set $\mathscr{P}$ of all primes $p \leqslant n^{1/(l+1)}$, where l is a suitably chosen natural number. Provided that, in each case, we can obtain a positive lower estimate for $S(\mathscr{A}, \mathscr{P}; n)$ when n is large, it will follow that:

(a') *there exists an infinity of pairs of numbers* q, $q+2$ *such that each is a product of at most* l *prime factors;*

(b') *every large even integer is the sum of two numbers, each being a product of at most* l *prime factors.*

As one would expect, the technical difficulties involved in obtaining a lower bound for $S(\mathscr{A}, \mathscr{P}; n)$ increase as $\mathscr{P}$ becomes more numerous, and this is the limiting factor in the choice of l.

† This remark is not intended to reflect a general preference for elementary over analytic methods. Our sole intention is to underline the contrast between the beautiful simplicity of Selberg's upper-bound method and the complexities implicit in his lower-bound approach; and to remark that no such difference exists in Brun's method.

‡ (*b*) is, of course, a slightly weaker form of Goldbach's famous conjecture.

Brun obtained the results (a'), (b') with $l = 9$. His method was later refined, notably by Rademacher[(4)], Estermann[(5)] and Buchstab[(6)], and correspondingly better results of type (a'), (b') were obtained.

We have already mentioned the possibility of applying sieve methods to sequences $\mathscr{A}$ of the form $\{h(q_j)\}$ where h is a polynomial and $\{q_j\}$ is the sequence of primes; and we have indicated the difficulties inherent in such a choice of $\mathscr{A}$. It is clear, however, that if these difficulties can be overcome, there is a considerable advantage in defining the sequence $\mathscr{A}$ in terms of the primes; in problems concerning representations involving two numbers, it is possible to arrange for one of these numbers to be automatically a prime, by virtue of the definition of $\mathscr{A}$. The following examples will make this explicit.

Take $\mathscr{A}$ to be the finite sequence

(I) $$q_1+2,\ q_2+2,\ldots,\ q_n+2$$

where n is a large positive integer and $q_1, q_2,\ldots, q_n$ are the first n primes; then if one can again obtain a positive lower estimate for the number of unsifted elements in a sieve which sifts out† all elements of $\mathscr{A}$ containing more than l prime factors, the following important result emerges.

(a'') *There exist infinitely many primes q such that $q+2$ is the product of at most l prime factors.*

Similarly, if G is a large natural number and $\mathscr{A}$ is taken to be

(II) $$2G-q_1,\ 2G-q_2,\ldots,\ 2G-q_n,$$

where the q's are all the primes less than $2G$, one may hope to obtain a result of the following type.

(b'') *Every large even integer is the sum of a prime and a number which is the product of at most l prime factors.*

The difficulties referred to above arise (as has already been indicated)

† The simplest such sieve removes from $\mathscr{A}$ all multiples of primes not exceeding $(q_n+2)^{1/(l+1)}$, but more complicated sieves have been used to obtain the result (a'') with a smaller (constant integer) value of l (cf. § 9).

(4) H. Rademacher, 'Beiträge zur Viggo Brunschen Methode in der Zahlentheorie', *Hamburger Abh.* **3** (1924) 12–30.

(5) T. Estermann, 'Eine neue Darstellung und neue Anwendungen der Viggo Brunschen Methode', *J. reine angew. Math.* **168** (1932) 106–16.

(6) A. A. Buchstab, 'Neue Verbesserungen in der Methode des Eratosthenischen Siebes', *Mat. Sb.* N.S. **4** (1938) 375–87; also

'On the representation of integers as sums of two numbers each the product of a finite number of primes', *Dokl. Akad. Nauk SSSR*, **29** (1940) 544–8. [Result (b') with $l = 4$.]

'On an additive representation of integers', *Mat. Sb.* N.S. **10** (52) 1–2 (1942) 87–91.

in connexion with the relation (1.9). If $\mathscr{A}$ is either of the sequences (I), (II), the left-hand side of (1.9) represents the number of primes lying in a certain (finite) arithmetic progression modulo d. Our knowledge concerning the distribution of primes among congruence classes is far from complete and, in particular, insufficient to enable one to obtain really good estimates for $|R_d|$ in this context. The analytic techniques for investigating the distribution of primes among congruence classes is based on the use of the so-called L-functions; and the limitations of existing results are due to corresponding limitations in our knowledge of the behaviour of L-functions. Results of type (a''), (b''), were first deduced from unproved hypotheses concerning L-functions. In 1947, however, Rényi[7] succeeded in obtaining, *without the use of any unproved hypothesis*, estimates, for sums of type $\sum_d |R_d|$, which were sufficiently effective to enable him to establish (a''), (b'') for a large (absolute constant) l. Curiously enough, it was again a sieve technique, namely an application of a 'large'† sieve, which, combined with the analytic apparatus, yielded this estimate for $\sum_d |R_d|$. It should be stressed that this very indirect application of the 'large' sieve takes place in a highly technical context.

It was also in 1947 that A. Selberg discovered his form of the sieve method. In the area of problems amenable to applications of the Brun or Selberg sieves, all the most precise results now known have been proved by means of a Selberg-type sieve. So far as (a''), (b'') are concerned, these are now known‡ with $l = 4$; whilst results of (essentially) the (a'), (b') type are known§ with one of the pair of numbers being the

† 'Large' sieves will be discussed later in this section and in § 10. We shall not, however, describe the actual sieve used by Rényi, as this involved technical complications of limited independent interest.

‡ (8) M. B. Barban, 'The density of zeros of Dirichlet L-series and the problem of sums of prime and "almost prime" numbers', *Mat. Sb.* N.S. **61** (103) (1963) 418–25. (This paper contains references to previous work.)

(9) B. V. Levin, 'Distribution of "almost prime" numbers in polynomial sequences', *Mat. Sb.* N.S. **61** (103) (1963) 389–407.

(10) Wang Yuan, 'On the representation of a large integer as a sum of a prime and an almost prime', *Acta math. sin.* **10**, No. 2 (1960), pp. 168–81. See also *Sci. sin.* **11**, No. 8 (1962) 1033–54 (in English), especially the supplement.

(Both Levin and Wang Yuan use the fundamental Theorem 1 of Barban.[8])

§ This result was announced by Selberg[11] in 1949. In 1957 A. I. Vinogradov[12] proved (a'), (b') with $l = 3$; at this time no proof of the 2–3 results had been published. Since then, however, several proofs [9,13] of these sharper results have appeared. Although

(7) A. Rényi, 'On the representation of an even number as the sum of a prime and of an almost prime', *Izv. Akad. Nauk SSSR*, Ser. mat. **12** (1948) 57–78; see also *Am. math. Soc. Trans.* ser. 2, **19** (1962) 299–321.

product of at most two primes and the other of at most three primes. (These are, of course, not the only results obtained[9] by the methods used.) Unfortunately, we have found no account (of this work) which is at the same time both self-contained and reasonably easy to check, and we have not studied in detail the proofs of the results we have just stated.

In § 3 we consider the very special case of the sifting problem, in which the sequence $\mathscr{A}$ consists simply of the natural numbers. We determine the order of magnitude of the number $\Phi(x, y)$ of unsifted elements remaining when the natural numbers not exceeding x are sifted by the primes $p < y$. We do this without the use of Brun's method which, whilst applicable, is dispensable in this context. Instead, we use a method which incorporates ideas of Buchstab† and de Bruijn.†

In the sieves so far considered, the sequence $\mathscr{A}$ is sifted, for each $p \in \mathscr{P}$, by the congruence class 0 (mod p) (by removing, from $\mathscr{A}$, all those elements which lie in such a congruence class). For some applications, however, one requires a larger type of sieve in which, for each $p \in \mathscr{P}$, the sequence is sifted by some given set of congruence classes

$$h_{p,1}, h_{p,2}, \ldots, h_{p,k(p)} \pmod{p}. \tag{1.18}$$

The Brun–Selberg methods can be generalized to apply to this case (cf. § 4), but are ineffective unless $k(p)$ is (on average) exceedingly small compared with p. In 1941 Linnik[14] devised a most remarkable method which yields an effective upper bound for the number of unsifted elements in 'large sieves' for which $k(p)$ is, on the average, large.

Suppose that the Z integers n_j satisfy

$$1 \leqslant n_1 < n_2 < \ldots < n_Z \leqslant N. \tag{1.19}$$

In the application to the 'large sieve', the n_j will be the *unsifted* elements (not exceeding N) of $\mathscr{A}$. However, the discussion will include results, concerning arbitrary sets of integers n_j satisfying (1.19), which are of great interest quite apart from this application.

For a given prime p, let $Z(p, h)$ be the number of n_j falling into the

Wang's proof was the first, the reader may find Levin's account, though not very accurate, less difficult to study.

(11) A. Selberg, 'On elementary methods in prime number theory and their limitations', *Den 11-te Skand. Mat.-Kongress Trondhjem* (1949), 13–22.

(12) A. I. Vinogradov, 'The application of $\zeta(s)$ to the sieve of Eratosthenes', *Mat. Sb.* N.S. **41** (83) (1957) 49–80; ibid. 415–6 (correction of the earlier paper); also *Am. math. Soc. Transl.* ser. 2, **13** (1960) 29–60.

(13) Wang Yuan, 'On the representation of a large even number as a sum of two almost primes', *Sci. Rec.* N.S. **1** (1957) 291–5. See also Levin[9] above.

† See references (19) and (20) in § 3.

(14) Ju. V. Linnik, 'The large sieve', *Dokl. Akad. Nauk SSSR*, N.S. **30** (1941) 292–4.

congruence class h (mod p). The variance

$$D(p) = \sum_{h=0}^{p-1} \left(Z(p,h) - \frac{1}{p}Z\right)^2 \tag{1.20}$$

provides a measure of the regularity of distribution of the n_j among the congruence classes (mod p). All known estimates for the number of unsifted elements in a 'large sieve' depend (although not necessarily explicitly) on estimates of type

$$\sum_{p \in \mathscr{P}} pD(p) \leqslant K(N, Z, \mathscr{P}), \tag{1.21}$$

where $K(N, Z, \mathscr{P})$ is completely determined by the 'sieving set' $\mathscr{P}$ together with the values of N and Z. In the particular case when $\mathscr{P}$ consists of the set of all primes p not exceeding some number X, we rewrite (1.21) in the form

$$\sum_{p \leqslant X} pD(p) \leqslant K(N, Z, X). \tag{1.21'}$$

It is clear that if the n_j are the unsifted elements in the large sieve described above, then

$$Z(p, h_{p,i}) = 0 \quad \text{for} \quad p \in \mathscr{P}, \qquad i = 1, 2, \ldots, k(p);$$

so that

$$\sum_{p \in \mathscr{P}} pD(p) \geqslant \sum_{p \in \mathscr{P}} pk(p)\left(\frac{Z}{p}\right)^2.$$

Given suitable conditions, this inequality, in conjunction with (1.21), provides an effective upper bound for the number Z of unsifted elements. ($K(N, Z, \mathscr{P})$ does not depend too heavily on Z; in fact, it will be of the form $Z^{\beta}K'(N, \mathscr{P})$ where $\beta \leqslant 1$). Linnik obtained the estimate

$$Z \leqslant 20\pi N/(\tau^2 r), \quad \text{where} \quad \tau = \min_{p \in \mathscr{P}} p^{-1}k(p).$$

(This and similar estimates do not depend on the sequence $\mathscr{A}$ in the sieve and therefore remain unaffected if all those integers among 1, 2,..., N, which are not already in $\mathscr{A}$, are placed in $\mathscr{A}$ before sifting.)

Linnik's paper(14), which was concerned solely with the 'large sieve' problem, did not include any estimate of type (1.21), although such an estimate was implicit in his method. This aspect of Linnik's work was made explicit by Rényi, who also refined Linnik's method in various respects.

In fact, the inequalities of type (1.21) which can be obtained by Linnik's method contain much more information than is required for the large sieve application. Clearly (1.21) implies that at most K/M of the primes p can be such that $pD(p) > M$. The estimates (1.21) are indeed so powerful that, provided *only* that Z is not too small compared with N (i.e. provided only the sequence $\{n_j\}$ is not too thin) the above observation suffices to tell us that for almost all $p \leqslant X$ the sequence $\{n_j\}$ is well-distributed modulo p in the sense that the variance (1.20) is small compared to Z^2/p. This is a result of surprising generality; for given N and Z, it is quite independent of the arithmetic character of the sequence $\{n_j\}$.

It is possible to generalize Linnik's method in various ways without encountering any new difficulties. The most natural extension is to 'weight' the elements of the sequence for the purpose of computing (1.20); so that, in (1.20), $Z(p,h)$ now measures the sum of the 'weights' falling into the congruence class $h \pmod p$, and Z is the total 'weight' of the sequence. The large sieve then leads (in suitable circumstances) to the conclusion that, apart from a small number of exceptional p, the total 'weight' falling into the congruence class $h \pmod p$ is approximately Z/p, for nearly all the residues h modulo p. Rényi's form[7] of the Linnik large sieve, which Rényi applied[7]† to the proofs of (a''), (b'') (cf. above), consisted of this generalization superimposed on another. It provided a valuable technical device in the treatment of the type of sum that arises in the investigation of L-functions; we have previously explained the relevance of L-functions to the proofs of (a''), (b''). We shall not, however, include a detailed account of any such indirect and technical (albeit important) application, and we now resume our discussion of the basic form of the large sieve.

As was pointed out by Rényi, inequalities of type (1.21′) are probabilistic statements, and suggest that such results are special cases of general theorems in probability theory. In 1949 Rényi[15] discovered a probabilistic method (based on principles altogether different from those of Linnik) which enabled him to prove a general theorem in probability theory and deduce an inequality of type (1.21) as a special application. He developed this method further in a series of papers.[16],[17]

† Linnik's large-sieve idea also plays an analogous role in the proofs of more recent results[8] of type (a''), (b'').

(15) A. Rényi, 'Un nouveau théorème concernant les fonctions indépendantes et ses applications à la théorie des nombres', *J. Math. pures appl.* **28** (1949) 137–49.

(16) A. Rényi, 'Sur un théorème général de probabilité', *Annls Inst. Fourier Univ. Grenoble*, **1** (1950) 43–52; 'On the large sieve of Ju. V. Linnik', *Compositio Math.* **8** (1950) 68–75; 'On a general theorem in probability theory and its application in the theory of numbers' (in Russian), *Zprávy o spolecnem 3 sjezdu matematiku československych a 7 sjezdu matematiku Polskych, Praha* (1950), 167–74; 'On the probabilistic generalization of the large sieve of Linnik', *Magy. tudom. Akad. mat. Kut. Intez. Közl.* **3** (1958) 199–206; 'New version of the probabilistic generalization of the large sieve', *Acta math. hung.* **10** (1959) 217–26.

(17) A. Rényi, 'Probabilistic methods in number theory' (in Chinese), *Prog. mat. Sci.* **4** (1958) 465–510.

The results Rényi obtained in this way, even when expressed in terms of inequalities of type (1.21′), differ from those obtained by applications of Linnik's method. Over the range $p \leqslant N^{1/3}$ (i.e. when $X = N^{1/3}$ above) his inequality is essentially best possible and very much more precise than the corresponding result obtainable by Linnik's method. Over the range $p \leqslant N^{1/2}$, however (and this is the most appropriate range in some applications), Rényi's method fails completely, whereas Linnik's is still powerful. For intermediate ranges, where $X = N^{\alpha}$, with $\frac{1}{3} < \alpha < \frac{1}{2}$, both methods are applicable; Rényi's method is more powerful when α is near to $\frac{1}{3}$ and Linnik's when α is near to $\frac{1}{2}$. A more detailed comparison will constitute part of our discussion of the two methods.

Our account of Rényi's method will not be in the context of probability theory. We are concerned only to prove the most precise number theoretic results (in the form of inequalities of type (1.21′)) obtained by Rényi, and our aim will be to do so by extracting the arithmetical content of his probabilistic method. This will enable us to give quite simple proofs of his arithmetical theorems.

The methods of Linnik and Rényi are the subject-matter of § 10.

We conclude by mentioning that there is one important development of the sieve idea that we do not touch on in this chapter. This arises in Vinogradov's proof[18] of his famous theorem which states that every large odd integer is the sum of three primes (and in later developments of his work). This proof is analytic and involves the estimation of exponential sums of the type

$$\sum_{p \leqslant N} e(\alpha p) = \sum_{N^{1/2} < p \leqslant N} e(\alpha p) + O(N^{1/2}), \tag{1.22}$$

where $e(\theta) = e^{2\pi i\theta}$.

It is required to show that such sums are small when α is not too near a rational number with small denominator. (There is a dominant term only in the case, here excluded, when α is near a rational number with small denominator.)

Vinogradov's method is a highly ingenious combination of ideas some of which bear some resemblance to the mechanism of the Viggo Brun sieve. His method is one of such remarkable power that the resulting estimate for the sum (1.22) (subject to the above-mentioned type of restriction on α) is far more precise than any known estimate for the

(18) I. M. Vinogradov, *The Method of Trigonometrical Sums in the Theory of Numbers* (1947), English translation by K. F. Roth and A. Davenport (Interscience, London, 1954).

error term in the prime number theorem (which estimates the sum (1.22) when $\alpha = 1$).

The relationship between Vinogradov's proof and sieve methods, will already be evident from the fact that his starting-point is the identity

$$\Phi(1)+\sum_{N^{1/2}<p\leqslant N}\Phi(p)=\sum_{d|\Pi}\mu(d)\sum_{\substack{n\leqslant N\\ d|n}}\Phi(n), \tag{1.23}$$

applied with $\Phi(n) = e(\alpha n)$. It must be stressed, however, that this is only the simplest of the sieve ideas featuring in his proof.

2. Notation and preliminaries

The notation introduced in this section will be in use throughout §§ 3–9. Some of this notation is rather elaborate, and could be dispensed with in any single application. There is, however, a marked relationship between the structures of the various sieve methods described, and it requires some form of notational unity to reveal this. For example, various decompositions of type (2.4) will be used, and the notation below will make the connexion explicit. Similar considerations motivate the use of $s^{(+)}$, $s^{(-)}$, and $s^{(0)}$.

As in the introduction, $\mathscr{P}$ will denote the set of r primes

$$p_1 < p_2 < \ldots < p_r, \tag{2.1}$$

and we again write

$$\Pi(\mathscr{P}) = \prod_{p\in\mathscr{P}} p. \tag{2.2}$$

$\mathscr{D}$ will denote a subset of the positive divisors d of $\Pi(\mathscr{P})$. We say that $\mathscr{D}$ is *divisor-closed* if all the positive divisors of each element d of $\mathscr{D}$ also lie in $\mathscr{D}$. We shall restrict our attention to divisor-closed sets $\mathscr{D}$.

Let $\mathscr{D}^{(p)}$ denote the set of those elements d of $\mathscr{D}$ which have p for their *greatest* prime factor. Then $\mathscr{D}$ consists of 1 and the mutually exclusive classes $\mathscr{D}^{(p_i)}$ $(i = 1, 2, \ldots, r)$; we express this by the relation

$$\mathscr{D} = \dot{\bigcup_{i=1}^{r}}{}^{*}\, \mathscr{D}^{(p_i)}, \tag{2.3}$$

where the asterisk signifies that 1 *is to be included in the union.* Writing $\mathscr{D}_{(p)}$ for the set of d having p as their *least* prime factor, we obtain the analogous decomposition

$$\mathscr{D} = \dot{\bigcup_{i=1}^{r}}{}^{*}\, \mathscr{D}_{(p_i)}. \tag{2.4}$$

The decompositions (2.3), (2.4) of $\mathscr{D}$ into mutually exclusive classes (one of which consists of the single element 1) will play an important part in this chapter.

Let y be a real number. We write

$$\Delta^{(y)} = \bigcup_{p_i \leqslant y}^{*} \mathscr{D}^{(p_i)}, \qquad \Delta_{(y)} = \bigcup_{y \leqslant p_i}^{*} \mathscr{D}_{(p_i)}; \tag{2.5}$$

so that $\Delta^{(y)} = \Delta^{(y)}(\mathscr{D})$ consists of those elements d of $\mathscr{D}$ which contain no prime factor exceeding y, and $\Delta_{(y)} = \Delta_{(y)}(\mathscr{D})$ consists of those d containing no prime factor less than y. We note that $\Delta^{(y)}$ and $\Delta_{(y)}$ are divisor-closed with $\mathscr{D}$ (although $\mathscr{D}^{(p)}$ and $\mathscr{D}_{(p)}$ are not).

We shall make use of the Moebius inversion formulae below, which are an immediate consequence of the characteristic property

$$\sum_{d|m} \mu(d) = \begin{cases} 1 & \text{if } m = 1 \\ 0 & \text{if } m > 1, \end{cases} \tag{2.6}$$

of the Moebius function.

If the function $g(d)$ is defined for natural numbers d, and $f(d)$ is defined in terms of $g(d)$ by

$$f(d) = \sum_{\delta|d} g(\delta), \tag{2.7}$$

then

$$g(d) = \sum_{\delta|d} \mu(d/\delta) f(\delta) = \sum_{\delta|d} \mu(\delta) f(d/\delta). \tag{2.8}$$

Furthermore, if for all elements d of a (finite) divisor-closed set $\mathscr{D}$ of natural numbers, the function $F(d)$ is defined in terms of the function $G(d)$ by

$$F(d) = \sum_{\delta \in \mathscr{D}, d|\delta} G(\delta), \tag{2.9}$$

then

$$G(d) = \sum_{\delta \in \mathscr{D}, d|\delta} \mu(\delta/d) F(\delta) = \sum_{\substack{t \\ td \in \mathscr{D}}} \mu(t) F(td). \tag{2.10}$$

In view of (2.6), these formulae are easily verified by substituting the right-hand sides of (2.7) and (2.9) (with the appropriate change of variable) for f and F in the right-hand sides of (2.8), (2.10) respectively. For example,

$$\sum_{\substack{t \\ td \in \mathscr{D}}} \mu(t) \sum_{\delta \in \mathscr{D}, td|\delta} G(\delta) = \sum_{\delta \in \mathscr{D}, d|\delta} G(\delta) \sum_{t|(\delta/d)} \mu(t) = G(d).$$

We shall also have occasion to refer to the elementary identity

$$\sum_{i=1}^{r} x_i \prod_{j=1}^{i-1} (1-x_j) + \prod_{j=1}^{r} (1-x_j) = 1, \tag{2.11}$$

which is very easily proved by induction on r.

Throughout the chapter, the symbol c (with or without suffixes) is reserved for positive absolute constants.

3. The number of natural numbers not exceeding x not divisible by any prime less than y

Let x, y be real numbers satisfying $x > y > 2$. In this section we discuss the sieve problem in the special case when $\mathscr{A}$ is simply the sequence of natural numbers, and $\mathscr{P}$ is the set $\mathscr{P}_y$ of all primes less than y. Let $\Phi(x, y)$ be the number referred to in the heading of this section, so that, in the notation of § 1,

$$(3.1) \qquad \Phi(x, y) = S(\mathscr{A}, \mathscr{P}_y; [x]);$$

where $\mathscr{A}$ and $\mathscr{P}_y$ are defined as above.

If we apply the sieve of Eratosthenes (cf. formula (1.12) of § 1), we arrive at

$$(3.2) \qquad \Phi(x, y) = x \prod_{p < y} \left(1 - \frac{1}{p}\right) + O(2^{\pi(y)}).$$

This formula leads to an effective estimate of $\Phi(x, y)$ only if y is very small compared to x; certainly only if $y/\log y \ll \log x$.

On the other hand, if $x^{\frac{1}{2}} \leqslant y \leqslant x$, $\Phi(x, y)$ is essentially the number of primes between x and y, and can be estimated by the use of Tchebycheff's theorem (Lemma 1, Appendix) or the prime number theorem (Lemma 5, Appendix) according to the degree of accuracy required. This leaves the problem of estimating $\Phi(x, y)$ when $y < x^{\frac{1}{2}}$ but (3.2) is ineffective.

Actually, analytic methods lead to asymptotic formulae for $\Phi(x, y)$. Before stating these, it is convenient to make a change of variable by writing

$$(3.3) \qquad x = y^u.$$

Buchstab[19] proved that, for fixed $u \geqslant 2$,

$$(3.4) \qquad \lim_{y \to \infty} \Phi(y^u, y) y^{-u} \log y = \omega(u),$$

where $\omega(u)$ satisfies the relation†

$$\frac{d}{du}\{u\omega(u)\} = \omega(u-1) \quad (u \geqslant 2).$$

The relation (3.4) has the (serious) disadvantage of not being uniform in u, but De Bruijn[20] later refined Buchstab's method to obtain an

---

† At $u = 2$, the right-hand derivative is taken.

(19) A. A. Buchstab, 'Asymptotische Abschätzungen einer allgemeinen zahlentheoretischen Funktion', *Mat. Sb.* n.s. **44** (1937) 1239–46.
(20) N. G. De Bruijn, 'On the number of uncancelled elements in the sieve of Eratosthenes', *Indag. math.* **12** (1950) 247–56.

analogous relation which holds uniformly in u, for $u \geqslant 1$. He also showed that

$$\lim_{u \to \infty} \omega(u) = e^{-\gamma}.$$

We shall not prove any results of this type here, and we refer the reader to Hua's survey in the *Enzyklopädie der Mathematischen Wissenschaften*[21] for references to this and related topics.

For many applications in number-theory, and in particular, for all applications elsewhere in this book, it is quite sufficient to know the order of magnitude of $\Phi(x, y)$. We shall therefore content ourselves with proving Theorem 1 below, which will amply suffice for our purposes. The proof is based on Tchebycheff's estimates for $\pi(x)$. (See Appendix, § 2.) If the prime number theorem were assumed, the assertion of Theorem 1 could be correspondingly strengthened (without any significant changes in the proof). The reader will note that the assumption $y > c_1$, which we shall find convenient in the proof, is quite irrelevant to the result. When $y \leqslant c_1$, (3.5) is trivially true for suitable c_2, c_3. We also remark that the product appearing on each side of (3.5) below may be replaced by $(\log y)^{-1}$ (see Appendix, Lemma 4).

THEOREM 1. *There exist positive absolute constants c_1, c_2, c_3, such that*

$$c_2 x \prod_{p<y} \left(1 - \frac{1}{p}\right) \leqslant \Phi(x, y) \leqslant c_3 x \prod_{p<y} \left(1 - \frac{1}{p}\right) \tag{3.5}$$

for all x, y satisfying $x \geqslant c_1 y > c_1^2$.

Proof. We begin by remarking that it suffices to prove† the theorem for the number $\Phi'(x, y)$ of square-free integers $\leqslant x$ divisible by no prime less than y; for

$$\Phi'(x, y) \leqslant \Phi(x, y) \leqslant \Phi'(x, y) + \sum_{p \geqslant y} \frac{x}{p^2} \leqslant \Phi'(x, y) + 2xy^{-1}.$$

We use the notation of § 2, taking $\mathscr{P}$ to be the set of primes not exceeding x and $\mathscr{D}$ to be the set of square-free numbers not exceeding x. (Note that $\mathscr{P}$ *is no longer the set occurring in* (3.1). This change of notation should cause no confusion, as we shall not refer to (3.1).) With this notation, $\Phi'(x, y)$ is simply the number of elements in $\Delta_{(y)}$.

† The argument below could be applied directly to Φ, but the change to Φ' enables us to retain the notation introduced in § 2.

(21) L. K. Hua, 'Die Abschätzung von Exponentialsummen und ihre Anwendung in der Zahlentheorie', *Enzyklopädie der Math. Wissenschaften* (Teubner, Leipzig, 1959), I2, book 13, part 1, Art. 29, § 24.

We note that $\Delta_{(y)}$ certainly includes 1 and all primes p in the interval $y \leqslant p \leqslant x$, so that

$$\Phi'(x,y) \geqslant \pi(x)-\pi(y). \tag{3.6}$$

On the other hand, if $x < y^m$, each element of $\Delta_{(y)}$ contains at most $m-1$ prime factors, so that, in particular,

$$\Phi'(x,y)-1 \leqslant \begin{cases} \pi(x) & \text{if } x < y^2 \\ \pi(x)+\{\pi(xy^{-1})\}^2 & \text{if } y^2 \leqslant x < y^3. \end{cases} \tag{3.7}$$

Now, by Tchebycheff's theorem (Lemma 1, Appendix) and Lemma 4, Appendix,

$$\pi(x) \ll x \prod_{p \leqslant x} \left(1-\frac{1}{p}\right) \tag{3.8}$$

and for sufficiently large c_1,

$$\pi(x)-\pi(x/c_1) \gg x \prod_{p \leqslant x} \left(1-\frac{1}{p}\right) \tag{3.9}$$

for $x \geqslant 2$. We choose such a c_1, which will clearly satisfy

$$c_1 > 1. \tag{3.10}$$

With c_1 thus fixed, we now choose c_2 and c_3 so that (3.5), with Φ replaced by Φ', is satisfied whenever

$$c_1 y^2 \geqslant x \geqslant c_1 y > c_1^2; \tag{3.11}$$

this is possible in view of (3.6), (3.7), (3.8), and (3.9).

The fact that (3.5) holds subject to (3.11) will provide a starting-point for an inductive argument.

It is now convenient to write

$$\Phi'(x,y) = \left\{x \prod_{p<y} \left(1-\frac{1}{p}\right)\right\} \psi(x,y) \tag{3.12}$$

and

$$x = c_1 y^v. \tag{3.13}$$

We shall show that, for every $v \geqslant 1$,

$$c_2 \leqslant \psi(c_1 y^v, y) \leqslant c_3 \tag{3.14}$$

for all $y > c_1$. The proof will be by induction. We consider the numbers α with the property that for every v satisfying $1 \leqslant v \leqslant \alpha$, (3.14) holds for all $y > c_1$. It will suffice to prove that if α_0 is any such α, then (3.14) remains valid, for all $y > c_1$, in the range $\alpha_0 < v \leqslant \alpha_0+1$. Furthermore, since (3.14) holds subject to (3.11), we may assume that

$$\alpha_0 \geqslant 2. \tag{3.15}$$

We require a formula of Buchstab (and some immediate consequences). For every real $h > 1$, we have

$$\Delta_{(y)} = \Big(\bigcup_{y \leqslant p < y^h} \mathscr{D}_{(p)} \Big) \cup \Delta_{(y^h)},$$

and on noting that $\mathscr{D}_{(p)}$ contains $\Phi'(xp^{-1}, p)$ elements, we arrive at Buchstab's formula

$$\Phi'(x,y) = \sum_{y \leqslant p < y^h} \Phi'\Big(\frac{x}{p}, p\Big) + \Phi'(x, y^h).$$

On making the substitutions (3.12) and (3.13), this takes the form

$$(3.16) \qquad \psi(c_1 y^v, y) = \sum_{y \leqslant p < y^h} A_p(y) \psi\Big(c_1 \frac{y^v}{p}, p\Big) + B(y,h) \psi(c_1 y^v, y^h),$$

where $A_p(y)$ and $B(y,h)$ are non-negative and independent of v. We shall not need to make use of the explicit formulae for A and B, which are

$$A_p(y) = \frac{1}{p} \prod_{y \leqslant p' < p} \Big(1 - \frac{1}{p'}\Big), \qquad B(y,h) = \prod_{y \leqslant p' < y^h} \Big(1 - \frac{1}{p'}\Big).$$

The asymptotic density of the sequence of all natural numbers not divisible by any prime less than y, is given by $\prod_{p<y} (1 - 1/p)$ so that, by (3.12), $\psi(x,y) \to 1$ as $x \to \infty$ (for every fixed y). Hence, on letting $v \to \infty$ in (3.16), we obtain the identity†

$$(3.17) \qquad \sum_{y \leqslant p < y^h} A_p(y) + B(y,h) = 1.$$

We are now ready to make the inductive step. We assume that $\alpha_0 < v \leqslant \alpha_0 + 1$, $y > c_1$, and must show that (3.14) is satisfied. In view of (3.17), it will suffice to show that for a suitably chosen h, the hypothesis of induction ensures that each term ψ on the right-hand side of (3.16) satisfies (3.14).

We write

$$(3.18) \qquad y^v p^{-1} = p^{v_p}.$$

For the above-mentioned applications of the hypothesis of induction to the terms ψ on the right-hand side of (3.16), we require

$$(3.19) \qquad 1 \leqslant v_p \leqslant \alpha_0 \quad \text{and} \quad 1 \leqslant v/h \leqslant \alpha_0;$$

we note that, since $v > \alpha_0$, the inequality $h > 1$ (and hence $y^h > c_1$) is implicit in (3.19). It therefore suffices to show that h can be chosen so that (3.19) holds, subject to

$$(3.20) \qquad y \leqslant p < y^h.$$

† (3.17) is in fact merely a special case of (2.11).

But for each p satisfying (3.20), we have

$$p^{v_p} = y^v p^{-1} \geqslant p^{(v/h)-1} \quad \text{and} \quad p^{v_p} \leqslant p^{v-1} \leqslant p^{\alpha_0},$$

so that

$$\frac{v}{h} - 1 \leqslant v_p \leqslant \alpha_0.$$

It is now clear that the first pair of inequalities of (3.19) is satisfied provided $v/h \geqslant 2$. Hence, by (3.15), both pairs of inequalities of (3.19) are satisfied when $h = v/\alpha_0$. With this value of h, the inductive process described above is valid, and the proof of the theorem is now complete.

4. The generalized sieve problem

We take $\mathscr{A}$ to be the finite sequence

$$\mathscr{A} = \{a_1\, a_2, \ldots, a_n\}, \tag{4.1}$$

where the integers a_ν are not necessarily distinct.

Suppose now that with each prime p_i of $\mathscr{P}$ there is associated a set of k_i incongruent residue classes $\mathscr{R}_{i,1}, \mathscr{R}_{i,2}, \ldots, \mathscr{R}_{i,k_i}$ modulo p_i. We define $\bar{S} = \bar{S}(\mathscr{A}, \mathscr{R})$ to be the number of elements of $\mathscr{A}$ which lie in *none* of the classes $\mathscr{R}_{i,j}$; in other words, the number of unsifted elements of $\mathscr{A}$ with respect to the sifting classes $\mathscr{R}_{i,j}$. We note that, when all the sifting classes are of the form 0 modulo p_i (so that, in particular, $k_i = 1$ for each i), $\bar{S}$ is simply the number $S(\mathscr{A}, \mathscr{P}; n)$ defined in § 1. In the following sections, we consider the problem of estimating $\bar{S}$ from above and below.

Let a be an integral variable, and suppose that a lies in exactly $t = t(a)$ of the sifting classes. The appropriate sifting function is now given by

$$s^{(0)}(a) = \begin{cases} 1 & \text{if } t = 0 \\ 0 & \text{if } t > 0, \end{cases}$$

and has the sifting property

$$\bar{S} = \sum_{\nu=1}^{n} s^{(0)}(a_\nu).$$

We now construct a formula analogous to (1.11). We write

$$\sigma(a) = \begin{cases} 1 & \text{if } t = 0 \\ p_{i_1} p_{i_2} \cdots p_{i_t} & \text{if } t > 0, \end{cases} \tag{4.2}$$

where $p_{i_1}, p_{i_2}, \ldots, p_{i_t}$ are the moduli of the sifting classes containing a. We note that $\sigma(a) \mid \Pi(\mathscr{P})$. In place of (1.11), we now have

$$s^{(0)}(a) = \sum_{d \mid \sigma(a)} \mu(d). \tag{4.3}$$

This is in fact a generalization of (1.11), for $\sigma(a) = (a, \Pi(\mathscr{P}))$ when all the sifting classes are of the form 0 (mod p_i).

We must, of course, make an assumption analogous to (1.9). As before, let $f(d)$ be a multiplicative function of d satisfying†

(4.4) $$1 < f(p) \leqslant p \quad \text{for } p \in \mathscr{P}.$$

In place of (1.9) we suppose that for each divisor d of $\Pi(\mathscr{P})$,

(4.5) $$\sum_{\substack{\nu \\ d|\sigma(a_\nu)}} 1 = \frac{n}{f(d)} + R_d.$$

Once again, (4.5) is merely a generalization‡ of (1.9). If d is the product of t primes, the left-hand side of (4.5) represents the number of a_ν lying in any intersection of exactly t sifting classes with the prime divisors of d as moduli.

Clearly, we cannot hope to obtain non-trivial estimates for $\bar{S}$ unless $f(d)$ and R_d satisfy some further conditions. Whilst we do not state such conditions explicitly at this stage, we remark that R_d plays the role of an error term and must be sufficiently small, at least when d lies in some suitable subset $\mathscr{D}^*$ of the divisors of $\Pi(\mathscr{P})$. In many important applications, $a_\nu = h(\nu)$ for some integral-valued polynomial h, and in this case there is a relation of type (4.5) with $f(d)$ and R_d satisfying suitable conditions.

We denote by $\mathscr{C}$ the class of all functions $s(a)$ representable in the form

(4.6) $$s(a) = \sum_{d|\sigma(a)} \lambda(d),$$

where $\lambda(d)$ is a real-valued function. The class $\mathscr{C}$ is closed with respect to multiplication, for if

(4.7) $$s_1(a) = \sum_{d|\sigma(a)} \lambda_1(d), \qquad s_2(a) = \sum_{d|\sigma(a)} \lambda_2(d),$$

then the function $s = s_1 s_2$ is representable in the form (4.6) with

(4.8) $$\lambda(d) = \sum_{\substack{\delta_1,\delta_2 \\ [\delta_1,\delta_2]=d}} \lambda_1(\delta_1)\lambda_2(\delta_2).$$

Let $\mathscr{C}^{(+)}$ be the subclass of $\mathscr{C}$ formed by those functions $s^{(+)}$ of $\mathscr{C}$ which satisfy, for all natural numbers a,

(4.9) $$s^{(+)}(a) \geqslant s^{(0)}(a), \quad \text{with equality if } \sigma(a) = 1;$$

and let $\mathscr{C}^{(-)}$ consist of those functions $s^{(-)}$ of $\mathscr{C}$ which satisfy

(4.10) $$s^{(-)}(a) \leqslant s^{(0)}(a), \quad \text{with equality if } \sigma(a) = 1.$$

It is important to note that $\mathscr{C}^{(+)}$ is closed with respect to multiplication, whereas $\mathscr{C}^{(-)}$ is not.

† Cf. (1.8). ‡ For the d under consideration, which are square-free.

Now suppose that $s_1 = s^{(+)}$ is any fixed element of $\mathscr{C}^{(+)}$ and $s_2 = s^{(-)}$ is any fixed element of $\mathscr{C}^{(-)}$. Then, by (4.9) and (4.10), we have

$$\sum_{\nu=1}^{n} s_2(a_\nu) \leqslant \bar{S} \leqslant \sum_{\nu=1}^{n} s_1(a_\nu). \tag{4.11}$$

Suppose that in the representation (4.7) of $s_1 = s^{(+)}$, we have $\lambda_1(d) = 0$ outside a subset $\mathscr{D}_1^*$ of the divisors d of $\Pi(\mathscr{P})$. Then, by (4.5), we have

$$\bar{S} \leqslant n \sum_{d \in \mathscr{D}_1^*} \frac{\lambda_1(d)}{f(d)} + E_1, \tag{4.12}$$

where

$$E_1 = \sum_{d \in \mathscr{D}_1^*} |\lambda_1(d) R_d|. \tag{4.13}$$

Similarly, if $\lambda_2(d) = 0$ outside the set $\mathscr{D}_2^*$, we have

$$\bar{S} \geqslant n \sum_{d \in \mathscr{D}_2^*} \frac{\lambda_2(d)}{f(d)} - E_2, \tag{4.14}$$

where

$$E_2 = \sum_{d \in \mathscr{D}_2^*} |\lambda_2(d) R_d|. \tag{4.15}$$

The inequalities (4.12), (4.14) represent the starting-point in the sieve methods of Brun and A. Selberg.

5. The Viggo Brun method

As we have already mentioned in § 1, Brun's method has been superseded by that of A. Selberg, which leads to the more precise results. Nevertheless, we give a very brief indication of the ideas underlying Brun's method. The discussion will be very sketchy and, in some respects, presents an over-simplified picture of the Brun method. We include this section solely to compare the Selberg method, which we treat in some detail, with that of Brun. We first discuss Brun's method for obtaining an upper estimate for $\bar{S}$; the method for obtaining a lower estimate being closely analogous. Brun worked exclusively with the function $\lambda(d) = \mu(d)$ (§ 4 already incorporates some of Selberg's ideas).

Accordingly, Brun's method is based on the construction of a subset $\mathscr{D}^*$ (which will be divisor-closed) of the divisors d of $\Pi(\mathscr{P})$, satisfying the following conditions.

(i) $s(a) = \sum_{d \in \mathscr{D}^*, d | \sigma(a)} \mu(d)$ *is an* $s^{(+)}$ (*i.e. lies in* $\mathscr{C}^{(+)}$).

(ii) $\sum_{d \in \mathscr{D}^*} \frac{\mu(d)}{f(d)}$ *is not too large.*

(iii) $\sum_{d \in \mathscr{D}^*} 1$ *is not too large.*

Of these, the need for (i) is obvious, (iii) is intended to ensure that the expression E_1 on the right-hand side of (4.12), with $\lambda_1(d) = \mu(d)$, is small compared with the leading term, and (ii) will then imply that (4.12) leads to a good upper estimate for $\bar{S}$. (In both the methods of Brun and Selberg, the error E_1 is estimated directly without reference to the sum occurring in (iii). Nevertheless, this sum may be used for descriptive purposes, since R_d will always be subject to some condition restricting its magnitude, so that the above-mentioned sum provides a sufficiently accurate measure of the error E_1 for a heuristic discussion.)

The following lemma provides a method of constructing sets $\mathscr{D}$ with the property (i). We use d' to denote an element of $\mathscr{D}$ with an odd number of prime factors, and d'' to denote an element with an even number of prime factors. We also make use of the notation of § 2.

LEMMA. *If $\mathscr{D}$ is divisor-closed and has the property*

(iv) *for each $i = 1, 2, \ldots, r-1$, $d' \in \Delta_{(p_{i+1})}$ implies $p_i d' \in \mathscr{D}_{(p_i)}$,*

then $\mathscr{D}$ also has the property (i).

Proof. Let a be an element of $\mathscr{A}$ satisfying $\sigma(a) > 1$. Let p_k be the least prime factor of $\sigma(a)$, so that $\sigma(a) \in \mathscr{D}_{(p_k)}$. Then, interpreting $\Delta_{(p_{r+1})}$ (which arises as $\Delta_{(p_{k+1})}$ if $k = r$) as consisting of the single element 1 we have $s(a) = \sum_1 \mu(d) + \sum_2 \mu(d)$, where $\sum_1$ and $\sum_2$ are subject to the conditions of summation

$$d \in \mathscr{D}_{(p_k)},\ d \mid \sigma(a) \quad \text{and} \quad d \in \Delta_{(p_{k+1})},\ d \mid \sigma(a)$$

respectively.

Now if $\mu(d')$ is a term in $\sum_1$, then $\mu(d'/p_k)$ is a term in $\sum_2$; and if $\mu(d')$ is a term in $\sum_2$, then $\mu(p_k d')$ is a term in $\sum_1$. Both d'/p_k and $p_k d'$ are of type d'' (we count 1 as being of type d''), and hence the sum $s(a)$ contains at least as many positive as negative terms. This proves the lemma.

Now let $\mathscr{Q}$ be a subset of $\mathscr{P}$, consisting of the t primes

$$p_1 = q_t < q_{t-1} < \ldots < q_1. \tag{5.1}$$

With a suitably chosen set $\mathscr{Q}$, Viggo Brun defines the set $\mathscr{D}^*$ to consist of all divisors d of $\Pi(\mathscr{P})$ satisfying the condition

(v) *for each $i = 1, 2, \ldots, t$, at most $2i$ of the prime factors of d lie in $\Delta_{(q_i)}$.*

This set $\mathscr{D}^*$ is clearly divisor-closed. Furthermore, if an element of type d' of $\mathscr{D}^*$ satisfies (v), and p_k is less than every prime factor of d', then $p_k d'$ also satisfies (v) (since $2i$ is even). Thus $\mathscr{D}^*$ satisfies (iv) and hence, by the lemma, (i).

Brun showed that, under suitable conditions, the set $\mathscr{Q}$ can be so chosen that $\mathscr{D}^*$ also satisfies (ii) and (iii). Actually (iii) is easily satisfied, this being merely a matter of choosing $\mathscr{Q}$ so that the number t of its elements is not too large. On the other hand, as might be expected in view of the complex structure of $\mathscr{D}^*$, the estimation of the sum occurring in (ii) is no simple matter. Nevertheless Brun obtained (under suitable conditions and with $\mathscr{Q}$ suitably chosen) a highly effective upper estimate by elementary (although very complicated) means.

The method for obtaining a lower estimate for $\bar{S}$ is the same in principle. Here the set $\mathscr{D}^*$ must satisfy the following analogous conditions.

(I) $s(a) = \sum_{d \in \mathscr{D}^*, d | \sigma(a)} \mu(d)$ *is an* $s^{(-)}$ (*i.e. lies in* $\mathscr{C}^{(-)}$).

(II) $\sum_{d \in \mathscr{D}^*} \frac{\mu(d)}{f(d)}$ *is not too small.*

(III) $\sum_{d \in \mathscr{D}^*} 1$ *is not too large.*

The lemma still holds good if (i) is replaced by (I) and (iv) is replaced by (IV) below.

(IV) *For each* $i = 1, 2, \ldots, r$, $d'' \in \Delta_{(p_{i+1})}$ *implies* $p_i d'' \in \mathscr{D}_{(p_i)}$; *where* 1 *is counted as a* d'' *and* $\Delta_{(p_r+1)}$ *is interpreted as consisting of the single element* 1.

$\mathscr{Q}$ is again chosen of the type (5.1), but now $\mathscr{D}^*$ is taken to consist of all d satisfying

(V) *for each* $i = 1, 2, \ldots, t$, *at most* $2i-1$ *of the prime factors of* d *lie in* $\Delta_{(q_i)}$.

As before, $\mathscr{D}^*$ is divisor-closed and (in this case since $2i-1$ is *odd*) satisfies (IV) and hence (I).

Once again the main difficulty lies in estimating the sum in (II) (this time from below). Brun's method for dealing with this is closely analogous to his upper-bound method.

We now turn to A. Selberg's method, which we give in detail. Here the methods of obtaining upper and lower estimates for $\bar{S}$ are not quite so closely analogous, and we deal with the two separately.

6. Selberg's upper-bound method: informal discussion

We begin with an informal description of Selberg's upper-bound method. As far as the statements and proofs of Selberg's theorems are concerned, the reader may ignore this section and turn to § 7.

As before, $\mathscr{D}$ denotes a divisor-closed subset of the divisors d of $\Pi(\mathscr{P})$. We denote by $\mathscr{C}(\mathscr{D})$ and $\mathscr{C}^{(+)}(\mathscr{D})$ the subclasses of $\mathscr{C}$ and $\mathscr{C}^{(+)}$ respectively, consisting of those functions s and $s^{(+)}$ representable in the form (cf. (4.6))

$$s(a) = \sum_{d\in\mathscr{D}, d|\sigma(a)} \lambda(d). \tag{6.1}$$

We note that if $\mathscr{D}$ is a subclass of $\mathscr{D}'$, then $\mathscr{C}(\mathscr{D})$ and $\mathscr{C}^{(+)}(\mathscr{D})$ are subclasses of $\mathscr{C}(\mathscr{D}')$ and $\mathscr{C}^{(+)}(\mathscr{D}')$ respectively (since $\lambda(d)$ may be taken to be 0 outside $\mathscr{D}$).

In accordance with the now familiar procedure, we seek a set $\mathscr{D}^*$ and a function $s^{(+)}$ of $\mathscr{C}^{(+)}(\mathscr{D}^*)$ for which

(i) $\sum_{d\in\mathscr{D}^*} 1$ is sufficiently small,

(ii) $\displaystyle\sum_{d\in\mathscr{D}^*} \frac{\lambda(d)}{f(d)}$ is sufficiently small.

In most applications $|\lambda(d)R_d|$ is not much worse than $O(1)$, and, in any case, a suitable condition of type (i) will ensure that the term E_1 in (4.12) (with $\lambda_1(d) = \lambda(d)$) is negligible. (The appropriate $s^{(+)}$ are in some sense approximations to $s^{(0)}$, and the $\lambda(d)$ approximations to $\mu(d)$; at least to the extent that $\lambda(d)$ is not too large.)

Now suppose we take $\mathscr{D}$ of the form

$$\mathscr{D} = \mathscr{D}(z) = \{d : d \mid \Pi(\mathscr{P}), d \leqslant z\} \tag{6.2}$$

where z is chosen just small enough to make the term E_1 in (4.12) (with $\mathscr{D}_1^* = \mathscr{D}$) negligible. The choice $z = n^{1-\epsilon}$, or even $z = n(\log n)^{-c}$, would usually be appropriate for the reasons given above.

$\mathscr{D} = \mathscr{D}(z)$ being thus fixed, we have by (4.12)

$$\bar{S} \leqslant nL_1 + E_1, \tag{6.3}$$

where

$$L_1 = \min_{s^{(+)}\in\mathscr{C}^{(+)}(z)} \sum_{d\in\mathscr{D}(z)} \frac{\lambda(d)}{f(d)}, \qquad \mathscr{C}^{(+)}(z) = \mathscr{C}^{(+)}(\mathscr{D}(z)), \tag{6.4}$$

and, for the minimizing function $\lambda(d)$ in (6.4),

$$E_1 = \sum_{d\in\mathscr{D}(z)} |\lambda(d)R_d|. \tag{6.5}$$

Roughly speaking, (6.3) represents the most precise† inequality we can hope to achieve by means of a sifting function $s^{(+)}$ of $\mathscr{C}^{(+)}$. At the same time, the task of computing L_1 is a hopeless one.

One of the major ideas in Selberg's method is to seek the sifting function $s^{(+)}$ in a suitably chosen subclass of $\mathscr{C}^{(+)}$. Let $\mathscr{C}^{(2)}$ consist of the squares s^2 of all the functions s in $\mathscr{C}$. Since $\mathscr{C}$ is closed with respect

† On the not unreasonable assumption that the set (6.2) is no less favourable than other sets $\mathscr{D}$ with the same number of elements.

to multiplication, $\mathscr{C}^{(2)}$ is a subclass of $\mathscr{C}^{(+)}$. We denote by $\mathscr{C}^{(2)}(\mathscr{D})$ the class of the squares s^2 of functions s in $\mathscr{C}(\mathscr{D})$. Then $\mathscr{C}^{(2)}(z^{\frac{1}{2}}) = \mathscr{C}^{(2)}\{\mathscr{D}(z^{\frac{1}{2}})\}$ is a subclass of $\mathscr{C}^{(+)}(z)$; for if in (4.8) each of δ_1, δ_2 is at most $z^{\frac{1}{2}}$, then $d = [\delta_1, \delta_2]$ is at most z.

Selberg uses, in place of (6.3), the inequality

$$\bar{S} \leqslant nL_2+E_2, \tag{6.6}$$

where ($\lambda(d)$ being 0 when the sum on the right of (4.8) is empty)

$$L_2 = \min_{s^{(+)}\in\mathscr{C}^{(2)}(z^{\frac{1}{2}})} \sum_{d\in\mathscr{D}(z)} \frac{\lambda(d)}{f(d)}, \tag{6.7}$$

and for the minimizing function $\lambda(d)$ in (6.7),

$$E_2 = \sum_{d\in\mathscr{D}(z)} |\lambda(d)R_d|. \tag{6.8}$$

Selberg shows that both L_2 and the corresponding minimizing function $\lambda(d)$ can be determined exactly. Indeed, the corresponding problem can be completely solved for *every* divisor-closed subset $\mathscr{D}$ of the divisors of $\Pi(\mathscr{P})$, irrespective of whether it is, or is not, of the type (6.2). This extremal problem is much simpler in the class $\mathscr{C}^{(2)}(\mathscr{D})$ because the functions (cf. (7.2), (7.3), and (7.4))

$$s^{(+)} = s^2 = \Big(\sum_{d\in\mathscr{D},d|\sigma(a)} \Lambda_d\Big)^2 = \sum_{d\in\mathscr{D}^*,d|\sigma(a)} \lambda(d)$$

satisfy (4.9) automatically provided $\Lambda_1 = 1$; and (for the purpose of computing the minimum) the Λ_d may thus be regarded as independent variables, subject to the single restriction, $\Lambda_1 = 1$. (For a more detailed explanation of this, see § 7.)

At first sight it may appear very wasteful to concentrate on the comparatively small subclass $\mathscr{C}^{(2)}(z^{\frac{1}{2}})$ of $\mathscr{C}^{(+)}(z)$, and one might fear that the resulting estimate for $\bar{S}$ will be a poor one. The following considerations, however, suggest that, on the contrary, the estimate (6.6) can reasonably be expected to be of the same order of magnitude as the estimate (6.3).

In most applications heuristic considerations suggest that the order of magnitude of L_1 is (very roughly speaking, indeed) on the scale of $(\log z)^{-c}$. At any rate, one might reasonably expect L_1 to be of the same order of magnitude (to within a constant factor) as

$$L_3 = \min_{s^{(+)}\in\mathscr{C}^{(+)}(z^{\frac{1}{2}})} \sum_{d\in\mathscr{D}(z^{\frac{1}{2}})} \frac{\lambda(d)}{f(d)}. \tag{6.9}$$

In other words, if s_1 and s_3 are the minimizing functions in (6.4) and (6.9) respectively, we would expect these to be about equally effective for the purpose of estimating $\bar{S}$. Let s_2 be the minimizing function in (6.7). Then, since $(s_3)^2$ is an element of $\mathscr{C}^{(2)}(z^{\frac{1}{2}})$, the function s_2 is at least as effective as $(s_3)^2$. So s_2 will be less effective than s_1 only if $(s_3)^2$ is less effective than s_3. This can only take place if $s_3(a)$ is often (i.e. for many elements a_ν of $\mathscr{A}$) substantially greater than 1.

It is reasonable to hope that this will not happen, however. For if sieve methods are to succeed at all, one would expect a minimizing function $s^{(+)}(a)$ to be small, on average, when $\sigma(a) > 1$.

In fact, in many important applications, the bounds achieved by Selberg's method are nearly (or quite) as good as one can hope to attain by a sieve method.(22)

7. Selberg's upper-bound method

We shall now prove two theorems. Theorem 2 will be required both in the proof of Theorem 3 and (in the following section) in the proof of Theorem 4. Theorem 3 provides an upper bound for the number $\bar{S}$ defined in § 4.

Let $\mathscr{D}$ be a divisor-closed subset of the divisors d of $\Pi(\mathscr{P})$. Suppose that with each element d of $\mathscr{D}$ there is associated a real variable Λ_d. We shall consider all sets of values

$$\Lambda = \{\Lambda_d : d \in \mathscr{D}\}$$

subject to the restriction

$$\Lambda_1 = 1. \tag{7.1}$$

To each Λ there corresponds a function $s^{(+)}$ of $\mathscr{C}^{(+)}$ defined by

$$s^{(+)}(a) = \Big(\sum_{d\in\mathscr{D}, d|\sigma(a)} \Lambda_d\Big)^2. \tag{7.2}$$

This function $s^{(+)}$ is representable in the form (4.6) with (cf.(4.8))

$$\lambda(d) = \sum_{\substack{\delta_1,\delta_2\in\mathscr{D}\\ [\delta_1,\delta_2]=d}} \Lambda_{\delta_1}\Lambda_{\delta_2}, \tag{7.3}$$

where $\lambda(d)$ is taken to be zero outside the set

$$\mathscr{D}^* = \{d : d = [\delta_1, \delta_2],\ \delta_1 \in \mathscr{D},\ \delta_2 \in \mathscr{D}\}. \tag{7.4}$$

Furthermore, $s^{(+)}(a)$ is never negative, and therefore satisfies (4.9) in view of (7.1).

Let $f(d)$ be the multiplicative function appearing in (4.5) and let $g(d)$ be defined by

$$g(d) = \sum_{\delta|d} \mu(\delta) f\Big(\frac{d}{\delta}\Big) = f(d) \prod_{p|d} \Big(1 - \frac{1}{f(p)}\Big), \tag{7.5}$$

(22) A. Selberg, 'The general sieve method and its place in prime number theory', *Proc. Intern. Congr. Math., Cambridge, Mass. 1950*, **1**, 286–92 (Am. math. Soc., 1952).

so that by (4.4),

$$g(d) > 0 \quad \text{for} \quad d \mid \Pi(\mathscr{P}). \tag{7.6}$$

By a Moebius inversion (cf. the closing paragraphs of § 4 of the Appendix) we have

$$f(d) = \sum_{\delta \mid d} g(\delta). \tag{7.7}$$

Furthermore, since f is multiplicative,

$$f((d_1, d_2))f([d_1, d_2]) = f(d_1)f(d_2) \tag{7.8}$$

so that

$$\frac{1}{f([d_1, d_2])} = \frac{1}{f(d_1)f(d_2)} \sum_{\delta \mid (d_1, d_2)} g(\delta). \tag{7.9}$$

THEOREM 2 (A. Selberg). *Let*

$$H(\Lambda) = \sum_{d \in \mathscr{D}^*} \frac{\lambda(d)}{f(d)}, \tag{7.10}$$

where $\lambda(d)$ is defined by (7.3); and let $L = L(\mathscr{D})$ be the lower bound, taken over all Λ subject to (7.1), of $H(\Lambda)$.

Then this lower bound is given by $L = Q^{-1}$, where

$$Q = Q(\mathscr{D}) = \sum_{d \in \mathscr{D}} \frac{1}{g(d)}; \tag{7.11}$$

and is attained when Λ is the set $\Lambda(\mathscr{D})$ given by

$$\Lambda_d = \frac{\mu(d)f(d)}{Q(\mathscr{D})} \sum_{\delta \in \mathscr{D}, d \mid \delta} \frac{1}{g(\delta)} \quad (d \in \mathscr{D}). \tag{7.12}$$

Proof. We have, by (7.3) and (7.9),

$$\begin{aligned} H(\Lambda) &= \sum_{d_1 \in \mathscr{D}} \sum_{d_2 \in \mathscr{D}} \frac{\Lambda_{d_1} \Lambda_{d_2}}{f([d_1, d_2])} \\ &= \sum_{d_1 \in \mathscr{D}} \sum_{d_2 \in \mathscr{D}} \frac{\Lambda_{d_1} \Lambda_{d_2}}{f(d_1)f(d_2)} \sum_{\delta \mid (d_1, d_2)} g(\delta) \\ &= \sum_{\delta \in \mathscr{D}} g(\delta) \Big(\sum_{d \in \mathscr{D}, \delta \mid d} \frac{\Lambda_d}{f(d)} \Big)^2. \end{aligned}$$

We now make a linear transformation of the variables, by writing

$$y_\delta = \sum_{d \in \mathscr{D}, \delta \mid d} \frac{\Lambda_d}{f(d)} \quad (\delta \in \mathscr{D}). \tag{7.13}$$

By the second Moebius inversion formula of § 2 (see (2.9) and (2.10)), this transformation is non-singular, the inverse transformation being given by

(7.14) $$\Lambda_d = f(d) \sum_{\delta\in\mathscr{D}, d|\delta} \mu(\delta/d) y_\delta.$$

Since f is multiplicative, $f(1) = 1$, so that the transform of condition (7.1) is

(7.15) $$\sum_{\delta\in\mathscr{D}} \mu(\delta) y_\delta = 1.$$

Hence, if Q is defined by (7.11), we have the identity

$$H(\Lambda) = \sum_{\delta\in\mathscr{D}} g(\delta) y_\delta^2 = \sum_{\delta\in\mathscr{D}} \frac{1}{g(\delta)} \{g(\delta)y_\delta - \mu(\delta)Q^{-1}\}^2 + Q^{-1}.$$

In view of (7.6), it is now obvious that

$$L = \min_{\Lambda} H(\Lambda) = Q^{-1};$$

for this value is attained when

(7.16) $$y_\delta = \frac{1}{Q} \frac{\mu(\delta)}{g(\delta)} \quad (\delta \in \mathscr{D}),$$

and this set $\{y_\delta\}$ satisfies (7.15). Furthermore, (7.16) is equivalent to (7.12), in view of (7.14).

THEOREM 3 (A. Selberg). *If* $Q = Q(\mathscr{D})$ *is defined by* (7.11) *and* $\Lambda = \Lambda(\mathscr{D})$ *is the minimal set* (7.12), *then*

(7.17) $$\bar{S} \leqslant nQ^{-1} + \sum_{d_1\in\mathscr{D}} \sum_{d_2\in\mathscr{D}} |\Lambda_{d_1}\Lambda_{d_2} R_{[d_1,d_2]}|.$$

In particular, if R_d *and* $f(d)$ *satisfy*†

(7.18) $$|R_d| \leqslant d/f(d) \quad \textit{for} \quad d \mid \Pi(\mathscr{P}),$$

we obtain, on choosing

(7.19) $$\mathscr{D} = \{d : d \mid \Pi(\mathscr{P}),\, d \leqslant z^{\frac{1}{2}}\},$$

the upper estimate

(7.20) $$\bar{S} \leqslant n\{Q(\mathscr{D})\}^{-1} + z \prod_{p\in\mathscr{P}} \left(1 - \frac{1}{f(p)}\right)^{-2}.$$

Proof. In view of Theorem 2, (7.17) is merely a restatement of (4.12), where $\lambda_1(d) = \lambda(d)$ is given by (7.3). It remains to prove

(7.21) $$E = \sum_{d_1\in\mathscr{D}} \sum_{d_2\in\mathscr{D}} |\Lambda_{d_1}\Lambda_{d_2} R_{[d_1,d_2]}| \leqslant z \prod_{p\in\mathscr{P}} \left(1 - \frac{1}{f(p)}\right)^{-2}$$

subject to (7.18) and (7.19); $\Lambda = \Lambda(\mathscr{D})$ being the minimal set (7.12).

† Cf. (1.10).

Now
$$|\Lambda_d| = \frac{f(d)}{Q} \sum_{\delta \in \mathscr{D}, d|\delta} \frac{1}{g(\delta)} \quad (d \in \mathscr{D}),$$
so that by (7.6) (g is multiplicative),
$$|\Lambda_d| \leqslant \frac{f(d)}{g(d)} \frac{1}{Q} \sum_{\delta \in \mathscr{D}} \frac{1}{g(\delta)} = \frac{f(d)}{g(d)}.$$
Furthermore, by (7.8), (7.18), and (4.4),
$$|R_{[d_1,d_2]}| \leqslant \frac{d_1 d_2}{f(d_1)f(d_2)} \frac{f((d_1, d_2))}{(d_1, d_2)} \leqslant \frac{d_1 d_2}{f(d_1)f(d_2)}.$$
Hence

(7.22)
$$E \leqslant \left(\sum_{d \in \mathscr{D}} \frac{f(d)}{g(d)} \frac{d}{f(d)}\right)^2 \leqslant z\left(\sum_{d|\Pi(\mathscr{P})} \frac{1}{g(d)}\right)^2.$$
Finally, since

(7.23)
$$1+\frac{1}{g(p)} = 1+\frac{1}{f(p)-1} = \left(1-\frac{1}{f(p)}\right)^{-1},$$
we have

(7.24)
$$\sum_{d|\Pi(\mathscr{P})} \frac{1}{g(d)} = \prod_{p \in \mathscr{P}} \left(1+\frac{1}{g(p)}\right) = \prod_{p \in \mathscr{P}} \left(1-\frac{1}{f(p)}\right)^{-1}.$$
The theorem is now proved, since (7.22) and (7.24) give (7.21).

In many applications of interest, it is not hard to show that if z is chosen of the form
$$z = c_4 n \prod_{p \in \mathscr{P}} \left(1-\frac{1}{f(p)}\right)^3,$$
then
$$Q = \sum_{d \in \mathscr{D}} \frac{1}{g(d)} \geqslant c_5 \sum_{d|\Pi(\mathscr{P})} \frac{1}{g(d)} = c_5 \prod_{p \in \mathscr{P}} \left(1-\frac{1}{f(p)}\right)^{-1}.$$
Under these circumstances, (7.20) yields

(7.25)
$$\bar{S} \leqslant c_6 n \prod_{p \in \mathscr{P}} \left(1-\frac{1}{f(p)}\right).$$

The reader will find many applications of Theorem 3 in K. Prachar's *Primzahlverteilung*, Chapter II, § 4.

8. Selberg's lower-bound method

To obtain a lower estimate for $\bar{S}$, we require a suitable sifting function $s^{(-)}$ of the class $\mathscr{C}^{(-)}$ (cf. § 4). We describe a method of Selberg, in which a subclass of $\mathscr{C}^{(-)}$ is so constructed as to make Theorem 2 again applicable to the problem of choosing the most favourable $s^{(-)}$ from this subclass. The construction of the subclass is based on the decomposition (2.3) of the divisors of $\Pi(\mathscr{P})$.

We write

(8.1) $$\Pi_i = \prod_{j=1}^{i} p_j \quad (i = 0, 1, \ldots, r),$$

where the empty product is taken to be 1. We note that, in particular, $\Pi_r = \Pi(\mathscr{P})$. In this section, we take $\mathscr{D}$ to be the set of *all* the divisors d of $\Pi(\mathscr{P})$. In accordance with the notation of § 2, $\mathscr{D}^{(p_i)}$ is the set of those d which have p_i as their largest prime factor, and $\Delta^{(p_i)}$ is the set of divisors of Π_i. We recall that $\mathscr{D}$ consists of 1 and the discrete sets $\mathscr{D}^{(p_i)}$, as expressed by (2.3).

Now, since $\sigma(a) \in \mathscr{D}$ for all a, we have

$$s^{(0)}(a) = \sum_{d|\sigma(a)} \mu(d) = 1 + \sum_{i=1}^{r} \sum_{\delta}^{(i)} \mu(p_i \delta),$$

where the inner sum $\sum^{(i)}$ on the right is subject to the conditions of summation

(8.2) $$p_i \delta \in \mathscr{D}^{(p_i)}, \qquad p_i \delta \mid \sigma(a).$$

We rewrite this expression for $s^{(0)}$ in the form

(8.3) $$s^{(0)}(a) = 1 - \sum_{i=1}^{r} s_i^{(0)}(a),$$

where $$s_i^{(0)}(a) = \sum_{\delta}^{(i)} \{-\mu(p_i \delta)\}.$$

We note that

$$s_i^{(0)}(a) = \begin{cases} \sum_{d|(\Pi_{i-1}, \sigma(a))} \mu(d) & \text{if } p_i \mid \sigma(a) \\ 0 & \text{if } p_i \nmid \sigma(a); \end{cases}$$

thus $s_i^{(0)}(a)$ is 1 if p_i is the least prime dividing $\sigma(a)$, and is 0 otherwise.

For each $i = 1, 2, \ldots, r$ we consider the class $\mathscr{C}_i$ of functions s_i representable in the form (with d replacing δ in the conditions (8.2))

(8.4) $$s_i(a) = \sum_{d}^{(i)} \lambda(p_i d);$$

and the subclass $\mathscr{C}_i^{(+)}$ of those functions $s_i^{(+)}$ of $\mathscr{C}_i$ which satisfy

(8.5) $$s_i^{(+)}(a) \geqslant s_i^{(0)}(a)$$

with equality if p_i is the least prime factor of $\sigma(a)$. In view of (8.3), to each set $\{s_i^{(+)}: i = 1, 2, \ldots, r\}$ there corresponds a function $s^{(-)}$ of the class $\mathscr{C}^{(-)}$, defined by

(8.6) $$s^{(-)}(a) = 1 - \sum_{i=1}^{r} s_i^{(+)}(a).$$

We now proceed as in § 7. For each i, let Ω_i be a divisor-closed subset of $\Delta^{(p_{i-1})}$, where $\Delta^{(p_0)}$ is taken to consist of the single element 1. We consider the functions $s_i^{(+)}(a)$ of the special form

$$s_i^{(+)}(a) = \Big(\sum_{d\in\Omega_i, p_i d|\sigma(a)} \Lambda_d^{(i)}\Big)^2, \tag{8.7}$$

where

$$\Lambda_1^{(i)} = 1. \tag{8.8}$$

Such $s_i^{(+)}$ satisfy (8.5) in view of (8.8), and are representable in the form (8.4) with

$$\lambda(p_i d) = \sum_{\substack{\delta_1,\delta_2\in\Omega_i\\ [\delta_1,\delta_2]=d}} \Lambda_{\delta_1}^{(i)}\Lambda_{\delta_2}^{(i)} \tag{8.9}$$

where $\lambda(p_i d)$ is taken to be zero when d is outside the set

$$\Omega_i^* = \{d: d = [\delta_1, \delta_2],\ \delta_1 \in \Omega_i,\ \delta_2 \in \Omega_i\}. \tag{8.10}$$

In place of (4.14), we now have

$$\bar{S} \geqslant n - n\sum_{i=1}^{r}\sum_{d\in\Omega_i^*}\frac{\lambda(p_i d)}{f(p_i d)} - E^*, \tag{8.11}$$

where

$$E^* = E^*(\Omega) = \sum_{i=1}^{r}\sum_{d_1\in\Omega_i}\sum_{d_2\in\Omega_i}|\Lambda_{d_1}^{(i)}\Lambda_{d_2}^{(i)}R_{p_i[d_1,d_2]}|. \tag{8.12}$$

THEOREM 4 (A. Selberg). *If $Q_i = Q(\Omega_i)$ is defined by* (7.11) *(with Ω_i in place of $\mathscr{D}$) and $\Lambda^{(i)} = \Lambda(\Omega_i)$ is the set* (7.12) *(again with Ω_i in place of $\mathscr{D}$), then*

$$\bar{S} \geqslant n\Big(1 - \sum_{i=1}^{r}\frac{1}{f(p_i)Q_i}\Big) - E^*, \tag{8.13}$$

where $E^* = E^*(\Omega)$ is defined by (8.12).

In particular, if R_d and $f(d)$ satisfy (7.18), *we obtain, on choosing*

$$\Omega_i = \{d: d \mid \Pi_{i-1},\ d \leqslant (z/p_i)^{\frac{1}{2}}\}, \tag{8.14}$$

the lower estimate

$$\bar{S} \geqslant n\Big(1 - \sum_{i=1}^{r}\frac{1}{f(p_i)Q(\Omega_i)}\Big) - z\sum_{i=1}^{r}\frac{1}{f(p_i)}\prod_{p|\Pi_{i-1}}\Big(1-\frac{1}{f(p)}\Big)^{-2}. \tag{8.15}$$

Proof. On applying Theorem 2 with Ω_i, $\Lambda^{(i)}$ in place of $\mathscr{D}$, Λ respectively (here (8.9) takes the place of (7.3)), we obtain

$$\sum_{d\in\Omega_i^*}\frac{\lambda(p_i d)}{f(d)} = L(\Omega_i) = \{Q(\Omega_i)\}^{-1}. \tag{8.16}$$

On noting that the d appearing in (8.16) are not divisible by p_i, so that $f(p_i d) = f(p_i)f(d)$, we obtain (8.13) on substituting (8.16) in (8.11).

It remains to prove

$$E^* \leqslant z \sum_{i=1}^{r} \frac{1}{f(p_i)} \prod_{p|\Pi_{i-1}} \left(1-\frac{1}{f(p)}\right)^{-2}, \tag{8.17}$$

subject to (7.18) and (8.14).

As in the proof of Theorem 3,

$$|\Lambda_d^{(i)}| \leqslant f(d)/g(d)$$

and by (7.8), (7.18), and (4.4),

$$|R_{p_i[d_1,d_2]}| \leqslant \frac{p_i}{f(p_i)} \frac{[d_1,d_2]}{f([d_1,d_2])} \leqslant \frac{p_i}{f(p_i)} \frac{d_1 d_2}{f(d_1)f(d_2)}.$$

Hence

$$E^* \leqslant \sum_{i=1}^{r} \frac{p_i}{f(p_i)} E_i^2, \tag{8.18}$$

where (again as in the proof of Theorem 3)

$$E_i = \sum_{d\in\Omega_i} \frac{d}{g(d)} \leqslant \left(\frac{z}{p_i}\right)^{\frac{1}{2}} \sum_{d|\Pi_{i-1}} \frac{1}{g(d)} = \left(\frac{z}{p_i}\right)^{\frac{1}{2}} \prod_{p|\Pi_{i-1}} \left(1-\frac{1}{f(p)}\right)^{-1}. \tag{8.19}$$

The theorem is now proved, since (8.18) and (8.19) give (8.17).

In actual applications it is convenient to have an inequality of the form

$$E^* \leqslant z \prod_{p\in\mathscr{P}} F(p)$$

in place of (8.17). In fact, (8.17) can be replaced by the weaker inequality

$$E^* \leqslant z \prod_{p\in\mathscr{P}} \left(1-\frac{1}{f(p)}\right)^{-3}; \tag{8.20}$$

for the sum on the right of (8.17) is less than

$$\left(\sum_{p\in\mathscr{P}} \frac{1}{f(p)}\right) \prod_{p\in\mathscr{P}} \left(1-\frac{1}{f(p)}\right)^{-2} < \prod_{p\in\mathscr{P}} \left\{\left(1-\frac{1}{f(p)}\right)^{-2}\left(1+\frac{1}{f(p)}\right)\right\}.$$

In many applications, the error term (8.20) is easily estimated, and is sufficiently small once z is suitably chosen. It is clear, however, that it is also necessary to estimate the first term on the right-hand side of (8.15) from below; and this is seldom a simple task. We shall say more about this in the following section.

9. Selberg's lower-bound method: further discussion

We shall consider the expression

$$J = 1 - \sum_{i=1}^{r} \frac{1}{f(p_i)Q(\Omega_i)} = 1 - \sum_{i=1}^{r} \frac{1}{f(p_i)Q_i} \tag{9.1}$$

occurring in the leading term on the right-hand side of (8.15).

Provided the order of magnitude of p_r is substantially less than that of z (and the method fails if this is not the case), the numbers

$$Q_i = \sum_{d \in \Omega_i} \frac{1}{g(d)} = \sum_{\substack{d|\Pi_{i-1} \\ d \leqslant (z/p_i)^{\frac{1}{2}}}} \frac{1}{g(d)} \tag{9.2}$$

may be regarded as approximations to the numbers†

$$Q'_i = \sum_{d|\Pi_{i-1}} \frac{1}{g(d)} = \prod_{p|\Pi_{i-1}} \left(1+\frac{1}{g(p)}\right). \tag{9.3}$$

We note that, in view of (7.23),

$$Q'_i = P_i^{-1} \quad \text{where} \quad P_i = \prod_{p|\Pi_{i-1}} \left(1-\frac{1}{f(p)}\right). \tag{9.4}$$

We introduce the numbers

$$\beta_i = P_i Q_i, \qquad \epsilon_i = 1-\beta_i, \tag{9.5}$$

so that, by (9.2) and (7.6), we have $\beta_i \leqslant 1$ (with equality when $i = 1$). These numbers may be regarded as a measure of the accuracy of the approximations Q'_i to Q_i; the smaller the ϵ_i, the better the approximation. We show that the numbers β_i have the convenient property

$$\beta_1 \geqslant \beta_2 \geqslant \ldots \geqslant \beta_r. \tag{9.6}$$

In view of (7.6), we have for $i = 1, 2,\ldots, r-1$,

$$Q_{i+1} \leqslant \sum_{\substack{d|\Pi_i \\ d \leqslant (z/p_i)^{\frac{1}{2}}}} \frac{1}{g(d)} = Q_i + \sum_{\substack{p_i|d|\Pi_i \\ d \leqslant (z/p_i)^{\frac{1}{2}}}} \frac{1}{g(d)} \leqslant \left(1+\frac{1}{g(p_i)}\right)Q_i;$$

so that (9.6) follows from the product form (9.3) since $P_i^{-1} = Q'_i$.

We turn to the expression (9.1). It is natural to consider the error when J is replaced by the approximation

$$J' = 1-\sum_{i=1}^{r} \frac{1}{f(p_i)Q'_i} = 1-\sum_{i=1}^{r} \frac{P_i}{f(p_i)}. \tag{9.7}$$

By (2.11) we have $J' = P_r$, so that

$$\begin{aligned} J &= J'-(J'-J) \\ &= P_r - \sum_{i=1}^{r} \frac{1}{f(p_i)}\left\{\frac{1}{Q_i}-P_i\right\} \\ &= P_r\left(1-\sum_{i=1}^{r} \frac{1}{f(p_i)}\frac{P_i}{P_r}\left\{\frac{1-P_i Q_i}{P_i Q_i}\right\}\right) \\ &= P_r\left(1-\sum_{i=1}^{r} \frac{1}{f(p_i)}\frac{P_i}{P_r}\frac{\epsilon_i}{\beta_i}\right). \end{aligned}$$

† As usual, empty products are interpreted to be 1.

In view of (9.6) we are entitled to replace the numbers β_i, in this last expression, by β_r. Furthermore, since the β_i are approximations to 1, this process introduces a proportionately small error. We are therefore not only entitled to replace (9.1) by the right hand side of

$$J \geqslant P_r\left(1-\frac{1}{\beta_r}\sum_{i=1}^{r}\frac{1}{f(p_i)}\frac{P_i}{P_r}\epsilon_i\right), \tag{9.8}$$

but may expect to lose little by so doing.

It is now clear that, for a successful application of Theorem 4, one must (after choosing z just small enough for (8.20) to be sufficiently effective) prove that

$$\sum_{i=1}^{r}\frac{1}{f(p_i)}\frac{P_i}{P_r}\epsilon_i < c < 1. \tag{9.9}$$

We note that on writing

$$J_s = 1-\sum_{i=1}^{s}\frac{1}{f(p_i)Q_i} \quad (1 \leqslant s \leqslant r),$$

we have in analogy to (9.8),

$$J_s \geqslant P_s\left(1-\frac{1}{\beta_s}\sum_{i=1}^{s}\frac{1}{f(p_i)}\frac{P_i}{P_s}\epsilon_i\right), \tag{9.10}$$

and consequently

$$J \geqslant P_s\left(1-\frac{1}{\beta_s}\sum_{i=1}^{s}\frac{1}{f(p_i)}\frac{P_i}{P_s}\epsilon_i-\frac{1}{P_s}\sum_{i=s+1}^{r}\frac{1}{f(p_i)Q_i}\right).$$

In practice it is sometimes more convenient to prove, in place of (9.9),

$$\frac{1}{\beta_s}\sum_{i=1}^{s}\frac{1}{f(p_i)}\frac{P_i}{P_s}\epsilon_i+\frac{1}{P_s}\sum_{i=s+1}^{r}\frac{1}{f(p_i)Q_i} < c < 1$$

for some suitably chosen s in the range $1 \leqslant s \leqslant r$. Whilst this can be proved, under suitable circumstances, one requires a really good estimate for ϵ_i (unless the number r of sifting classes is very small). In other words, a really good asymptotic formula for Q_i is needed. At present, such formulae can usually be obtained only by the use of analytic methods. Indeed, analytic methods (or profound elementary methods) would be needed even to deduce from Theorem 4 the lower estimate in Theorem 1. In this (unimportant) respect, Selberg's lower-bound method is inferior to that of Brun. Whilst Selberg's method leads to the sharper estimates, it succeeds only with the aid of analysis.

It is therefore desirable to examine the proof of Theorem 4 for possible weaknesses. The (cumulative) effect of seeking the functions

$s_i^{(+)}$ in the classes $\mathscr{C}_i^{(2)}$ of functions of type (8.7), rather than in the wider classes $\mathscr{C}_i^{(+)}$, is probably not too serious. The real weakness of the method may lie in the fact that already the class of functions (8.6) represents a heavily restricted subclass of $\mathscr{C}^{(-)}$. It may be that one must seek an $s^{(-)}$ outside the class (8.6) to obtain a really substantial improvement on Theorem 4.

Selberg† has investigated various alternative sifting functions, but the results of these investigations have not been published. The task of finding effective functions $s^{(-)}$ is more difficult than the corresponding task relating to functions $s^{(+)}$, because the class $\mathscr{C}^{(-)}$ does not appear to possess a subclass which is both as extensive and simple in structure as the subclass $\mathscr{C}^{(2)}$ of $\mathscr{C}^{(+)}$. Whilst it is true that if $s^{(-)}$ is any element of $\mathscr{C}^{(-)}$, then

$$s^{(-)}(a)\Big\{\sum_{d|\sigma(a)} \lambda(d)\Big\}^2 \quad (\text{where } \lambda(1) = 1)$$

is another element of $\mathscr{C}^{(-)}$, this fact is by no means easy to utilize.

Selberg himself has written:† 'We have not yet found the right approach to the problem of the lower bound. It would be very desirable if one could find a method of the same simplicity and quality as in the case of the upper bound.'

In conclusion, we make some remarks concerning two important applications of the sieve idea. In fact, the various devices we shall describe are equally useful in a wide variety of other applications. We have already mentioned in § 1 that the Selberg sieve, in conjunction with analytic methods, can be used to prove results of type (a''), (b'') (stated in § 1, but repeated below).

(a'') *There exist infinitely many primes q such that $q+2$ is the product of at most l prime factors.*

(b'') *Every large even integer is the sum of a prime and a number which is the product of at most l prime factors.*

Let $\mathscr{A}$ be the finite sequence (cf. (4.1))

$$\mathscr{A} = \{a_1, a_2,..., a_n\}.$$

In the application to (a''), above, we let n be a large integer and take

$$a_j = q_j+2 \quad (j = 1, 2,..., n),$$

where q_j denotes the jth prime. In the application to (b''), on the other hand, we take

$$a_j = 2G-q_j \quad (j = 1, 2,..., n),$$

where G is a large integer and n is the number of primes less than $2G$. In each case, the objective is to apply a sieve method to show that there exists an element a of $\mathscr{A}$ containing at most l prime factors; preferably for a small (constant) value of l.

† See ref. (11) in § 1.

We first describe a device, due to P. Kuhn,(23) which enables one to use, for this purpose, a combination of the upper-bound method and the lower-bound method which is more effective than the direct application of the lower-bound method alone.

Suppose that $2 < y < x$, and that b is a non-negative integer. We denote by $\psi^{(b)}(x, y)$ the number of elements a of $\mathscr{A}$ such that

(i) a has no prime factors in the range $p < y$,

(ii) a has at most b prime factors in the range $y \leqslant p < x$, and none of these is repeated.†

Lemma A, below, shows how a combination of the upper and lower bound methods can be used to estimate $\psi^{(b)}(x, y)$ from below, whilst Lemma B shows how such an estimate can lead to our objective.

As before, we use $\mathscr{P}$ to denote a finite set of primes, and $\mathscr{P}_t$ to denote the set of all primes less than the real number t. We use

$$S(\mathscr{A}, \mathscr{P})$$

to denote the number of elements of $\mathscr{A}$ which lie in none of the sifting classes 0 modulo p, with $p \in \mathscr{P}$.

Lemma A. *For each natural number d, let $\mathscr{A}'_d$ be the set consisting of all those natural numbers a' for which da' lies in $\mathscr{A}$: or symbolically,*

$$\mathscr{A}'_d = \{a' \colon da' \in \mathscr{A}\}.$$

Then

$$\psi^{(b)}(x, y) \geqslant S(\mathscr{A}, \mathscr{P}_y) - \frac{1}{b+1} \sum_{y \leqslant p < x} S(\mathscr{A}'_p, \mathscr{P}_y). \tag{9.11}$$

Remark. It is clear that, subject to appropriate conditions, we can obtain a lower estimate for $\psi^{(b)}(x, y)$ by the following combination of the upper and lower bound methods. We apply Theorem 4 (cf. (8.13)) to obtain a lower bound for $S(\mathscr{A}, \mathscr{P}_y)$; and, for each p in the range $y \leqslant p < x$, we apply Theorem 3 ((7.17) being interpreted with respect to the sequence $\mathscr{A}'_p$ in place of the sequence $\mathscr{A}$) to obtain an upper bound for $S(\mathscr{A}'_p, \mathscr{P}_y)$.

We must stress, however, that the two sequences $\mathscr{A}$ chosen for the applications to (a'') and (b'') respectively (see immediately below the statements of (a'') and (b'')) lead to a serious difficulty in these applications of Theorems 3 and 4. Whilst for each choice, the sequence $\mathscr{A}$ satisfies a relation of type (4.5), the error term R_d is very difficult to

(23) P. Kuhn, 'Zur Viggo Brun'schen Siebmethode' 1, *K. norske Vidensk. Selsk. Forh.* **39** (1942) 145–7; also *Tolfte Skandinaviska Matematikerkongressen, Lund* (1953), 160–8.

† The requirement that a should be squarefree with respect to the primes between y and x is not severe, for, in practice, one can usually show that there are relatively few elements in $\mathscr{A}$ having large prime square factors.

estimate and certainly does not satisfy (7.18). Actually, in order to estimate the error term E^* in the application of (8.13) (and the corresponding error term arising in (7.17), appropriately interpreted after $\mathscr{A}$ has been replaced by $\mathscr{A}'_p$) it suffices to estimate certain sums of type $\sum_d |R_d|$. But even this is a major task (cf. ref. (8)) and involves the use of analytic techniques in conjunction with the large sieve. (See also the discussion immediately preceding Lemma C below.)

Proof of Lemma A. We subdivide $\mathscr{A}$ into disjoint classes $\mathscr{K}_d$ as follows: for each $d \mid \Pi(\mathscr{P}_x)$, we place into $\mathscr{K}_d$ those elements a of $\mathscr{A}$ for which $(a, \Pi(\mathscr{P}_x)) = d$. We use $|\mathscr{K}|$ to denote the number of elements of $\mathscr{K}$. Then, writing

$$\pi(x,y) = \Pi(\mathscr{P}_x)/\Pi(\mathscr{P}_y),$$

we have

$$\psi^{(b)}(x,y) = \sum_{\substack{d \mid \pi(x,y) \\ \nu(d) \leqslant b}} |\mathscr{K}_d|$$

and

$$S(\mathscr{A}, \mathscr{P}_y) = \sum_{d \mid \pi(x,y)} |\mathscr{K}_d|;$$

so that

$$\psi^{(b)}(x,y) = S(\mathscr{A}, \mathscr{P}_y) - \sum_{\substack{d \mid \pi(x,y) \\ \nu(d) \geqslant b+1}} |\mathscr{K}_d|.$$

Now suppose that $a \in \mathscr{K}_d$, where $d \mid \pi(x,y)$ and $\nu(d) \geqslant b+1$. Then $a \in \mathscr{A}^*$, where $\mathscr{A}^*$ consists of those elements of $\mathscr{A}$ counted in $S(\mathscr{A}, \mathscr{P}_y)$; furthermore, $p \mid a$ for each of the $\nu(d)$ prime factors p of d, and hence for at least $b+1$ primes in the range $y \leqslant p < x$. Thus

$$\sum_{\substack{d \mid \pi(x,y) \\ \nu(d) \geqslant b+1}} |\mathscr{K}_d| \leqslant \frac{1}{b+1} \sum_{y \leqslant p < x} \sum_{\substack{a \in \mathscr{A}^* \\ p \mid a}} 1 = \frac{1}{b+1} \sum_{y \leqslant p < x} S(\mathscr{A}'_p, \mathscr{P}_y),$$

as required. This completes the proof of the lemma.

LEMMA B. *Let*

(9.12) $$0 < \theta_1 < \theta_2$$

and let l_1, l_2 be non-negative integers such that

(9.13) $$l_1 \theta_1 + (l_2+1)\theta_2 \geqslant 1.$$

Write

(9.14) $$N = 1 + \max_{a \in \mathscr{A}} a.$$

Then if

(9.15) $$\psi^{(l_1)}(N^{\theta_2}, N^{\theta_1}) > 0,$$

there exists an element a of $\mathscr{A}$ containing at most $l_1 + l_2$ prime factors.

Remark. Needless to say, in practice θ_1, θ_2, l_1, l_2 are chosen so as to lead to the most favourable (minimal) value of $l = l_1+l_2$ consistent with the requirements for the proof of (9.15).

Proof of Lemma B. We show that only those elements of $\mathscr{A}$ containing at most l_1+l_2 prime factors are counted in

$$\psi^{(l_1)}(N^{\theta_2}, N^{\theta_1}).$$

Suppose that, on the contrary, there exists an element a of $\mathscr{A}$ such that:

(i) a has no prime factor in the range $p < N^{\theta_1}$,

(ii) a has at most l_1 prime factors in the range $N^{\theta_1} \leqslant p < N^{\theta_2}$,

(iii) $\nu(a) \geqslant l_1+l_2+1$.

Then, by (9.12), if a has exactly l_1' prime factors in the range $N^{\theta_1} \leqslant p < N^{\theta_2}$,

$$a \geqslant (N^{\theta_1})^{l_1'}(N^{\theta_2})^{l_1+l_2+1-l_1'} = N^{\theta_1 l_1'+\theta_2(l_1-l_1')}N^{\theta_2(l_2+1)} \geqslant N^{\theta_1 l_1}N^{\theta_2(l_2+1)},$$

which contradicts (9.14) in view of (9.13). This completes the proof of the lemma.

The procedure indicated above (by Lemmas A, B and the associated remarks) gives rise to only one further serious difficulty. This difficulty occurs when Theorem 4 is applied to obtain a lower estimate for the number $S(\mathscr{A}, \mathscr{P}_y)$ occurring in (9.11). For this purpose it is necessary (after the various parameters have been suitably chosen in terms of n) to obtain an effective lower estimate for (cf. (9.1) and (9.2))

$$J = 1-\sum_{p<y} \frac{1}{f(p)Q\{(z/p)^{\frac{1}{2}}, p\}},$$

where, for real numbers u, v,

$$Q(u,v) = \sum_{\substack{d|\Pi(\mathscr{P}_v)\\ d\leqslant u}} \frac{1}{g(d)}. \tag{9.16}$$

Note that the first condition of summation is inoperative when $v > u$ (and essentially inoperative in the term $Q(u,u)$ appearing on the left-hand side of (9.17) below).

We have already explained that an asymptotic formula for the function Q is required in order to achieve this. To describe the analytic techniques needed to obtain such an asymptotic formula would be well beyond the scope of these remarks. In Lemma C below, however, we give a reduction formula which can be made to play an important role in the analysis. There is a close analogy between this reduction formula and the formula of Buchstab (cf. § 3, (3.16) and above) used in the proof of Theorem 1, both in structure and in the manner in which each is proved and applied.

LEMMA C. *Let $u > v > 2$. Then*

$$Q(u,u) = Q(u,v) + \sum_{v \leqslant p < u} \frac{1}{g(p)} Q\left(\frac{u}{p}, p\right). \tag{9.17}$$

Proof. We use the notation of § 2, with $\mathscr{P} = \mathscr{P}_u$ and

$$\mathscr{D} = \{d : d \mid \Pi(\mathscr{P}_u),\ d \leqslant u\}.$$

Then (cf. (2.5)) writing $u' = u-1$ or u according as u is or is not an integer, and defining v' in the same way in terms of v, we have

$$\Delta^{(u')} = (\Delta^{(v')}) \cup \Big(\bigcup_{v \leqslant p < u} \mathscr{D}^{(p)}\Big). \tag{9.18}$$

But
$$\sum_{d \in \Delta^{(u')}} \frac{1}{g(d)} = Q(u,u), \qquad \sum_{d \in \Delta^{(v')}} \frac{1}{g(d)} = Q(u,v),$$

whilst, on writing $d = pd^*$, we obtain

$$\sum_{d \in \mathscr{D}^{(p)}} \frac{1}{g(d)} = \sum_{\substack{d^* \mid \Pi(\mathscr{P}_p) \\ pd^* \leqslant u}} \frac{1}{g(pd^*)} = \frac{1}{g(p)} Q\left(\frac{u}{p}, p\right).$$

In view of (9.18), these relations yield the desired identity (9.17).

10. The 'large' sieves of Linnik and Rényi

In this section we shall be concerned with finite sequences of distinct natural numbers. Accordingly, we suppose that $\mathscr{A}$ is a subsequence of the sequence 1, 2,..., N. We write $A(N) = Z$; thus $\mathscr{A}$ contains exactly Z elements.

It is here convenient to introduce the characteristic function $\kappa(n)$ of $\mathscr{A}$, defined to be 1 or 0 according as n does or does not lie in $\mathscr{A}$. We shall require, for primes p, a measure of the regularity of distribution of the elements of $\mathscr{A}$ among the congruence classes modulo p. The number $Z(p,l)$ of elements of $\mathscr{A}$ falling into the congruence class l modulo p is given by

$$Z(p,l) = \sum_{\substack{n=1 \\ n \equiv l \pmod p}}^{N} \kappa(n). \tag{10.1}$$

The corresponding 'expectation' is

$$Z^*(p,l) = \sum_{\substack{n=1 \\ n \equiv l \pmod p}}^{N} \kappa^*, \tag{10.2}$$

where κ^* is the probability for an n, chosen at random from among $1, 2, \ldots, N$, to lie in $\mathscr{A}$; this probability κ^* is determined by

$$\sum_{n=1}^{N} \{\kappa(n) - \kappa^*\} = 0, \tag{10.3}$$

so that

$$\kappa^* = ZN^{-1}. \tag{10.4}$$

The variance

$$V(p) = \sum_{l=1}^{p} \{Z(p,l) - Z^*(p,l)\}^2 \tag{10.5}$$

provides a natural measure of the regularity of distribution of the elements of $\mathscr{A}$ among the congruence classes modulo p.

We shall be working with the actual variance $V(p)$, rather than the approximate variance $D(p)$ discussed in the introduction in § 1 (cf. (1.20)). The approximation is sufficiently good for the two expressions to be interchangeable in the statements of all the theorems in this section.

We shall now prove a number of theorems and discuss their relationship to each other. The setting of these theorems in the wider context of sieve methods in general, and the special point of view adopted here, have already been explained in the introduction to the chapter (§ 1).

Theorem 5 below, which we ascribe to Linnik–Rényi, is not to be found in the literature in the exact form stated here. Its source is the proof, based on Linnik's method, of Lemma 1 in Rényi's paper,(7) although our modification of Linnik's method is slightly different from that of Rényi. We should also mention that Rényi's work was more general in the sense that the elements of $\mathscr{A}$ were 'weighted' and the distribution modulo p was, for each p, measured relative to the congruence classes corresponding to a *fixed* modulus q. This additional generality was introduced for a specific purpose and would not be appropriate here. In other respects, our Theorem 5 is, if anything, slightly more general.

As regards Rényi's own method, we have already explained in § 1 that we do not treat this in the context of probability theory. The quality of the arithmetic results will not be affected, however; the statements of Theorems 6 and 6′ are as they appear in the literature,(15),(17) apart from negligible differences arising solely from our use of the $\ll$ notation.

THEOREM 5 (Linnik–Rényi). *Let $X \geqslant 2$, and suppose that for each $p \leqslant X$, δ_p is a number satisfying*

$$0 \leqslant \delta_p < \tfrac{1}{2}p^{-1}X^{-1}. \tag{10.6}$$

Then

$$\sum_{p \leqslant X} p\delta_p V(p) \ll N^4 \sum_{p \leqslant X} \delta_p^3 + Z, \tag{10.7}$$

where the implied constant is absolute.

Before proving the theorem, we discuss its content. For the purpose of this discussion we shall suppose that (10.7) remains valid when $V(p)$ is replaced by

$$D(p) = \sum_{l=1}^{p} \left\{Z(p,l) - \frac{Z}{p}\right\}^2.$$

For

$$Z^*(p,l) = \frac{Z}{p} + O(1),$$

and hence

$$D(p) \ll V(p) + p.$$

Thus $V(p)$ may be replaced by $D(p)$ in (10.7) provided the term

$$\sum_{p \leqslant X} p^2 \delta_p$$

is absorbed in the right-hand side of (10.7). This can easily be shown to be so, but we require the result for informal discussion only (and not for any proof). It therefore suffices to remark that, intuitively, one expects (10.7) to remain true when $V(p)$ is replaced by p, in view of the definition of $V(p)$, in which $Z(p,l)$ is an integer whilst $Z^*(p,l)$ usually is not.

Now if

$$p \sum_{l=1}^{p} \left\{Z(p,l) - \frac{Z}{p}\right\}^2 \quad \text{is small compared with} \quad p \sum_{l=1}^{p} \left(\frac{Z}{p}\right)^2,$$

i.e. if $pD(p)$ is small compared with Z^2, the elements of $\mathscr{A}$ are (in a sense) well distributed among the congruence classes modulo p; certainly Z/p would then be a good approximation to $Z(p,l)$ for almost all l. Thus an inequality of type

$$\sum p\delta_p V(p) \ll R \tag{10.8}$$

will provide information, for almost all p, concerning the well-distribution of the sequence $\mathscr{A}$ among the congruence classes modulo p, provided

$$R/(Z^2 \sum \delta_p) \tag{10.9}$$

is small. We shall refer to (10.9) as the 'effectiveness' factor of (10.8).

It is clear that the inequality (10.7) becomes most powerful when the δ_p are chosen (subject to (10.6)) to minimize the 'effectiveness factor'. For this purpose, nothing is lost by choosing the δ_p so that they all have the same value δ. However, since† for fixed $\beta < 1$,

$$\sum_{n \leqslant X} n^{-\beta} \ll \sum_{n \leqslant X} X^{-\beta},$$

there is the additional possibility of attaching greater 'weight' to irregularities of distribution corresponding to the smaller p, without any loss of information with respect to the larger p; this can be done by choosing the δ_p to be of the general form

$$\delta_p = Yp^{-\nu} \quad \text{for } p \leqslant X, \tag{10.10}$$

where Y and ν are independent of p, and

$$0 < \nu \leqslant \tfrac{1}{3}. \tag{10.11}$$

Such a choice might well be appropriate for application to a problem in which well-distribution corresponding to the smaller primes is of comparatively greater importance.

We should also mention that for some applications (see the proof of (10.15′) below) it is appropriate to choose the δ_p to be 0 unless p lies in some suitable subset $\mathscr{P}$ of the set of primes $p \leqslant X$.

It will be useful, for comparison with subsequent theorems, to state the most precise inequality for $\sum pV(p)$ inherent in (10.7). On choosing the same value δ for all the δ_p, (10.7) yields

$$\sum_{p \leqslant X} pV(p) \ll N^4X\delta^2 + Z\delta^{-1}, \tag{10.12}$$

subject only to (cf. (10.6))

$$0 < \delta < \tfrac{1}{2}X^{-2}. \tag{10.13}$$

To minimize the right-hand side of (10.12), we choose δ of the order of magnitude $Z^{1/3}N^{-4/3}X^{-1/3}$ so that the two terms on the right-hand side of (10.12) are both of the same order of magnitude, namely $Z^{2/3}N^{4/3}X^{1/3}$. Writing

$$\delta = \tfrac{1}{8}Z^{1/3}N^{-4/3}X^{-1/3},$$

we see that (10.13) is satisfied provided

$$X \leqslant N^{4/5}Z^{-1/5} = N^{3/5}(\kappa^*)^{-1/5}. \tag{10.14}$$

Theorem 5 thus implies the following result.

THEOREM 5′ (Linnik–Rényi). *Let* $2 \leqslant X \leqslant N^{3/5}(\kappa^*)^{-1/5}$. *Then*

$$\sum_{p \leqslant X} pV(p) \ll Z^{2/3}N^{4/3}X^{1/3}, \tag{10.15}$$

where the implied constant is absolute.

† In the actual sums under consideration, n is restricted to prime values, but this is of no importance here.

We note that $\kappa^* \leqslant 1$, so that (10.15) certainly holds for $X \leqslant N^{3/5}$. The effectiveness factor of (10.15) is roughly (ignoring a factor $\log X$) $Z^{-4/3}N^{4/3}X^{-2/3} = (\kappa^*)^{-4/3}X^{-2/3}$; this is roughly $X^{-2/3}$ for fairly 'dense' sequences (such as the sequence of primes not exceeding N).

We remark that Theorem 5 also implies the following generalization of Theorem 5′.

If $\mathscr{P}$ is any subset of the set of primes $p \leqslant X$, and

$$2 \leqslant X \leqslant N^{1/2}|\mathscr{P}|^{1/6}(\kappa^*)^{-1/6},$$

then

$$\sum_{p\in\mathscr{P}} pV(p) \ll Z^{2/3}N^{4/3}|\mathscr{P}|^{1/3}. \tag{10.15′}$$

To obtain (10.15′) we need only write

$$\delta = \tfrac{1}{3}Z^{1/3}N^{-4/3}|\mathscr{P}|^{-1/3},$$

and apply (10.7) with $\delta_p = \delta$ or 0 according as $p \in \mathscr{P}$ or $p \notin \mathscr{P}$.

This generalized form of Theorem 5′ is required for some applications; for example, the result stated as Theorem 2 in Rényi, 'On the large sieve of Ju. V. Linnik', *Compositio Math.* **8** (1950), 68–75, follows† from (10.15′) on taking $\mathscr{P}$ to be the set of 'exceptional' primes featuring in this theorem of Rényi.

Proof of Theorem 5. Let

$$e(\alpha) = e^{2\pi i\alpha} \tag{10.16}$$

where (in this definition only) i stands for a square root of -1. We write

$$f(n) = \kappa(n) - \kappa^*, \tag{10.17}$$

and

$$F(\alpha) = \sum_{n=1}^{N} f(n)e(n\alpha). \tag{10.18}$$

Lemma 1.

$$pV(p) = \sum_{h=0}^{p-1} \left|F\left(\frac{h}{p}\right)\right|^2. \tag{10.19}$$

Proof. Since

$$|F(\alpha)|^2 = F(\alpha)F(-\alpha) = \sum_{n=1}^{N}\sum_{n'=1}^{N} f(n)f(n')e\{(n-n')\alpha\}, \tag{10.20}$$

we have

$$\sum_{h=0}^{p-1} \left|F\left(\frac{h}{p}\right)\right|^2 = \sum_{n=1}^{N}\sum_{n'=1}^{N} t_p(n-n')f(n)f(n'), \tag{10.21}$$

† Apart from the fact that the value of the constant, given explicitly by Rényi, is here concealed by our use of the $\ll$ notation.

where the function t_p is defined by

$$(10.22)\qquad t_p(a) = \sum_{h=0}^{p-1} e\left(\frac{ha}{p}\right) = \begin{cases} p & \text{if } a \equiv 0 \pmod p \\ 0 & \text{if } a \not\equiv 0 \pmod p. \end{cases}$$

Finally, in view of (10.22), the right-hand side of (10.21) can be written in the form

$$p \sum_{l=1}^{p} \Big(\sum_{\substack{n=1 \\ n \equiv l \pmod p}}^{N} f(n) \Big)^2 = pV(p).$$

Lemma 2. *For every* $\delta > 0$,

$$(10.23)\qquad \sum_{h=0}^{p-1} \left| F\left(\frac{h}{p}\right) \right|^2 = \sum_{h=0}^{p-1} \frac{1}{2\delta} \int_{-\delta}^{\delta} \left| F\left(\frac{h}{p}+\beta\right) \right|^2 d\beta + O(N^4\delta^2).$$

Proof. On writing $\alpha = h/p+\beta$ in (10.20) and summing over h, we obtain (cf. (10.21))

$$(10.24)\qquad \sum_{h=0}^{p-1} \left| F\left(\frac{h}{p}+\beta\right) \right|^2 = \sum_{n=1}^{N} \sum_{n'=1}^{N} t_p(n-n') f(n) f(n') e\{\beta(n-n')\}.$$

Now (interpreting $(\sin x)/x$ to be 1 when $x = 0$) we have

$$(10.25)\qquad \frac{1}{2\delta} \int_{-\delta}^{\delta} e(a\beta)\, d\beta = \frac{\sin 2\pi a\delta}{2\pi a\delta} = 1 + O(|a\delta|^2)$$

for all integers a. We integrate each side of (10.24) over the range $-\delta \leqslant \beta \leqslant \delta$ and divide by 2δ. On substituting (10.25) in the various terms of the resulting expression for

$$\sum_{h=0}^{p-1} \frac{1}{2\delta} \int_{-\delta}^{\delta} \left| F\left(\frac{h}{p}+\beta\right) \right|^2 d\beta,$$

we obtain a main term equal to (cf. (10.21))

$$\sum_{h=0}^{p-1} \left| F\left(\frac{h}{p}\right) \right|^2$$

and an error term

$$O\Big(N^2\delta^2 \sum_{n=1}^{N} \sum_{n'=1}^{N} |t_p(n-n') f(n) f(n')|\Big).$$

Since $|f(n)| \leqslant 1$ for all n, it follows from the definition (10.22) of the function t_p that this error term is in fact $O(N^4\delta^2)$. This concludes the proof of the lemma.

By Lemmas 1 and 2, we have for every set of non-negative numbers δ_p,

(10.26) $$\sum_{p\leqslant X} p\delta_p V(p) = \frac{1}{2}\sum_{p\leqslant X}\sum_{h=0}^{p-1}\int_{-\delta_p}^{\delta_p}\left|F\left(\frac{h}{p}+\beta\right)\right|^2 d\beta+O(T_3)$$
$$= \tfrac{1}{2}(T_1+T_2)+O(T_3),$$

where

(10.27) $$T_1 = \sum_{p\leqslant X}\int_{-\delta_p}^{\delta_p}|F(\beta)|^2\,d\beta,$$

(10.28) $$T_2 = \sum_{p\leqslant X}\sum_{h=1}^{p-1}\int_{-\delta_p}^{\delta_p}\left|F\left(\frac{h}{p}+\beta\right)\right|^2 d\beta,$$

and

(10.29) $$T_3 = N^4\sum_{p\leqslant X}\delta_p^3.$$

The error term $O(T_3)$ already appears on the right-hand side of (10.7). Lemmas 3 and 4 below provide the required estimates for T_1 and T_2.

Lemma 3. *For every* $\delta > 0$,

(10.30) $$\int_{-\delta}^{\delta}|F(\beta)|^2\,d\beta \ll N^4\delta^3.$$

Proof. In view of (10.3) we have $F(0) = 0$, so that

$$F(\beta) = F(\beta)-F(0) = \sum_{n=1}^{N} f(n)\{e(n\beta)-1\}.$$

Hence, since $|f(n)| \leqslant 1$ and $|e(\alpha)-1| \ll |\alpha|$, we have

(10.31) $$|F(\beta)| \ll N^2\delta \quad \text{provided } |\beta| \leqslant \delta.$$

Clearly (10.31) implies (10.30).

Lemma 4. *Suppose* $X \geqslant 2$, *and the numbers* δ_p *satisfy* (10.6). *Then*

(10.32) $$T_2 = \sum_{p\leqslant X}\sum_{h=1}^{p-1}\int_{(h/p)-\delta_p}^{(h/p)+\delta_p}|F(\alpha)|^2\,d\alpha \leqslant Z.$$

Proof. The condition (10.6) ensures that the intervals of integration corresponding to the various integrands constituting the double sum in (10.32) are disjoint. For we have (subject to the summation conditions),

$$\left|\frac{h}{p}-\frac{h'}{p'}\right| \geqslant \frac{1}{pp'} \geqslant \frac{1}{2}\left(\frac{1}{p}+\frac{1}{p'}\right)\frac{1}{X} > \delta_p+\delta_{p'} \quad \text{if } p \neq p',$$

and $$\left|\frac{h}{p}-\frac{h'}{p}\right| \geqslant \frac{1}{p} > 2\delta_p \quad \text{if } h \neq h'.$$

Thus
$$T_2 \leqslant \int_0^1 |F(\alpha)|^2\,d\alpha = \sum_{n=1}^{N} f^2(n)$$

in view of (10.20). Finally, using (10.3) and (10.4), we obtain

$$\sum_{n=1}^{N} \{\kappa(n)-\kappa^*\}^2 = \sum_{n=1}^{N} \kappa^2(n)-2N(\kappa^*)^2+N(\kappa^*)^2 = (1-\kappa^*)Z,$$

so that

(10.33) $$\sum_{n=1}^{N} f^2(n) \leqslant Z.$$

The proof of the lemma is now complete. So also is the proof of the theorem, for on using Lemmas 3 and 4 to estimate T_1 and T_2 in (10.26), we obtain (10.7).

THEOREM 6 (Rényi[15]).

(10.34) $$\sum_{p\leqslant N^{1/3}} pV(p) \ll NZ$$

where the implied constant is absolute.

Remarks. The 'effectiveness factor' of (10.34) is roughly (ignoring a factor $\log N$) $Z^{-1}N^{2/3} = (\kappa^*)^{-1}N^{-1/3}$, as compared with the 'effectiveness factor' of roughly $Z^{-4/3}N^{10/9}$ of (10.15) with $X = N^{1/3}$. The right-hand side of (10.34) is smaller than the right-hand side of (10.15), with $X = N^{1/3}$, by the factor $(ZN^{-1})^{1/3}N^{-1/9} = (\kappa^*)^{1/3}N^{-1/9}$; this represents a very substantial improvement.

A simple example† shows that (10.34) is exceedingly powerful, and indeed essentially best possible (at least for fairly 'dense' sequences). Consider the special case when $\mathscr{A}$ consists of the primes not exceeding N. In this case $Z \sim N(\log N)^{-1}$ and, for each $p \leqslant N^{1/3}$, $Z(p,0) = 1$. Hence

$$\sum_{p\leqslant N^{1/3}} pD(p) \geqslant \sum_{p\leqslant N^{1/3}} p\left\{Z(p,0)-\frac{Z}{p}\right\}^2 \gg Z^2 \sum_{p\leqslant N^{1/3}} \frac{1}{p}$$
$$\gg N^2(\log\log N)/(\log N)^2,$$

whilst
$$NZ \ll N^2/\log N.$$

Thus, for this sequence $\mathscr{A}$, the right-hand side of (10.34) exceeds the left-hand side at most by a factor of order $(\log N)/(\log\log N)$.

Proof of Theorem 6. We again write (cf. 10.17))

(10.35) $$f(n) = \kappa(n)-\kappa^*.$$

† If $\mathscr{A}$ is taken to consist of the odd integers not exceeding N, then the term $p = 2$ contributes at least $\frac{1}{4}N^2$ to the left-hand side of (10.34). This shows that (10.34) is best possible to within a constant factor, but the example given in the text is also of interest.

We shall use L to denote the operator $\sum_{n=1}^{N}$; for example, we write

$$L(f) = 0 \tag{10.36}$$

in place of (10.3).

Let $f_p(n)$ be the function which is periodic modulo p and satisfies, for each l modulo p,

$$\sum_{\substack{n=1 \\ n \equiv l (\mathrm{mod}\, p)}}^{N} \{f(n) - f_p(n)\} = 0; \tag{10.37}$$

so that $f_p(l)$ is defined by

$$(\kappa^*)^{-1} Z^*(p, l) f_p(l) = Z(p, l) - Z^*(p, l). \tag{10.38}$$

We note that (in view of (10.36)) on summing (10.37) over l we obtain

$$L(f_p) = 0; \tag{10.39}$$

and on multiplying (10.37) by $f_p(l)$ before summing over l,

$$L(ff_p) = L(f_p^2). \tag{10.40}$$

Before embarking on the formal proof, we give a brief indication of the ideas on which it is based. In view of (10.39) and the periodicity of f_p, $L(f_p f_{p'})$ is small compared to $L(|f_p f_{p'}|)$ if p, p' are relatively prime (as is the case for distinct primes p, p') and each sufficiently small compared to N; in other words the functions are 'quasi-orthogonal' provided p is restricted to a sufficiently small range. An extension of Bessel's inequality to 'quasi-orthogonal' functions will enable us to estimate $\sum_{p \leqslant X} L(f_p^2)$ in terms of $L(f^2)$, provided only that X is sufficiently small for the functions f_p to be quasi-orthogonal in an appropriate sense. In this way we shall prove that

$$\sum_{p \leqslant N^{1/3}} L(f_p^2) \ll L(f^2), \tag{10.41}$$

which, as we now show, implies (10.34).

We have

$$L(f_p^2) = \sum_{l=1}^{p} (\kappa^*)^{-1} Z^*(p, l) f_p^2(l)$$

$$= \kappa^* \sum_{l=1}^{p} \frac{\{Z(p, l) - Z^*(p, l)\}^2}{Z^*(p, l)}$$

in view of (10.38). Thus, to see that (10.41) implies (10.34), we need only recall (10.33) and note that, for $p \leqslant N^{1/3}$,

$$Z^*(p, l) = \kappa^* \{Np^{-1} + O(1)\} \gg \kappa^* N p^{-1}.$$

We now proceed with the proof of (10.41). Lemma 5 below is, in essence, Bessel's inequality.

Lemma 5. *Let $\mathscr{S}$ be a set of elements x. Let $\phi(x)$ denote a real function of x, so that ϕ assigns a real number to each element x of $\mathscr{S}$. Let L be a linear operator which assigns, to each† function ϕ, a real number $L(\phi)$; and such that the numbers assigned to non-negative functions are non-negative.*

Suppose that ϕ; $\phi_1,\ldots,\phi_k$ is a finite set of functions satisfying (10.42) *and* (10.43) *below.*

(10.42) $$L(\phi\phi_i) = L(\phi_i^2) \quad \text{for } 1 \leqslant i \leqslant k.$$

(10.43) $$L(\phi_i\phi_j) = 0 \quad \text{for } 1 \leqslant i < j \leqslant k,$$

Then

(10.44) $$\sum_{i=1}^{k} L(\phi_i^2) \leqslant L(\phi^2).$$

To prove the lemma, we need only note that

(10.45) $$0 \leqslant L\Big(\Big\{\phi - \sum_{i=1}^{k}\phi_i\Big\}^2\Big)$$
$$= L(\phi^2) - 2\sum_{i=1}^{k} L(\phi\phi_i) + \sum_{i=1}^{k}\sum_{j=1}^{k} L(\phi_i\phi_j) = L(\phi^2) - \sum_{i=1}^{k} L(\phi_i^2).$$

We now obtain an analogous result for quasi-orthogonal functions.

Lemma 6. *Suppose the hypotheses of Lemma 5 are modified by replacing* (10.43) *by*

(10.46) $$|L(\phi_i\phi_j)| \leqslant \delta_i\delta_j\{L(\phi_i^2)L(\phi_j^2)\}^{\frac{1}{2}} \quad \text{for } i \neq j;$$

where $\delta_1, \delta_2,\ldots, \delta_k$ are real numbers satisfying

(10.47) $$\sum_{i=1}^{k} \delta_i^2 \leqslant \tfrac{1}{2}.$$

Then

(10.48) $$\sum_{i=1}^{k} L(\phi_i^2) \leqslant 2L(\phi^2).$$

Proof. This differs from the proof of Lemma 5 only in that there is an extra term
$$\sum_{\substack{i=1 \\ i\neq j}}^{k}\sum_{j=1}^{k} L(\phi_i\phi_j)$$
at the end of (10.45). Since the absolute value of this term is at most
$$\Big(\sum_{i=1}^{k} \delta_i\{L(\phi_i^2)\}^{\frac{1}{2}}\Big)^2 \leqslant \Big(\sum_{i=1}^{k}\delta_i^2\Big)\Big(\sum_{i=1}^{k} L(\phi_i^2)\Big) \leqslant \tfrac{1}{2}\sum_{i=1}^{k} L(\phi_i^2),$$
the result follows.

† In the application, $L(\psi)$ will be meaningful for all functions ψ; but the lemma remains valid if $L(\psi)$ is defined only when ψ lies in the ring generated by the functions ϕ; $\phi_1,\ldots,\phi_k$ appearing in (10.44).

To prove (10.41) we apply Lemma 6 with f in place of ϕ and the set of functions $\{f_p : p \leqslant N^{1/3}\}$ in place of the set $\{\phi_i : i = 1, 2, \ldots, k\}$; of course L will once more stand for $\sum_{n=1}^{N}$. But, for this purpose, we must first obtain an appropriate inequality of type (10.46) for the set of functions f_p. Assuming that

(10.49) $$p' < p \leqslant N^{1/2},$$

we shall prove

(10.50) $$|L(f_p f_{p'})| \leqslant (cpN^{-\frac{1}{2}})(cp'N^{-\frac{1}{2}})\{L(f_p^2)L(f_{p'}^2)\}^{\frac{1}{2}},$$

where c is an absolute constant.

Since (by Lemma 1, Appendix)

(10.51) $$\sum_{p \leqslant N^{1/3}} (cpN^{-\frac{1}{2}})^2 \leqslant \tfrac{1}{2}$$

for large N, it will follow† that the requirement (10.46) of Lemma 6 is satisfied for large N.

We shall use $C^{(q)}(\psi)$ to denote the sum of $\psi(n)$ over a complete set of residues modulo q. We note that if ψ is periodic modulo q (as will be the case when this notation is used), we have

(10.52) $$L(\psi) = \left[\frac{N}{q}\right] C^{(q)}(\psi) + O\{C^{(q)}(|\psi|)\}.$$

Now, in view of the periodicity of f_p,

(10.53) $$\begin{aligned} |L(f_p f_{p'})| &\leqslant \frac{N}{pp'} |C^{(pp')}(f_p f_{p'})| + C^{(pp')}(|f_p f_{p'}|) \\ &= \frac{N}{pp'} |C^{(p)}(f_p) C^{(p')}(f_{p'})| + C^{(p)}(|f_p|) C^{(p')}(|f_{p'}|) \\ &\ll C^{(p)}(|f_p|) C^{(p')}(|f_{p'}|); \end{aligned}$$

to verify this last step we need only apply (10.52) with $\psi = f_p$ and use (10.39) (and repeat this procedure with p' in place of p). Furthermore, on using Cauchy's inequality and again applying (10.52), we obtain

(10.54) $$C^{(p)}(|f_p|) \leqslant p^{\frac{1}{2}}\{C^{(p)}(f_p^2)\}^{\frac{1}{2}} \ll pN^{-\frac{1}{2}}\{L(f_p^2)\}^{\frac{1}{2}}.$$

On substituting (10.54) (and the corresponding inequality with p' in place of p) in (10.53), we obtain (10.50).

Lemma 6 is now applicable, since (10.40) and (10.50) correspond to (10.42) and (10.46) respectively. We thus obtain (10.41), and the theorem is proved.

† In this connexion, (10.50) is required only for $p' < p \leqslant N^{1/3}$. But the fact that (10.50) holds under the weaker hypothesis (10.49) will be used in the proof of Theorem 6′.

Extension of the range $p \leqslant N^{1/3}$ in Rényi's method

We suppose that

(10.55) $$X = N^{\alpha}, \quad \text{where} \quad \tfrac{1}{3} \leqslant \alpha \leqslant \tfrac{1}{2}.$$

Rényi's method is easily adapted to permit the range $p \leqslant N^{1/3}$ to be extended (as Rényi himself showed), but only at the expense of replacing Lemma 6 by a modification which, whilst very similar in nature, is far less economical in its application. In consequence, the method becomes rapidly less effective as α increases, until it is entirely ineffective when $\alpha = \frac{1}{2}$.

THEOREM 6′ (Rényi[17]). *If X is of the form* (10.55)

(10.56) $$\sum_{p \leqslant X} pV(p) \ll X^3 Z,$$

where the implied constant is absolute.

Remarks. The effectiveness factor of (10.56) is† $X^2Z^{-1} = X^2N^{-1}(\kappa^*)^{-1}$ so that (10.56) is entirely ineffective when $X = N^{1/2}$. The ratio between the respective right-hand sides of (10.56) and (10.15) is

$$(X^3Z)/(Z^{2/3}N^{4/3}X^{1/3}) = (ZN^{-1})^{1/3}N^{-1}X^{8/3}.$$

Suppose, for example, that Z is roughly of the order N (say to within a power of $\log N$). Then of the two estimates (10.56) and (10.15), the former is the more effective in the range $\frac{1}{3} \leqslant \alpha < \frac{3}{8}$, and the latter in the range $\frac{3}{8} < \alpha \leqslant \frac{1}{2}$.

Proof of Theorem 6′. We replace Lemma 6 by the following result.

LEMMA 7. *Let θ satisfy $0 < \theta \leqslant 1$. Suppose the hypotheses of Lemma 5 are modified by replacing (10.43) by (10.46), where $\delta_1,\ldots,\delta_k$ are real numbers satisfying*

(10.57) $$\sum_{i=1}^{k} \delta_i^2 \leqslant \tfrac{1}{2}\theta^{-1}.$$

Then

(10.58) $$\sum_{i=1}^{k} L(\phi_i^2) \leqslant 2\theta^{-1}L(\phi^2).$$

Proof. We have (cf. the proofs of Lemmas 5 and 6)

$$\begin{aligned} 0 &\leqslant L\Big(\Big\{\phi-\theta\sum_{i=1}^{k}\phi_i\Big\}^2\Big) \\ &= L(\phi^2)-2\theta\sum_{i=1}^{k}L(\phi\phi_i)+\theta^2\sum_{i=1}^{k}\sum_{j=1}^{k}L(\phi_i\phi_j) \\ &\leqslant L(\phi^2)-\theta\sum_{i=1}^{k}L(\phi_i^2)+\theta^2\sum_{i=1}^{k}\sum_{\substack{j=1\\ i\neq j}}^{k}L(\phi_i\phi_j) \\ &\leqslant L(\phi^2)-\tfrac{1}{2}\theta\sum_{i=1}^{k}L(\phi_i^2). \end{aligned}$$

† Strictly speaking, the factor is $X^2Z^{-1}\log X$.

Remark. The above proof suggests that, in applications to given sets of functions, the lemma may often prove uneconomical when θ is small. For, in view of the normalization (10.42), $\phi - \sum \lambda_i \phi_i$ is orthogonal to all the functions ϕ_i when $\lambda_i = 1$ $(i = 1, 2, ..., k)$. Consequently, if the set of functions ϕ_i is in some sense 'nearly complete',

$$L\Big(\Big|\phi - \sum_{i=1}^{k} \phi_i\Big|^2\Big)$$

may be expected to be small. It is thus the inequality

$$L\Big(\Big|\phi - \sum_{i=1}^{k} \phi_i\Big|^2\Big) \geqslant 0$$

(on which the proof of Lemma 6 is based) and not the inequality

$$\theta^{-1} L\Big(\Big|\phi - \theta \sum_{i=1}^{k} \phi_i\Big|^2\Big) \geqslant 0$$

(on which the proof of Lemma 7 is based) which may be expected to be economical when applied to the set of functions f, f_p. (This concludes the remark.)

We now apply Lemma 7 with f in place of ϕ and the set of functions $\{f_p; p \leqslant X\}$ in place of the set $\{\phi_i; i = 1, 2, ..., k\}$. Since

$$\sum_{p \leqslant X} (cpN^{-\frac{1}{2}})^2 \leqslant \tfrac{1}{2} X^3 N^{-1}$$

for large N, it follows from (10.50) that the hypotheses of Lemma 7 are satisfied (for large N) with $\theta = NX^{-3}$. On applying the lemma, we obtain

$$\sum_{p \leqslant X} L(f_p^2) \ll N^{-1} X^3 L(f^2). \tag{10.59}$$

The inequality (10.56) now follows from (10.59) by the argument following (10.41).

Addendum (November 1964). K. F. Roth has recently proved the following result:

THEOREM 7. *If* $N \geqslant 2$ *and* $X \geqslant N^{1/2}(\log N)^{-1/2}$, *then*

$$\sum_{p \leqslant X} pV(p) \ll ZX^2 \log X. \tag{10.60}$$

This result is an equivalent form of Corollary 2 to the theorem proved in K. F. Roth, 'On the large sieves of Linnik and Rényi', *Mathematika* **12** (1965), 1–9.

Theorem 7 represents an improvement on Theorems 5′ and 6′. The nature of this improvement is borne out by comparing the estimate

$$\sum_{p \leqslant N^{1/2}(\log N)^{-1/2}} pV(p) \ll ZN,$$

obtained by writing $X = N^{1/2}(\log N)^{-1/2}$ in (10.60), with the estimate

$$\sum_{p \leqslant N^{1/3}} pV(p) \ll ZN$$

of Theorem 6.

The above proof of Theorem 6 is, however, simpler than the proof of Theorem 7.

V

PRIMITIVE SEQUENCES AND SETS OF MULTIPLES

1. Introduction

THROUGHOUT this chapter $\mathscr{A} = \{a_i\}$ will denote a subsequence of the sequence of natural numbers. We consider the set $\mathscr{B} = \mathscr{B}(\mathscr{A})$ consisting of all the distinct positive multiples of elements of $\mathscr{A}$. We note that $\mathscr{B}$ is the positive part of the union, taken over all elements a_i of $\mathscr{A}$, of the congruence classes $0 \pmod{a_i}$. Whilst none of our arguments will depend on ordering $\mathscr{B}$, it will often be convenient (to facilitate description and for reasons of notation) to imagine $\mathscr{B}$ to be ordered according to magnitude and to refer to it as a 'sequence'.

The present chapter is, in part, a study of the manner in which a general set $\mathscr{B} = \mathscr{B}(\mathscr{A})$ is distributed among the natural numbers.

The same sequence $\mathscr{B} = \mathscr{B}(\mathscr{A})$ may, of course, be generated by more than one sequence $\mathscr{A}$. For example, if $\mathscr{A}$ contains two distinct elements a_j, a_k such that a_j divides a_k, then the omission of a_k from $\mathscr{A}$ will not affect $\mathscr{B}$. This obvious remark leads us to the fact that among all the sequences $\mathscr{A}$ generating the same $\mathscr{B} = \mathscr{B}(\mathscr{A})$, there is one which is 'minimal' in the sense that it is contained in all the others. In other words, the intersection of all generating sequences of $\mathscr{B}$ is itself a generating sequence of $\mathscr{B}$.

To see this, let $\mathscr{A}'$ be the sequence consisting of each of those elements of $\mathscr{A}$ not divisible by any other element of $\mathscr{A}$. Then

(i) $$\mathscr{B}(\mathscr{A}) = \mathscr{B}(\mathscr{A}'),$$

since every element a of $\mathscr{A}$ has a divisor in $\mathscr{A}'$, namely the least among those divisors of a which lie in $\mathscr{A}$.

Furthermore, $\mathscr{A}'$ is the only sequence satisfying both (i) and the condition (ii) below.

(ii) No element of $\mathscr{A}'$ divides any other.

For let $\mathscr{A}''$ be the result of applying the same reduction procedure to another generating sequence of $\mathscr{B}$; then $\mathscr{A}''$ is also a generating sequence of $\mathscr{B}$ and no element of $\mathscr{A}''$ divides another. We shall see that $\mathscr{A}'$ and $\mathscr{A}''$ coincide. If, on the contrary, r is the least number such that the rth elements a'_r and a''_r of $\mathscr{A}'$ and $\mathscr{A}''$ are distinct and if $a'_r < a''_r$, then

a'_r cannot appear in $\mathscr{B}(\mathscr{A}'')$ and we arrive at a contradiction. Thus conditions (i) and (ii) determine $\mathscr{A}'$ uniquely and it follows that $\mathscr{A}'$ is the intersection of all generating sequences of $\mathscr{B}$.

The preceding discussion makes it clear that the structure of $\mathscr{B} = \mathscr{B}(\mathscr{A})$ depends on $\mathscr{A}'$ rather than on $\mathscr{A}$. In general $\mathscr{A}'$ is infinite, as is the case when $\mathscr{A}$ is the sequence of all natural numbers greater than 1. It may happen, however, that $\mathscr{A}'$ is finite; for instance, whenever $\mathscr{A}$ contains 1, $\mathscr{A}'$ consists of the single element 1, and if $\mathscr{A}$ is the sequence 2, 4, 6, 8,..., $\mathscr{A}'$ again contains only one element, namely 2. If $\mathscr{A}'$ is finite, $\mathscr{B}(\mathscr{A}) = \mathscr{B}(\mathscr{A}')$ is (the positive part of) a union of a finite number of congruence classes and therefore clearly possesses asymptotic density (in the usual sense; see § 2). It was conjectured at one time that $\mathscr{B}(\mathscr{A})$ possesses asymptotic density for any sequence $\mathscr{A}$. Although we shall find this to be false we shall prove that the corresponding result is true for a weaker type of density. The later part of the chapter is devoted to the investigation of the type of condition under which $\mathscr{B}(\mathscr{A})$ does, in fact, possess asymptotic density. The only known necessary and sufficient condition is somewhat cumbersome, but will be seen to imply that $\mathscr{B}(\mathscr{A})$ possesses asymptotic density provided that $\mathscr{A}$ is not, in a certain sense, too numerous. A best possible result of this type will be obtained.

We have seen above that the minimal generating sequence of $\mathscr{B}(\mathscr{A})$ is characterized by property (ii). Sequences having this property present a topic of independent interest. We shall call any sequence $\mathscr{A}'$ satisfying (ii) a *primitive* sequence. The set of all integers m in the interval $n < m \leqslant 2n$ (where $n > 0$ is fixed) provides an example of a finite primitive sequence, and such sets will play an important role in our investigations. An example of an infinite primitive sequence (which, again, will prove useful later) is the sequence of all those natural numbers composed of exactly r prime factors, where r is fixed. Several interesting results are known concerning the densities of primitive sequences, and these will be described in the opening part of the chapter.

It is natural to ask whether the condition (ii) is sufficiently restrictive to ensure that every primitive sequence must have zero asymptotic density. We shall see that this is not the case, insofar as a primitive sequence need not possess asymptotic density. On the other hand, we shall prove that the lower asymptotic density of a primitive sequence is always zero.

As a matter of historical interest, the investigation of general primitive sequences received its first impetus from the study of abundant numbers (that is, natural numbers n for which $\sigma(n) \geqslant 2n$; $\sigma(n)$ being the sum of the divisors of n) where certain special primitive sequences play an important part. The topic of abundant numbers is not included in our book and we refer the reader to the literature.(1)

Notation. Whilst the various notations are explained in the text, we draw special attention to the fact that in *this* chapter, we use $u\mathscr{A}$ to denote the sequence obtained from $\mathscr{A}$ by multiplying each element of $\mathscr{A}$ by u.

2. Density

We recall the definitions

$$\underline{\mathbf{d}}\mathscr{A} = \liminf_{n\to\infty} \frac{A(n)}{n}, \qquad \bar{\mathbf{d}}\mathscr{A} = \limsup_{n\to\infty} \frac{A(n)}{n}$$

of the *lower* and *upper asymptotic density* of a sequence $\mathscr{A}$; and that, if the two are equal, we refer to their common value as *the* (asymptotic) density $\mathbf{d}\mathscr{A}$. (For the definition of $A(n)$, see § 4 of the 'Notation' section at the beginning of the book.)

Where sequences of natural numbers are defined as *finite* unions of (the positive parts of) congruence classes, it is sometimes convenient to think in terms of the unions of the corresponding entire congruence classes (e.g. in the application of Rogers's theorem, proved in § 3, to Lemma 5). Whilst, technically, the asymptotic density of such a union of entire congruence classes is defined in terms of the counting number, which counts only the *positive* elements, the resulting density is clearly not affected by this restriction; if all the elements were taken into account, the same density would result. Where infinite unions are concerned (e.g. sequences of multiples), we shall work *exclusively* in terms of their positive parts.

Among the obvious results that will frequently be used are the following:

(i) The density of the sequence of all positive multiples of a given natural number a_1 is $1/a_1$.

(ii) The density of the sequence of all those positive multiples of a_1 which are not divisible by a_2 is

$$\frac{1}{a_1} - \frac{1}{[a_1, a_2]}.$$

(1) P. Erdös, 'On primitive abundant numbers', *J. Lond. math. Soc.* **10** (1935) 49–58, is an early paper on this (extensive) subject.

(iii) The density of the sequence of all those natural numbers which are not divisible by any one of a given set of distinct primes $p_1, p_2,..., p_r$, is

$$\prod_{\nu=1}^{r}\left(1-\frac{1}{p_\nu}\right).$$

It may happen that the distribution of a sequence is regular to some degree, but not sufficiently so for the sequence to possess asymptotic density. It is natural, therefore, to define 'weaker' densities, which exist for a wider class of sequences. Among the many 'weaker' types of density that have proved useful in the investigation of sequences, the concept of *logarithmic density* is especially appropriate to some of the problems considered in this chapter.

We define

$$\underline{\delta}\mathscr{A} = \liminf_{n\to\infty}\frac{1}{\log n}\sum_{a_i\leqslant n}\frac{1}{a_i}, \qquad \bar{\delta}\mathscr{A} = \limsup_{n\to\infty}\frac{1}{\log n}\sum_{a_i\leqslant n}\frac{1}{a_i}$$

to be respectively the lower and upper logarithmic densities of $\mathscr{A}$, and if the two are equal we say that $\mathscr{A}$ possesses logarithmic density $\delta\mathscr{A}$ given by

$$\delta\mathscr{A} = \lim_{n\to\infty}\frac{1}{\log n}\sum_{a_i\leqslant n}\frac{1}{a_i}.$$

As is an immediate consequence of the lemma below, logarithmic density is 'weaker' than asymptotic density in the sense that the existence of the latter implies the existence of the former.

LEMMA 1. $$0 \leqslant \underline{\mathbf{d}}\mathscr{A} \leqslant \underline{\delta}\mathscr{A} \leqslant \bar{\delta}\mathscr{A} \leqslant \bar{\mathbf{d}}\mathscr{A} \leqslant 1.$$

Proof. The lower and upper densities concerned obviously lie between 0 and 1.

For any integers $n > N \geqslant 1$, we have

$$\frac{1}{\log n}\sum_{a_i\leqslant n}\frac{1}{a_i} = \frac{1}{\log n}\sum_{a_i\leqslant N}\frac{1}{a_i}+\frac{1}{\log n}\sum_{N<a_i\leqslant n}\frac{1}{a_i}. \tag{2.1}$$

We apply an Abel summation to the second sum to the right, obtaining

$$\sum_{N<a_i\leqslant n}\frac{1}{a_i} = \sum_{m=N+1}^{n}\frac{A(m)-A(m-1)}{m}$$
$$= \sum_{m=N+1}^{n-1}A(m)\left(\frac{1}{m}-\frac{1}{m+1}\right)-\frac{A(N)}{N+1}+\frac{A(n)}{n}$$
$$= \sum_{m=N+1}^{n-1}\left(\frac{A(m)}{m}\right)\frac{1}{m+1}+O(1).$$

Now $$\sum_{m=N+1}^{n-1} \frac{1}{m+1} \sim \log n \quad \text{as } n \to \infty,$$

so that substituting in (2.1) and letting $n \to \infty$, we obtain

$$\inf_{m>N}\left(\frac{A(m)}{m}\right) \leqslant \underline{\delta}\mathscr{A} \leqslant \bar{\delta}\mathscr{A} \leqslant \sup_{m>N}\left(\frac{A(m)}{m}\right).$$

On letting $N \to \infty$, the result follows.

3. An inequality concerning densities of unions of congruence classes

In § 6 of this chapter, we shall require a result of Behrend (see Lemma 5) for the proof of a fundamental result of Erdös (Theorem 13). Instead of giving Behrend's own proof, we shall deduce his result from a theorem of C. A. Rogers, concerning unions of arithmetic progressions. Professor Rogers's theorem has not been published previously, and we are grateful to him for allowing us to include it in our book.

The notation introduced in the definition below is for use in this section (and the application to Lemma 5) only.

Definitions. $\mathscr{R}(q,h)$ denotes the congruence class $h \pmod q$;

$$\mathscr{R}(q,h)+c = \mathscr{R}(q,h+c), \qquad c\mathscr{R}(q,h) = \mathscr{R}(cq,ch).$$

THEOREM (C. A. Rogers†). *For any fixed system $\mathscr{R}_1, \mathscr{R}_2, \ldots, \mathscr{R}_l$ of congruence classes, the density of the union of the translates*

$$\mathscr{R}_1+t_1, \quad \mathscr{R}_2+t_2, \quad \ldots, \quad \mathscr{R}_l+t_l$$

is minimal when these translates have a common element. In other words, if

$$\mathscr{R}_\nu = \mathscr{R}(q_\nu, h_\nu) \quad (\nu = 1, 2, \ldots, l),$$

then

$$\mathbf{d} \bigcup_{\nu=1}^{l} \mathscr{R}_\nu \geqslant \mathbf{d} \bigcup_{\nu=1}^{l} \mathscr{R}(q_\nu, 0). \tag{3.1}$$

Proof. We write $$Q = q_1 q_2 \ldots q_l.$$

Suppose that Q contains exactly s distinct prime factors. The truth of (3.1) is obvious when $s = 1$; in this case each q_ν is a power of the same prime p, and consequently

$$\mathscr{R}(q_\mu, 0) \supset \bigcup_{\nu=1}^{l} \mathscr{R}(q_\nu, 0)$$

† Unpublished; communicated to the authors by Professor Rogers. The proof has been modified, but is essentially the same as that of Professor Rogers.

for at least one μ $(1 \leqslant \mu \leqslant l)$.

We proceed by induction on s and, accordingly, assume that $s > 1$ and that the theorem is true for all systems $\mathscr{R}_1^*,\ldots, \mathscr{R}_{l^*}^*$ whose corresponding s^* is at most $s-1$. Since $s > 1$, there exists a factorization

$$Q = uv, \qquad (u, v) = 1, \tag{3.2}$$

of Q, in which each of u, v contains at most $s-1$ distinct prime factors. Subject to the induction hypothesis, we shall prove the following result for such u, v.

LEMMA. *Suppose the numbers* $m_1,\ldots, m_l$ *are such that* (i) *each of the congruence classes* $\mathscr{R}_1+um_1,\ldots, \mathscr{R}_l+um_l$ *contains an element* $\equiv 0 \pmod v$. *Then*

$$\mathbf{d} \bigcup_{\nu=1}^{l} \mathscr{R}_\nu \geqslant \mathbf{d} \bigcup_{\nu=1}^{l} (\mathscr{R}_\nu+um_\nu). \tag{3.3}$$

The theorem then follows at once. For if the number pairs m_ν, n_ν are chosen to satisfy

$$h_\nu+um_\nu+vn_\nu \equiv 0 \pmod Q \quad (\nu = 1, 2,\ldots, l),$$

we have (3.3), and on applying the lemma once more (with u, v interchanged and $\mathscr{R}_\nu$ replaced by $\mathscr{R}_\nu+um_\nu$),

$$\mathbf{d} \bigcup_{\nu=1}^{l} (\mathscr{R}_\nu+um_\nu) \geqslant \mathbf{d} \bigcup_{\nu=1}^{l} \{(\mathscr{R}_\nu+um_\nu)+vn_\nu\}. \tag{3.4}$$

(3.3) and (3.4) combined now yield (3.1) as required.

It remains only to verify the lemma, subject to the induction hypothesis. We write

$$\mathscr{S}_\nu = \mathscr{R}_\nu+um_\nu, \tag{3.5}$$

$$\mathscr{R}_\nu^{(a)} = \mathscr{R}(u, a) \cap \mathscr{R}_\nu, \qquad \mathscr{S}_\nu^{(a)} = \mathscr{R}(u, a) \cap \mathscr{S}_\nu. \tag{3.6}$$

It will clearly suffice to prove that, for each a,

$$\mathbf{d} \bigcup_{\nu=1}^{l} \mathscr{R}_\nu^{(a)} \geqslant \mathbf{d} \bigcup_{\nu=1}^{l} \mathscr{S}_\nu^{(a)}. \tag{3.7}$$

We concentrate our attention on a *fixed* value a. Write $q_\nu = q'_\nu q''_\nu$ where $q'_\nu | u$, $q''_\nu | v$. Now $\mathscr{R}_\nu^{(a)}$, $\mathscr{S}_\nu^{(a)}$ are both non-empty or both empty according as

$$h_\nu \equiv a \pmod{q'_\nu} \tag{3.8}$$

is or is not satisfied. Suppose (3.8) is satisfied for exactly l^* values of ν. Renumbering the q_ν if necessary, we may take these values to be

$\nu = 1, 2,..., l^*$. On omitting the empty sets, (3.7) takes the form

(3.9) $$\mathbf{d}\bigcup_{\nu=1}^{l^*}\mathscr{R}_\nu^{(a)} \geqslant \mathbf{d}\bigcup_{\nu=1}^{l^*}\mathscr{S}_\nu^{(a)}.$$

Suppose ν has one of the values $1, 2,..., l^*$ (if $l^* = 0$, there is nothing to prove). Then $\mathscr{R}_\nu^{(a)}$ is the intersection of the congruence classes $h_\nu \pmod{q_\nu''}$ and $a \pmod{u}$, so that

(3.10) $$\mathscr{R}_\nu^{(a)} = u\mathscr{R}(q_\nu'', k_\nu)+a,$$

where k_ν is any number satisfying $uk_\nu+a \equiv h_\nu \pmod{q_\nu''}$. Furthermore, $h_\nu+um_\nu \equiv 0 \pmod{q_\nu''}$ by (i), so that $\mathscr{S}_\nu^{(a)}$ is the intersection of the congruence classes $0 \pmod{q_\nu''}$ and $a \pmod{u}$. Thus, if a number c is chosen to satisfy $uc+a \equiv 0 \pmod{v}$, we have

(3.11) $$\mathscr{S}_\nu^{(a)} = u\mathscr{R}(q_\nu'', c)+a \quad (\nu = 1, 2,..., l^*).$$

Finally, by the induction hypothesis,

$$\mathbf{d}\bigcup_{\nu=1}^{l^*}\mathscr{R}(q_\nu'', k_\nu) \geqslant \mathbf{d}\bigcup_{\nu=1}^{l^*}\mathscr{R}(q_\nu'', c);$$

for the q_ν'' are divisors of v, and v contains at most $s-1$ distinct prime factors. Thus (3.9) follows from (3.10), (3.11), and (3.3) is established. This completes the proof of the theorem.

4. Primitive sequences

We begin with an investigation of infinite primitive sequences.

THEOREM 1. *If $\mathscr{A}$ is an infinite primitive sequence, then $\bar{\mathbf{d}}\mathscr{A} \leqslant \frac{1}{2}$.*

Proof. If m_i denotes the greatest odd divisor of a_i, then the numbers m_i are all distinct; for $m_i = m_j$ implies either $a_i \mid a_j$ or $a_j \mid a_i$, and is possible only if $i = j$. Hence $A(2n) \leqslant n$, since there are exactly n distinct odd natural numbers not exceeding $2n$.

We shall see later that Theorem 1 is essentially best possible.† On the other hand, much more can be said concerning the lower density of a primitive sequence. Indeed, the logarithmic density (and therefore the lower asymptotic density) of a primitive sequence must be 0. The truth of this assertion (cf. Theorem 3) is not quite as obvious as that of Theorem 1, but will be shown to be an immediate consequence of the following theorem.

† It will be shown in Theorem 4 that $\bar{\mathbf{d}}\mathscr{A}$ can lie arbitrarily close to $\frac{1}{2}$. On the other hand $\bar{\mathbf{d}}\mathscr{A} = \frac{1}{2}$ is impossible, as can be seen from the result given in P. Erdös, Aufgabe 395, of *Elem. Math., Basel* **16** (1961) 21. This states that if u, v, n are natural numbers such that $a' = 2^u(2v+1) \leqslant n$, and $\mathscr{A}$ is an infinite primitive sequence containing a', then

$$A(n) \leqslant n-[\tfrac{1}{2}n]-\left[\frac{1}{2}\left(\frac{n}{3^u(2v+1)}-1\right)\right].$$

THEOREM 2 (Erdös[2]). *If $\mathscr{A}$ is an infinite primitive sequence, then*

$$\sum_{i=1}^{\infty} \frac{1}{a_i \log a_i}$$

converges.

Proof. We denote by p_i the greatest prime factor of a_i. We shall prove that

$$\sum_{i=1}^{\infty} \frac{1}{a_i} \prod_{p \leqslant p_i} \left(1 - \frac{1}{p}\right) \leqslant 1, \tag{4.1}$$

which, by Lemma 4 of the Appendix, implies Theorem 2. Let $\mathscr{S}^{(i)}$ denote the sequence of those natural numbers composed entirely of primes greater than p_i. Let $\mathscr{R}^{(i)} = a_i \mathscr{S}^{(i)}$ be the sequence obtained by multiplying each member of $\mathscr{S}^{(i)}$ by a_i. Then the sequences $\mathscr{R}^{(i)}$ are discrete; for

$$a_i s_1^{(i)} = a_j s_2^{(j)}, \qquad p_i \leqslant p_j,$$

would imply $a_i \mid a_j$. Hence, for every n,

$$\sum_{i=1}^{n} \mathbf{d}(\mathscr{R}^{(i)}) \leqslant 1. \tag{4.2}$$

Since (see § 2) $\quad \mathbf{d}\mathscr{R}^{(i)} = \dfrac{1}{a_i} \mathbf{d}\mathscr{S}^{(i)} = \dfrac{1}{a_i} \displaystyle\prod_{p \leqslant p_i} \left(1 - \frac{1}{p}\right),$

we obtain (4.1) by letting $n \to \infty$ in (4.2).

THEOREM 3. *For every primitive sequence $\mathscr{A}$, $\underline{\mathbf{d}}\mathscr{A} = \delta\mathscr{A} = 0$.*

Proof. In view of Lemma 1, it suffices to prove $\delta\mathscr{A} = 0$; and to see that this is a consequence of Theorem 2, we need only note that

$$\frac{1}{\log n} \sum_{N < a_i \leqslant n} \frac{1}{a_i} \leqslant \sum_{N < a_i \leqslant n} \frac{1}{a_i \log a_i}$$

in (2.1).

Theorem 3 might, at first sight, suggest that perhaps $\mathbf{d}\mathscr{A} = 0$ also. This is not always the case, however, and, as was remarked earlier, Theorem 1 is essentially best possible. We express this more precisely in the following theorem.

THEOREM 4 (Besicovitch[3]). *Corresponding to every $\epsilon > 0$, there exists a primitive sequence $\mathscr{A}$, depending on ϵ, such that $\bar{\mathbf{d}}\mathscr{A} > \frac{1}{2} - \epsilon$.*

Theorem 5 below represents a qualitative improvement on Theorem 3. The subsequence whose existence is asserted satisfies a condition which

(2) P. Erdös, 'Note on sequences of integers no one of which is divisible by any other', *J. Lond. math. Soc.* **10** (1935) 126–8.

(3) A. S. Besicovitch, 'On the density of certain sequences of integers', *Math. Annln* **110** (1934) 336–41.

is in the strongest possible contrast to the defining property (ii) (cf. § 1) of a primitive sequence. In other words, $\delta\mathscr{A} > 0$ implies much more than the mere 'non-primitiveness' of $\mathscr{A}$.

THEOREM 5 (Davenport–Erdös[4]). *Any sequence $\mathscr{A}$ of positive upper logarithmic density contains a subsequence $a_{i_1}, a_{i_2},...$ such that $a_{i_r} \mid a_{i_{r+1}}$ $(r = 1, 2, 3,...)$.*

The proofs of Theorems 4 and 5 require some preparation and are deferred to § 5.

Theorem 3 is equivalent to

$$\frac{1}{\log n} \sum_{a_i \leqslant n} \frac{1}{a_i} = o(1) \quad \text{as } n \to \infty, \tag{4.3}$$

and it is reasonable to seek also a quantitative improvement in which the right-hand side of (4.3) is replaced by $O\{\phi(n)\}$ for some specific function $\phi(n)$ which tends to zero as n tends to infinity. If we re-examine the proof of Theorem 2 with a view to obtaining such an improvement on (4.3), we observe that its effectiveness is limited by the need to choose $\mathscr{S}^{(i)}$ rather thin in order to ensure the discreteness of the sequences $\mathscr{R}^{(i)} = a_i \mathscr{S}^{(i)}$. This discreteness, which ensures that every integer u lies in at most one of the sequences $\mathscr{R}^{(i)}$, is not essential to the method, however. It would suffice instead to know that 'on average' an integer u does not lie in too many sequences $\mathscr{R}^{(i)}$. This remark suggests a line of approach which proves successful. Even when $\mathscr{R}^{(i)}$ is taken to consist of *all* the multiples of a_i, the 'primitiveness' of $\mathscr{A}$ restricts the average overlap of the sequences $\mathscr{R}^{(i)}$; and, as we shall now show, sufficiently so to enable us to obtain a much improved form of (4.3).

Suppose then that $\mathscr{R}^{(i)}$ is the sequence of all positive multiples of a_i, and denote by $r(u)$ the number of these sequences which contain u. Then, writing

$$\rho(n) = \sum_{u \leqslant n} r(u) \tag{4.4}$$

(this function is, in a sense, a measure of the 'overlapping' of the sequences $\mathscr{R}^{(i)}$), we have

$$\rho(n) = \sum_{ma_i \leqslant n} 1 = \sum_{a_i \leqslant n} \left[\frac{n}{a_i}\right] = n \sum_{a_i \leqslant n} \frac{1}{a_i} + O(n),$$

(4) H. Davenport and P. Erdös, 'On sequences of positive integers', *Acta arith.* **2** (1937) 147–51; *J. Indian math. Soc.* **15** (1951) 19–24.

so that

$$\sum_{a_i \leqslant n} \frac{1}{a_i} = \frac{1}{n}\rho(n) + O(1). \tag{4.5}$$

To prove Theorem 3 it would suffice to show that $\rho(n) = o(n \log n)$. Actually, we shall prove that

$$\rho(n) \ll n \frac{\log n}{(\log\log n)^{\frac{1}{2}}} \quad (n \geqslant 3), \tag{4.6}$$

which yields at once:

Theorem 6 (Behrend[(5)]). *If $\mathscr{A}$ is a primitive sequence,*

$$\frac{1}{\log n} \sum_{a_i \leqslant n} \frac{1}{a_i} \ll \frac{1}{(\log\log n)^{\frac{1}{2}}} \quad (n \geqslant 3), \tag{4.7}$$

where the implied constant is absolute.

The fact that the constant implied by the symbol is absolute represents an important feature of this result, for it means that exactly the same estimate is obtained for every primitive sequence $\mathscr{A}$.

To prove (4.6) we shall estimate $r(u)$, the number of divisors of u belonging to $\mathscr{A}$. We have

$$r(u) \leqslant s(u), \tag{4.8}$$

where $s(u)$ denotes the order (number of elements) of a maximal primitive set of divisors of u. Furthermore, if all the elements of $\mathscr{A}$ are square-free, we may replace (4.8) by

$$r(u) \leqslant s(u'), \tag{4.8$'$}$$

where u' denotes the square-free part of u (i.e. the product of all the distinct prime factors of u).

We shall see later that it suffices to establish (4.7) in the special case when $\mathscr{A}$ consists entirely of square-free integers, since the general result is easily deduced from this special case. We shall therefore be in a position to apply (4.8′), and need only estimate $s(v)$ for square-free v.

Accordingly, we now suppose that v is square-free (but see the remarks immediately below formula (4.18) later in the section). The number $s(v)$ is fully determined by the prime decomposition of v and is entirely independent of the sequence $\mathscr{A}$. The problem of estimating $s(v)$ is of a combinatorial nature; we shall first express it in a much more general form.

We note that the prime factors of a divisor d of the square-free number v form a subset M of the set $\mathfrak{M}$ of all distinct prime factors of v;

(5) F. Behrend, 'On sequences of numbers not divisible one by another', *J. Lond. math. Soc.* **10** (1935) 42–4.

and if d_1, d_2 are two such divisors, the relations $d_1 \mid d_2$ and $M_1 \subset M_2$ are equivalent. Our problem thus takes the following simple form.

Let $\mathfrak{M}$ be a set of n elements, and denote by Σ a collection of subsets of $\mathfrak{M}$. We say that Σ is primitive if no member of Σ is contained in another. What is the order of a maximal primitive collection Σ?

A more searching question would be to ask for the complete characterization of all maximal primitive collections. This question, one of independent interest, was settled by Sperner(6), who proved that a primitive Σ is maximal if and only if either Σ or Σ^c, the complement of Σ with respect to $\mathfrak{M}$, is the collection of all the distinct subsets of order $[\frac{1}{2}n]$ of $\mathfrak{M}$. The order of a maximal Σ is thus $\binom{n}{[\frac{1}{2}n]}$.

Since we are interested in the order (number of elements), and not the structure, of a maximal Σ, we shall content ourselves with a weaker form of Sperner's result.

THEOREM 7 (Sperner). *If Σ is primitive it consists of at most $\binom{n}{[\frac{1}{2}n]}$ subsets of $\mathfrak{M}$.*

We shall require the following lemma.

LEMMA 2. *If $k > [\frac{1}{2}n]$, then any m distinct subsets of $\mathfrak{M}$, each of order k, contain at least m distinct subsets of $\mathfrak{M}$ of order $k-1$.*

Proof. We denote by $M^{(k)}$ a subset of $\mathfrak{M}$ containing k elements. Every $M^{(k)}$ contains exactly k different subsets $M^{(k-1)}$. Thus m subsets of order k contain between them mk subsets of order $k-1$, although some of these may coincide. However, an $M^{(k-1)}$ can occur at most $(n-k+1)$ times as a subset of sets $M^{(k)}$ and so, if r among the mk subsets $M^{(k-1)}$ are distinct,

$$mk \leqslant r(n-k+1).$$

Now if $k > [\frac{1}{2}n]$, then $k \geqslant \frac{1}{2}(n+1)$ so that $n-k+1 \leqslant k$. Hence $r \geqslant m$ and the lemma is proved.

We are now in a position to prove Theorem 7. Suppose that Σ is a primitive collection containing $N(\Sigma)$ subsets of $\mathfrak{M}$. Suppose further that no subset in Σ has order greater than k and that exactly m of the subsets have order k. If $k > [\frac{1}{2}n]$, we replace these m subsets, in accordance with Lemma 2, by m subsets of order $k-1$. The resulting collection will again be primitive, since $k-1$ is the new maximal order of subsets and each new subset is contained in one of the discarded ones. Repeating this process if necessary, we arrive at a primitive

(6) E. Sperner, 'Ein Satz über Untermengen einer endlichen Menge', *Math. Z.* **27** (1928) 544–8.

collection Σ_1, with $N(\Sigma_1) = N(\Sigma)$, containing no subset of order exceeding $[\frac{1}{2}n]$. The complement Σ_1^c of Σ_1 with respect to $\mathfrak{M}$ is also primitive and contains no subset of order less than $[\frac{1}{2}n]$. Applying to Σ_1^c the same process previously applied to Σ, we arrive at a primitive collection Σ_2, with $N(\Sigma_2) = N(\Sigma_1^c)$, consisting entirely of subsets of order $[\frac{1}{2}n]$. Thus

$$N(\Sigma) = N(\Sigma_1) = N(\Sigma_1^c) = N(\Sigma_2) \leqslant \binom{n}{[\frac{1}{2}n]}$$

and Theorem 7 is proved.

If the set $\mathfrak{M}$ is now taken to consist of the $\omega(v)$ prime factors† of *a square-free* number v, Theorem 7 yields

$$s(v) \leqslant \binom{\omega(v)}{[\frac{1}{2}\omega(v)]}.$$

Assuming (temporarily) that $\mathscr{A}$ is a primitive sequence of *square-free* numbers, we obtain, by (4.8′),

$$r(u) \leqslant \binom{\omega(u')}{[\frac{1}{2}\omega(u')]} = \binom{\omega(u)}{[\frac{1}{2}\omega(u)]} \tag{4.9}$$

for every natural number u.

From Stirling's formula

$$n! \sim (2\pi)^{\frac{1}{2}} n^{n+\frac{1}{2}} e^{-n},$$

it follows that

$$\binom{n}{[\frac{1}{2}n]} \ll \frac{2^n}{n^{\frac{1}{2}}}.$$

Hence, by (4.4) and (4.9), we have

$$\begin{aligned} \rho(n) &\ll \Big(\sum_{\substack{u \leqslant n \\ \omega(u) \leqslant k}} + \sum_{\substack{u \leqslant n \\ \omega(u) > k}} \Big) \frac{2^{\omega(u)}}{\{\omega(u)\}^{\frac{1}{2}}} \\ &\ll \frac{2^k}{k^{\frac{1}{2}}} n + \frac{1}{k^{\frac{1}{2}}} \sum_{u \leqslant n} 2^{\omega(u)} \\ &\ll \frac{2^k}{k^{\frac{1}{2}}} n + \frac{1}{k^{\frac{1}{2}}} \sum_{u \leqslant n} d(u) \\ &\ll n \frac{\log n}{(\log\log n)^{\frac{1}{2}}} \end{aligned}$$

on applying‡ Lemma 6 of the Appendix, and choosing $k = \log\log n$. This proves (4.6), and hence also Theorem 6, in the case when the integers of $\mathscr{A}$ are square-free. To remove this restriction, suppose now

† As is explained in § 4 of the Appendix, $\omega(u)$ denotes the number of distinct prime factors of an integer u. We note, in connexion with (4.9) below, that $\omega(u) = \omega(u')$, where u' is the square-free part of u (in the sense explained immediately below (4.8′)).

‡ Obviously $2^{\omega(u)} \leqslant d(u)$, where $d(u)$ denotes the number of divisors of u (see § 4 of the Appendix).

that $\mathscr{A} = \{a_i\}$ is an unrestricted primitive sequence. Let $\{a_i^{(k)}\}$ be the subsequence of $\mathscr{A}$ consisting of all those integers of $\mathscr{A}$ which have k^2 as their greatest square factor. Writing $a_i^{(k)} = k^2 q_i^{(k)}$ $(i = 1, 2, 3, \ldots)$, so that $q_i^{(k)}$ is square-free, we have

$$\sum_{a_i \leqslant n} \frac{1}{a_i} = \sum_{k=1}^{\infty} \sum_{a_i^{(k)} \leqslant n} \frac{1}{a_i^{(k)}} = \sum_{k=1}^{\infty} \frac{1}{k^2} \sum_{q_i^{(k)} \leqslant n/k^2} \frac{1}{q_i^{(k)}} \leqslant \sum_{k=1}^{\infty} \frac{1}{k^2} \sum_{q_i^{(k)} \leqslant n} \frac{1}{q_i^{(k)}}.$$

But $\{q_i^{(k)}\}$ is a primitive sequence of square-free integers, so that we may apply to it the restricted form of Theorem 7 and obtain

$$\sum_{q_i^{(k)} \leqslant n} \frac{1}{q_i^{(k)}} \ll \frac{\log n}{(\log\log n)^{\frac{1}{2}}}.$$

Since $\sum k^{-2}$ converges, the proof of Theorem 6 is now complete.

Theorem 6 is best possible in the sense that (4.7) cannot be improved if the constant implied by the symbol $\ll$ is to be absolute. This is an immediate consequence of the following result.

THEOREM 8 (S. Pillai[7]). *There exists a positive absolute constant* c, *such that corresponding to every number* $x \geqslant 3$, *there exists a primitive sequence* $\mathscr{U}$, *depending on* x, *for which*

$$\frac{1}{\log x} \sum_{u_i \leqslant x} \frac{1}{u_i} > \frac{c}{(\log\log x)^{\frac{1}{2}}}. \tag{4.10}$$

Proof. It clearly suffices to prove the result for large x and, accordingly, we shall suppose that x is sufficiently large. We shall prove the result by taking $\mathscr{U}$ to be the sequence of all integers $u_i = u_i^{(r)}$ which are composed of exactly r prime factors, counted according to their multiplicities, with r chosen appropriately as a function of x.

Let the constant c_1 be chosen so large that each of the inequalities

$$\sum_{p \leqslant x} \frac{1}{p} \geqslant \log\log x, \tag{4.11}$$

$$\sum_{p \leqslant x} \frac{\log p}{p} \leqslant 2\log x \tag{4.12}$$

and

$$\log\log x > 8 \tag{4.13}$$

is true for $x > c_1$; this is possible by virtue of Lemmas 3 and 2 of the Appendix. Assume also that

$$r \leqslant 2(\log\log x - 4). \tag{4.14}$$

(7) S. Pillai, 'On numbers which are not multiples of any other in the set', *Proc. Indian Acad. Sci.* A **10** (1939) 392–4.

Then:

LEMMA 3. *If* $x > c_1^{2^{r-1}}$,

$$\lambda_r(x) = \sum_{u_i^{(r)} \leqslant x} \frac{1}{u_i^{(r)}} \geqslant \frac{(\log\log x - 4)^r}{r!}.$$

Proof. By (4.11) the lemma is true when $r = 1$. We proceed by induction on r. If $x > c_1^{2^r}$ and we assume the result to hold for $r \geqslant 1$, we have

$$(r+1)\lambda_{r+1}(x) \geqslant \sum_{pu_i^{(r)} \leqslant x} \frac{1}{pu_i^{(r)}} \geqslant \sum_{p \leqslant x^{\frac{1}{2}}} \frac{1}{p} \lambda_r\left(\frac{x}{p}\right).$$

Since $p \leqslant x^{\frac{1}{2}}$, $(x/p) > c_1^{2^{r-1}}$ and it follows from the induction hypothesis that

$$(r+1)\lambda_{r+1}(x) \geqslant \frac{1}{r!} \sum_{p \leqslant x^{\frac{1}{2}}} \frac{1}{p}\left(\log\log\frac{x}{p} - 4\right)^r$$

$$= \frac{(\log\log x - 4)^r}{r!} \tau_r(x),$$

where
$$\tau_r(x) = \sum_{p \leqslant x^{\frac{1}{2}}} \frac{1}{p}\left\{1 + \frac{\log(1 - \log p/\log x)}{\log\log x - 4}\right\}^r.$$

To complete the inductive step it suffices to prove that if $x > c_1^{2^r}$,

(4.15) $$\tau_r(x) \geqslant \log\log x - 4.$$

Since $p \leqslant x^{\frac{1}{2}}$ it is easily seen that

$$0 < -\log\left(1 - \frac{\log p}{\log x}\right) \leqslant (2\log 2)\frac{\log p}{\log x},$$

and it follows that

$$\left\{1 + \frac{\log(1 - \log p/\log x)}{\log\log x - 4}\right\}^r \geqslant \left\{1 - \frac{(2\log 2)\log p/\log x}{\log\log x - 4}\right\}^r$$

$$\geqslant 1 - \frac{(2\log 2) r \log p}{(\log\log x - 4)\log x}$$

$$\geqslant 1 - (4\log 2)\frac{\log p}{\log x};$$

the last step being justified by (4.14). Hence

$$\tau_r(x) \geqslant \sum_{p \leqslant x^{\frac{1}{2}}} \frac{1}{p} - \frac{(4\log 2)}{\log x} \sum_{p \leqslant x^{\frac{1}{2}}} \frac{\log p}{p},$$

and since $x > c_1^{2^r} \geqslant c_1^2$, we conclude from (4.11) and (4.12) that

$$\tau_r(x) \geqslant \log\log x - \log 2 - 4\log 2 > \log\log x - 4;$$

thus (4.15) is true, and the proof of the lemma is now complete.

The proof of Theorem 8 follows almost at once. We choose x so large that

$$x > c_1^{2^{\log\log x - 1}},$$

and then take $r = [\log\log x]$. The condition $x > c_1^{2^{r-1}}$ is obviously satisfied, and (4.13) ensures that (4.14) remains true. Hence, by Lemma 3,

$$\lambda_r(x) \geqslant \frac{(\log\log x - 4)^{[\log\log x]}}{([\log\log x])!}, \qquad r = [\log\log x],$$

and it follows by Stirling's formula that there exists an absolute constant c such that, when $r = [\log\log x]$,

$$\lambda_r(x) \geqslant c\frac{\log x}{(\log\log x)^{\frac{1}{2}}},$$

the desired result.

It is clear that each of Theorems 6 and 8 is really a result about finite sets of integers. Suppose we write, for each natural number n,

$$L(n) = \sup_{\substack{\mathscr{A} \subset [1,n] \\ \mathscr{A}\text{ primitive}}} \sum \frac{1}{a_i}$$

where the supremum is taken over all primitive subsequences $\mathscr{A} = \{a_i\}$ of the (finite) sequence

$$1, 2, 3, \ldots, n.$$

Then Theorem 6 is equivalent to

$$\frac{(\log\log n)^{\frac{1}{2}}}{\log n} L(n) \leqslant c' \quad \text{for large } n; \tag{4.16}$$

whilst Theorem 8 is equivalent to

$$\frac{(\log\log n)^{\frac{1}{2}}}{\log n} L(n) \geqslant c'' \quad \text{for large } n. \tag{4.17}$$

Erdös has stated[9] that it is possible to show that (4.16) holds for every number c' satisfying $c' > 1/\surd(2\pi)$, and that (4.17) holds for every number c'' satisfying $c'' < 1/\surd(2\pi)$: thus claiming to establish

$$\lim_{n\to\infty} \frac{(\log\log n)^{\frac{1}{2}}}{\log n} L(n) = \frac{1}{\surd(2\pi)}. \tag{4.18}$$

In connexion with (4.16), Erdös had in mind the fact that the maximal primitive sub-sets of the divisors of u can be characterized[8] even when u is not square-free. However, in the absence of a new idea, this does not yield† (4.16) for an arbitrary $c' > 1/\surd(2\pi)$ as claimed.

† (*Added June* 1965.) This has recently been confirmed by I. Anderson, who has shown that Erdös's approach yields (4.16) only when $c' > 1/\surd\pi$.

(8) N. G. De Bruijn, C. van E. Tengbergen, and D. Kruyswijk, 'On the set of divisors of a number', *Nieuw Arch. f. Wisk.* Ser. II, **23** (1949–51) 191–3.

It is worth remarking that whilst we have followed Sperner's original approach in our proof of Theorem 7, the De Bruijn–Tengbergen–Kruyswijk method[8] is almost as simple and gives much greater insight. (In fact the latter method is simpler, even in the case considered by Sperner, if the maximal sets are to be characterized.)

To verify that (4.17) holds for an arbitrary $c'' < 1/\sqrt{(2\pi)}$, one needs to obtain a suitable refinement of Lemma 3 (appearing in the proof of Theorem 8). Erdös[9] has proved an asymptotic formula,† which suffices for this purpose.

Finally, we remark that it requires only a straightforward adaptation of the proof of Theorem 8 to obtain the following result, which relates to *infinite* primitive sequences in the true sense.

Corresponding to every function $\psi(x)$ which is monotonic decreasing and tends to zero as x tends to infinity, there exists a primitive sequence‡ $\mathscr{V}$ such that

$$\frac{1}{\log x}\sum_{v_i \leqslant x}\frac{1}{v_i} > \frac{\psi(x)}{(\log\log x)^{\frac{1}{2}}}$$

for infinitely many x.

We shall give only an outline of the proof of this result. Let

$$x_1 < x_2 < x_3 < \dots$$

be an infinite sequence which increases sufficiently rapidly (in a manner depending on the function $\psi(x)$) and with x_1 sufficiently large.

Let $\mathscr{V}^{(s)}$ be the set of those integers v not exceeding x_s composed of exactly $r_s = [\log\log x_s]$ prime factors, each greater than x_{s-1}. Provided x_s is sufficiently large as a function of x_{s-1}, it can be shown (by an adaptation of the method used to prove Theorem 8) that

$$\frac{1}{\log x_s}\sum_{v_i \in \mathscr{V}^{(s)}}\frac{1}{v_i} > \frac{\psi(x_s)}{(\log\log x_s)^{\frac{1}{2}}}.$$

The union of all the sets $\mathscr{V}^{(s)}$ now provides the required primitive sequence $\mathscr{V}$.

It may also be worth remarking that for every infinite primitive sequence $\mathscr{V}$, even the comparatively weak inequality

$$\frac{1}{\log x}\sum_{v_i \leqslant x}\frac{1}{v_i} > \frac{1}{\log\log x}$$

† See also, A. Selberg, 'Note on a paper by L. G. Sathe', *J. Indian math. Soc.*, N.S. **18** (1954) 83–7.

‡ $\mathscr{V}$ depends on the function ψ but not, of course, on x.

(9) P. Erdös, 'Integers with exactly k prime factors', *Ann. Math.* II **49** (1948) 53–66.

must be *false* for infinitely many x; for otherwise it would require only an Abel summation to obtain a contradiction to Theorem 2.

Our discussion of primitive sequences remains incomplete in that we have yet to give proofs of Theorems 4 and 5. But, to prepare the ground for these, it is convenient to turn now to the other main topic of this chapter.

Addendum (September 1964). Just as the book was due to go to press, we received a letter from Dr. Erdös informing us of a number of new results, due to Sárkozy, Szmerédi, and himself. Of these, one is particularly closely related to the work of this section, and we state it here. This result is analogous to Theorem 6, but relates to *infinite* primitive sequences in the true sense.

If $\mathscr{A}$ is an infinite primitive sequence, then

$$\sum_{a_i \leqslant n} \frac{1}{a_i} = o\left(\frac{\log n}{(\log\log n)^{\frac{1}{2}}}\right).$$

This result is best possible (cf. the final result of the section).

5. The set of multiples of a sequence: applications including the proofs of Theorems 4 and 5

We propose to investigate the distribution of the set, $\mathscr{B}(\mathscr{A})$, of (positive) multiples of the elements of a given sequence $\mathscr{A}$. We have shown in § 1 that for the purpose of such an investigation there would be no loss of generality in restricting $\mathscr{A}$ to be primitive. We shall not need to impose this restriction, however, as the methods used will automatically take advantage of the situation described in § 1.

We denote by $\mathscr{B}_m = \mathscr{B}_m(\mathscr{A})$ the set of multiples of the first m elements of $\mathscr{A}$, namely $a_1, a_2, \ldots, a_m$. As was remarked in the introduction, $\mathscr{B}_m(\mathscr{A})$ can be represented as the union of a finite number of congruence classes, and therefore possesses asymptotic density. If we denote by $B^{(i)}(n)$ the number of natural numbers, not exceeding n, which are divisible by a_i but not divisible by any one of $a_1, a_2, \ldots, a_{i-1}$, we have

$$B_m(n) = \sum_{i=1}^{m} B^{(i)}(n). \tag{5.1}$$

Now it is easily seen that for each $i = 1, 2, 3, \ldots$

$$B^{(i)}(n) = \left[\frac{n}{a_i}\right] - \sum_{j<i}\left[\frac{n}{[a_j, a_i]}\right] + \sum_{k<j<i}\left[\frac{n}{[a_k, a_j, a_i]}\right] - \cdots$$

so that

$$\mathbf{b}^{(i)} = \lim_{n\to\infty} \frac{B^{(i)}(n)}{n} = \frac{1}{a_i} - \sum_{j<i} \frac{1}{[a_j, a_i]} + \sum_{k<j<i} \frac{1}{[a_k, a_j, a_i]} - \cdots; \tag{5.2}$$

it follows from (5.1) that

(5.3) $$\mathbf{d}\mathscr{B}_m(\mathscr{A}) = \sum_{i=1}^{m} \mathbf{b}^{(i)}.$$

The right-hand side of (5.2) has rather a complicated appearance, but in the special case when $a_1, a_2, a_3, \dots, a_m$ are coprime in pairs we find that

(5.4) $$\mathbf{d}\mathscr{B}_m(\mathscr{A}) = 1 - \prod_{i=1}^{m} \left(1 - \frac{1}{a_i}\right).$$

It is evident from (5.2) that $\mathbf{b}^{(i)} \geqslant 0$, and from (5.3) that for every m,

$$0 < \sum_{i=1}^{m} \mathbf{b}^{(i)} < 1.$$

Hence $\sum \mathbf{b}^{(i)}$ is a convergent series; if we write

(5.5) $$\mathbf{b} = \sum_{i=1}^{\infty} \mathbf{b}^{(i)},$$

we have

(5.6) $$0 < \mathbf{b} \leqslant 1,$$

and, by (5.3),

(5.7) $$\mathbf{b} = \lim_{m \to \infty} \mathbf{d}\mathscr{B}_m(\mathscr{A}).$$

The question as to whether or not $\mathscr{B}(\mathscr{A})$ possesses asymptotic density is more profound. Since $\mathscr{B}_m(\mathscr{A}) \subset \mathscr{B}(\mathscr{A})$ we can say at once that

(5.8) $$\mathbf{b} \leqslant \underline{\mathbf{d}}\mathscr{B}(\mathscr{A}).$$

On the other hand, if $\sum a_i^{-1}$ converges we have

$$\bar{\mathbf{d}}\mathscr{B}(\mathscr{A}) \leqslant \mathbf{d}\mathscr{B}_m(\mathscr{A}) + \sum_{i=m+1}^{\infty} \frac{1}{a_i},$$

so that on letting $m \to \infty$ we obtain the following theorem.

THEOREM 9. *If $\sum a_i^{-1}$ converges, $\mathbf{d}\mathscr{B}(\mathscr{A})$ exists and is equal to* $\mathbf{b}$.

It was conjectured at one time that the conclusion of this theorem remains valid even without the restriction $\sum a_i^{-1} < \infty$. This conjecture was disproved by Besicovitch,† who constructed a sequence $\mathscr{A}$ for which $\mathbf{d}\mathscr{B}(\mathscr{A})$ does not exist. We first prove the following theorem, which will be required in connexion with Besicovitch's construction.‡

† A. S. Besicovitch, loc. cit., see ref. (3).

‡ Besicovitch based his construction on a weaker form of Theorem 10; it was this which led Erdös to discover the stronger result.

THEOREM 10 (Erdös).† *Let $\mathscr{I}_T$ denote the set of integers lying in the interval $T < t \leqslant 2T$. Then*

$$\lim_{T\to\infty} \mathbf{d}\mathscr{B}(\mathscr{I}_T) = 0.$$

The theorem is a fairly straightforward deduction from the following special case of Lemma 7 of the Appendix. (This special case is obtained by taking $\mathscr{P}$ to be the set of primes $p < T$ in the corollary to the lemma, choosing $k = \frac{1}{4}$, and using also Lemma 3 of the Appendix.)

Let $\nu_T^*(u)$ denote the number of those prime factors of u (counted according to their multiplicities) which are less than T. Let $\delta > 0$ be freely chosen. Then there exists a number $T_0 = T_0(\delta)$ such that, if $n > T > T_0$, the number of natural numbers $u \leqslant n$ for which

$$|\nu_T^*(u) - \log\log T| \geqslant \tfrac{1}{3}\log\log T$$

is less than δn.

We subdivide the set $\{x\}$ of all natural numbers into two classes $\{x^{(1)}\}$ and $\{x^{(2)}\}$; denoting x by $x^{(1)}$ or $x^{(2)}$ according as

$$\nu_T^*(x) \leqslant \tfrac{2}{3}\log\log T \quad \text{or} \quad \nu_T^*(x) > \tfrac{2}{3}\log\log T.$$

Let t denote an integer of $\mathscr{I}_T$ and let $E_n(tm)$ be the number of numbers, not exceeding n, representable in the form tm. Since every integer of the form tm is representable in one of the forms $t^{(1)}m$, $t^{(2)}m^{(1)}$, $t^{(2)}m^{(2)}$, we have

$$E_n(tm) \leqslant E_n(t^{(1)}m) + E_n(t^{(2)}m^{(1)}) + E_n(t^{(2)}m^{(2)}).$$

We shall prove that if $n > 2T^2$ and $T > T_0(\delta)$,

$$E_n(tm) < 4\delta n. \tag{5.9}$$

In fact,

$$E_n(t^{(1)}m) \leqslant \sum_{t^{(1)}} \frac{n}{t^{(1)}} \leqslant \frac{n}{T}\sum_{t^{(1)}} 1 < \frac{n}{T}(2T\delta) = 2\delta n,$$

$$E_n(t^{(2)}m^{(1)}) \leqslant \sum_{t}\ \sum_{m^{(1)}\leqslant(n/t)} 1 < \sum_{t}\left(\frac{n}{t}\right)\delta < \delta n$$

since $(n/t) \geqslant (n/2T) > T$, and finally

$$E_n(t^{(2)}m^{(2)}) < \delta n,$$

since $\nu_T^*(t^{(2)}m^{(2)}) > \frac{4}{3}\log\log T$ for any integer of the form $t^{(2)}m^{(2)}$. This proves (5.9), thus establishing Theorem 10.

We are now in a position to prove the following theorem.

† P. Erdös, loc. cit., see ref. (2).

THEOREM 11 (Besicovitch). *There exists an infinite sequence of natural numbers whose set of multiples does not possess asymptotic density.*

Proof. As in Theorem 10, we denote by $\mathscr{I}_T$ the set of integers t satisfying $T < t \leqslant 2T$. We observe that $\mathscr{B}(\mathscr{I}_T)$ can be represented as the union of a number of congruence classes to the modulus $(2T)!$; thus if $l > (2T)!$, any sequence of l consecutive integers contains at most $\{\mathbf{d}\mathscr{B}(\mathscr{I}_T)\}2l$ members of $\mathscr{B}(\mathscr{I}_T)$.

Let ϵ and $\epsilon_k = (\frac{1}{2})^{k+1}\epsilon$ $(k = 1, 2, 3,...)$ be positive numbers satisfying

$$(5.10) \qquad 0 < \epsilon < \tfrac{1}{2}, \qquad \sum_{k=1}^{\infty} \epsilon_k = \tfrac{1}{2}\epsilon.$$

In view of Theorem 10, we may now define an infinite sequence $T_1, T_2,...$ of natural numbers satisfying both

$$(5.11) \qquad \mathbf{d}_k = \mathbf{d}\mathscr{B}(\mathscr{I}_{T_k}) < \epsilon_k$$

and

$$(5.12) \qquad T_{k+1} > (2T_k)!$$

for $k = 1, 2, 3,...$.

Denote by $\mathscr{G}$ the union of the sets $\mathscr{I}_{T_k}$. The number of integers in $\mathscr{B}(\mathscr{G})$ which do not exceed $2T_k$ is at least T_k (the number of integers in $\mathscr{I}_{T_k}$), so that

$$\bar{\mathbf{d}}\mathscr{B}(\mathscr{G}) \geqslant \tfrac{1}{2}.$$

Moreover, the number of integers in $\mathscr{B}(\mathscr{G})$ which do not exceed T_k is at most

$$\sum_{r=1}^{k-1} (2\mathbf{d}_r T_k) < \epsilon T_k,$$

so that

$$\underline{\mathbf{d}}\mathscr{B}(\mathscr{G}) \leqslant \epsilon.$$

Since $\epsilon < \frac{1}{2}$, the proof of Theorem 11 is complete.

The sequence $\mathscr{G}$ introduced in the proof of Theorem 11 is not necessarily primitive. Let $\mathscr{G}'$ be the minimal (primitive) generating sequence of $\mathscr{B}(\mathscr{G})$ in the sense of § 1; so that $\mathscr{G}'$ is obtained from $\mathscr{G}$ by removing from each interval $\mathscr{I}_{T_k}$ $(k = 2, 3,...)$ all those integers which belong to $\mathscr{B}(\mathscr{I}_{T_1})$, $\mathscr{B}(\mathscr{I}_{T_2}),...$, $\mathscr{B}(\mathscr{I}_{T_{k-1}})$. As is to be expected in view of the fact that $\mathscr{B}(\mathscr{G})$ does not possess asymptotic density, the distribution of $\mathscr{G}'$ is highly irregular. In fact, the sequence $\mathscr{G}'$ provides the example required to prove Theorem 4. More precisely, we have the following result.

COROLLARY. *The minimal (primitive) generating sequence* $\mathscr{G}'$ of $\mathscr{B}(\mathscr{G})$ *satisfies*

$$(5.13) \qquad \bar{\mathbf{d}}(\mathscr{G}') \geqslant \tfrac{1}{2}(1-\epsilon)$$

so that, in particular, Theorem 4 is true.

Proof. By (5.12), (5.11), and (5.10) the number of elements of $\mathscr{G}'$ which lie in $\mathscr{I}_{T_k}$ is at least

$$T_k - \sum_{r=1}^{k-1} (2\mathbf{d}_r T_k) > T_k(1-\epsilon),$$

so that (5.13) is true.

We now come to a theorem which shows that the distribution of the set of multiples of any infinite sequence is regular to some degree, and clarifies the relationship between this distribution and the value of $\mathbf{b}$.

THEOREM 12 (Davenport–Erdös).† *For any infinite sequence $\mathscr{A}$ the set of multiples, $\mathscr{B}(\mathscr{A})$, possesses logarithmic density, and*

$$\delta\mathscr{B}(\mathscr{A}) = \underline{\mathbf{d}}\mathscr{B}(\mathscr{A}) = \mathbf{b}.$$

Proof. In view of Lemma 1 and (5.8) it suffices to prove that

$$\delta\mathscr{B}(\mathscr{A}) \leqslant \mathbf{b}. \tag{5.14}$$

Let $p_1, p_2,..., p_k$ be the first k prime numbers, and let $\mathscr{A}^{(k)}$ denote the sequence of those integers of $\mathscr{A}$ which are composed entirely of $p_1, p_2,..., p_k$. Also let $\mathscr{N}^{(k)}$ denote the sequence of all natural numbers composed entirely of $p_1, p_2,..., p_k$, so that $\mathscr{A}^{(k)}$ is the intersection of $\mathscr{A}$ and $\mathscr{N}^{(k)}$. It is clear that

$$\sum_{i=1}^{\infty} \frac{1}{n_i^{(k)}} = \prod_{i=1}^{k} \left(1-\frac{1}{p_i}\right)^{-1} = \Pi_k,$$

say, and, in particular, that

$$\sum_{i=1}^{\infty} \frac{1}{a_i^{(k)}}$$

converges. It follows from Theorem 9 that $\mathbf{d}\mathscr{B}(\mathscr{A}^{(k)})$ exists. The proof of Theorem 12 will be based on the following result.

LEMMA 4. $$\lim_{k\to\infty} \mathbf{d}\mathscr{B}(\mathscr{A}^{(k)}) = \mathbf{b}.$$

Proof. Let $\mathscr{M}^{(k)}$ denote the intersection of $\mathscr{B}(\mathscr{A}^{(k)})$ and $\mathscr{N}^{(k)}$. In other words, $\mathscr{M}^{(k)}$ consists of those members of $\mathscr{B}(\mathscr{A})$ which are composed entirely of $p_1, p_2,..., p_k$. The integers $m_1^{(k)}, m_2^{(k)},...$ of $\mathscr{M}^{(k)}$ are thus numbers of the form $a^{(k)}n^{(k)}$, and it is easy to see that

$$\sum_{i=1}^{\infty} \frac{1}{m_i^{(k)}} = \frac{1}{a_1^{(k)}} \sum_{i=1}^{\infty} \frac{1}{n_i^{(k)}} + \left(\frac{1}{a_2^{(k)}} - \frac{1}{[a_1^{(k)}, a_2^{(k)}]}\right) \sum_{i=1}^{\infty} \frac{1}{n_i^{(k)}} + \cdots \tag{5.15}$$
$$= \Pi_k \, \mathbf{d}\mathscr{B}(\mathscr{A}^{(k)}).$$

† Davenport and Erdös, loc. cit., ref. (4). Their earlier proof of this theorem was based on a deep Tauberian theorem of Hardy and Littlewood.

The next step is justified by virtue of Theorem 9 applied to $\mathscr{A}^{(k)}$. If we write

$$\mathbf{m}_k = \frac{1}{\Pi_k} \sum_{i=1}^{\infty} \frac{1}{m_i^{(k)}},$$

so that, in a certain sense, $\mathbf{m}_k$ measures the proportion of members of $\mathscr{M}^{(k)}$ lying in $\mathscr{N}^{(k)}$, we have, by (5.15), that

$$\mathbf{m}_k = \mathbf{d}\mathscr{B}(\mathscr{A}^{(k)}); \tag{5.16}$$

in particular, it is plain that $\mathbf{m}_k$ increases with k and is always less than 1. Hence

$$\mathbf{m} = \lim_{k\to\infty} \mathbf{m}_k = \lim_{k\to\infty} \mathbf{d}\mathscr{B}(\mathscr{A}^{(k)}) \tag{5.17}$$

exists. It remains to prove that

$$\mathbf{m} = \mathbf{b}. \tag{5.18}$$

We observe first of all that if an integer r is assigned, $k = k(r)$ can be chosen large enough to ensure that $a_1, a_2,..., a_r$ occur in $\mathscr{A}^{(k)}$. Then $\mathbf{d}\mathscr{B}(\mathscr{A}^{(k)}) \geqslant \mathbf{d}\mathscr{B}_r(\mathscr{A})$. If now we let $r \to \infty$, we have by (5.7) and (5.17) that

$$\mathbf{m} \geqslant \mathbf{b}. \tag{5.19}$$

On the other hand, given any $\epsilon > 0$ and a fixed k, a sufficiently large integer $r_0 = r_0(\epsilon, k)$ can be found so that

$$\sum_{i=r+1}^{\infty} \frac{1}{a_i^{(k)}} < \epsilon \quad (r \geqslant r_0).$$

As in the proof of Theorem 9,

$$\mathbf{d}\mathscr{B}(\mathscr{A}^{(k)}) \leqslant \mathbf{d}\mathscr{B}_r(\mathscr{A}^{(k)}) + \sum_{i=r+1}^{\infty} \frac{1}{a_i^{(k)}} < \mathbf{d}\mathscr{B}_r(\mathscr{A}^{(k)}) + \epsilon;$$

if now we choose s so large that $a_1^{(k)}, a_2^{(k)},..., a_r^{(k)}$ are included among a_1, $a_2,..., a_s$, we have

$$\mathbf{d}\mathscr{B}(\mathscr{A}^{(k)}) < \mathbf{d}\mathscr{B}_s(\mathscr{A}) + \epsilon,$$

so that

$$\mathbf{d}\mathscr{B}(\mathscr{A}^{(k)}) < \mathbf{b} + \epsilon$$

for every $\epsilon > 0$ and every k. Hence $\mathbf{m} \leqslant \mathbf{b} + \epsilon$ for every $\epsilon > 0$, and therefore $\mathbf{m} \leqslant \mathbf{b}$. This, together with (5.19), proves (5.18), and completes the proof of the lemma.

We are now in a position to prove (5.14). For a fixed k, we have

$$\beta(n) = \sum_{b_i \leqslant n} \frac{1}{b_i} = \beta_1(n) + \beta_2(n) \tag{5.20}$$

where

$$\beta_1(n) = \sum_{\substack{b_i \leqslant n \\ b_i \in \mathscr{B}(\mathscr{A}^{(k)})}} \frac{1}{b_i} \quad \text{and} \quad \beta_2(n) = \sum_{\substack{b_i \leqslant n \\ b_i \notin \mathscr{B}(\mathscr{A}^{(k)})}} \frac{1}{b_i}. \tag{5.21}$$

We have already noted that since $\sum 1/a^{(k)}$ converges, $\mathbf{d}\mathscr{B}(\mathscr{A}^{(k)})$ exists. Hence, by Lemma 1,

$$\lim_{n\to\infty} \frac{\beta_1(n)}{\log n} = \mathbf{d}\mathscr{B}(\mathscr{A}^{(k)}) = \mathbf{m}_k. \tag{5.22}$$

We now consider $\beta_2(n)$. Let h be the unique positive integer defined by $p_h \leqslant n < p_{h+1}$. Then

$$\beta_2(n) \leqslant \sum_{\substack{i=1 \\ m_i^{(h)} \notin \mathscr{B}(\mathscr{A}^{(k)})}}^{\infty} \frac{1}{m_i^{(h)}}.$$

The integers $m_i^{(h)}$ counted on the right can be obtained by removing from $\mathscr{M}^{(h)}$ the sets $u_i \mathscr{M}^{(k)}$ $(i = 1, 2, 3,...)$ where u_1, u_2, u_3,... are all the distinct integers composed entirely of $p_{k+1}, p_{k+2},..., p_h$. Thus

$$\beta_2(n) \leqslant \sum_{i=1}^{\infty} \frac{1}{m_i^{(h)}} - \left(\sum_{i=1}^{\infty} \frac{1}{u_i}\right) \sum_{i=1}^{\infty} \frac{1}{m_i^{(k)}}.$$

But evidently

$$\sum_{i=1}^{\infty} \frac{1}{u_i} = \frac{\Pi_h}{\Pi_k},$$

so that it follows from (5.15) and (5.16) that

$$\begin{aligned} \beta_2(n) &\leqslant \Pi_h\{\mathbf{d}\mathscr{B}(\mathscr{A}^{(h)}) - \mathbf{d}\mathscr{B}(\mathscr{A}^{(k)})\} \\ &\leqslant (c_2 \log n)(\mathbf{m}_h - \mathbf{m}_k) \end{aligned}$$

by Lemma 4 of the Appendix. Letting $n \to \infty$, so that $h \to \infty$ also, we apply (5.17) to arrive at

$$\limsup_{n\to\infty} \frac{\beta_2(n)}{\log n} \leqslant c_2(\mathbf{m} - \mathbf{m}_k).$$

Hence, by (5.20) and (5.22),

$$\limsup_{n\to\infty} \frac{\beta(n)}{\log n} \leqslant \mathbf{m}_k + c_2(\mathbf{m} - \mathbf{m}_k);$$

the left-hand side is independent of k, so that, by letting $k \to \infty$ and applying (5.17), we obtain

$$\delta\mathscr{B}(\mathscr{A}) = \limsup_{n\to\infty} \frac{\beta(n)}{\log n} \leqslant \mathbf{m}.$$

Since $\mathbf{m} = \mathbf{b}$ by Lemma 4, the proof of (5.14) is now complete.

Theorem 12 leads us at last to a proof of Theorem 5. It will be recalled that Theorem 5 states: if $\mathscr{A}$ is an infinite sequence of positive upper logarithmic density, then $\mathscr{A}$ contains a subsequence $a_{i_1}, a_{i_2},\ldots$ with the property that $a_{i_k} \mid a_{i_{k+1}}$ $(k = 1, 2, 3,\ldots)$.

We may assume then that

$$\alpha = \limsup_{n\to\infty}(\log n)^{-1} \sum_{a_i \leqslant n} \frac{1}{a_i} > 0.$$

To prove Theorem 5 it suffices to show that there exists a number a_{i_1} such that

$$\limsup_{n\to\infty}(\log n)^{-1} \sum_{\substack{a_i \leqslant n \\ a_{i_1} | a_i}} \frac{1}{a_i} > 0. \tag{5.23}$$

For the same argument applied to the subsequence $\mathscr{A}' = \{a_i : a_{i_1} \mid a_i\}$ of $\mathscr{A}$ will then establish the existence of a number a_{i_2}, which belongs to $\mathscr{A}'$ and so is divisible by a_{i_1}, such that

$$\limsup_{n\to\infty}(\log n)^{-1} \sum_{\substack{a_i \leqslant n, a_i \in \mathscr{A}' \\ a_{i_2} | a_i}} \frac{1}{a_i} > 0;$$

it is clear that this procedure may be continued indefinitely. We need therefore justify only the first step.

Let r be chosen so large that

$$\sum_{i>r} \mathbf{b}^{(i)} < \alpha;$$

here $\mathbf{b}^{(i)}$, defined by (5.2), is the asymptotic density of the sequence of all natural numbers divisible by a_i but not by any of $a_1, a_2,\ldots, a_{i-1}$. We shall see that an integer a_{i_1} for which (5.23) is true may be found among $a_1, a_2,\ldots, a_r$. For if we suppose on the contrary that the left-hand side of (5.23) is 0 when a_{i_1} is replaced by each in turn of $a_1, a_2,\ldots, a_r$, then

$$\alpha = \limsup_{n\to\infty}(\log n)^{-1} \sum_{\substack{a_i \leqslant n \\ a_1 \nmid a_i, a_2 \nmid a_i,\ldots, a_r \nmid a_i}} \frac{1}{a_i}$$

$$\leqslant \limsup_{n\to\infty}(\log n)^{-1} \sum_{\substack{b_i \leqslant n \\ a_1 \nmid b_i, a_2 \nmid b_i,\ldots, a_r \nmid b_i}} \frac{1}{b_i}$$

where, as usual, $b_1, b_2,\ldots$ are the elements of $\mathscr{B}(\mathscr{A})$; but by Theorem 12 the last expression is equal to $\mathbf{b} - \sum_{i=1}^{r} \mathbf{b}^{(i)} = \sum_{i>r} \mathbf{b}^{(i)} < \alpha$ by our choice of r. We have thus been led to the contradiction $\alpha < \alpha$, showing that a_{i_1} satisfying (5.23) can be chosen from among $a_1, a_2,\ldots, a_r$. This completes the proof of Theorem 5.

It was proved earlier, in Theorem 9, that if $\sum a_i^{-1}$ converges, $\mathbf{d}\mathscr{B}(\mathscr{A})$ exists, and it is natural to ask whether there exist weaker conditions on $\mathscr{A}$ which ensure that $\mathscr{B}(\mathscr{A})$ possesses asymptotic density. This question is completely answered in the next section.

6. A necessary and sufficient condition for the set of multiples of a given sequence to possess asymptotic density

The main objective of this section is the following theorem.

THEOREM 13 (Erdös[10]). *A necessary and sufficient condition for* $\mathbf{d}\mathscr{B}(\mathscr{A})$ *to exist is*

$$\lim_{\epsilon\to+0}\limsup_{n\to\infty}\frac{1}{n}\sum_{n^{1-\epsilon}<a_i\leqslant n} B^{(i)}(n) = 0. \tag{6.1}$$

We recall that $B^{(i)}(n)$ was defined at the beginning of § 5 to be the number of natural numbers, not exceeding n, which are divisible by a_i but not by any one of $a_1, a_2,..., a_{i-1}$.

Condition (6.1) has a somewhat cumbersome appearance. Nevertheless, as we shall now show, a very simple sufficient condition, in terms of $A(n)$, can be obtained as an immediate consequence.

Since
$$B^{(i)}(n) \leqslant \left[\frac{n}{a_i}\right] \leqslant \frac{n}{a_i},$$
we have
$$\frac{1}{n}\sum_{n^{1-\epsilon}<a_i\leqslant n} B^{(i)}(n) \leqslant \sum_{n^{1-\epsilon}<m\leqslant n}\frac{1}{m}\{A(m)-A(m-1)\}$$
$$< \sum_{n^{1-\epsilon}<m\leqslant n}\left(\frac{1}{m}-\frac{1}{m+1}\right)A(m)+\frac{A(n)}{n+1}.$$

Hence, if there exists a constant c such that
$$A(m) \leqslant cm/(\log m) \quad \text{for large } m,$$
$$\frac{1}{n}\sum_{n^{1-\epsilon}<a_i\leqslant n} B^{(i)}(n) \ll \sum_{n^{1-\epsilon}<m\leqslant n}\frac{1}{m\log m}+\frac{1}{\log n} \ll \epsilon+\frac{1}{\log n},$$
and consequently (6.1) is satisfied. Thus we have deduced that:

THEOREM 14 (Erdös[10]). *If there exists a constant c such that*

$$A(m) \leqslant cm/(\log m) \quad \textit{for large } m, \tag{6.2}$$

then $\mathbf{d}\mathscr{B}(\mathscr{A})$ *exists.*

Furthermore, Theorem 14 is best possible, as we shall show by proving the following theorem.

(10) P. Erdös, 'On the density of some sequences of integers', *Bull. Am. math. Soc.* **54** (1948) 685–92.

THEOREM 15 (Erdös[10]). *If $\psi(m)$ is any monotonically increasing function which tends to infinity with m, there exists a sequence $\mathscr{A}$ such that $\mathbf{d}\mathscr{B}(\mathscr{A})$ does not exist, and yet*

$$A(m) \leqslant m\psi(m)/(\log m) \quad \textit{for large } m. \tag{6.3}$$

In the proofs of Theorems 13 and 15 we shall require the following three lemmas, of which the first is a consequence of the theorem proved in § 3, and the second is a straightforward application of Theorem 1 of Chapter IV (Sieve Methods).

LEMMA 5 (Behrend[11]). *Let $\mathbf{t}(q_1,\dots, q_l)$ denote the density of the sequence consisting of all those integers not divisible by any q_ν $(\nu = 1, 2,\dots, l)$. Then*

$$\mathbf{t}(q_1,\dots, q_r)\mathbf{t}(q_{r+1},\dots, q_{r+s}) \leqslant \mathbf{t}(q_1,\dots, q_{r+s}) \tag{6.4}$$

always.

Proof. We write

$$Q = \prod_{\nu=1}^{r+s} q_\nu; \tag{6.5}$$

$$\mathscr{V}^{(1)} = \bigcup_{\nu=1}^{r} \mathscr{R}(q_\nu, 0), \qquad \mathscr{V}^{(2)} = \bigcup_{\nu=r+1}^{r+s} \mathscr{R}(q_\nu, 0). \tag{6.6}$$

By the theorem of § 3, we have

$$\mathbf{d}\{\mathscr{V}^{(1)} \cup (\mathscr{V}^{(2)}+h)\} \geqslant \mathbf{d}(\mathscr{V}^{(1)} \cup \mathscr{V}^{(2)}),$$

and hence
$$\mathbf{d}\{\mathscr{V}^{(1)} \cap (\mathscr{V}^{(2)}+h)\} \leqslant \mathbf{d}(\mathscr{V}^{(1)} \cap \mathscr{V}^{(2)})$$

for every h. In particular,

$$\frac{1}{Q}\sum_{h=1}^{Q} \mathbf{d}\{\mathscr{V}^{(1)} \cap (\mathscr{V}^{(2)}+h)\} \leqslant \mathbf{d}(\mathscr{V}^{(1)} \cap \mathscr{V}^{(2)}). \tag{6.7}$$

But if $\mathscr{R}^{(1)}$, $\mathscr{R}^{(2)}$ are any two congruence classes modulo Q, we have

$$\frac{1}{Q}\sum_{h=1}^{Q} \mathbf{d}\{\mathscr{R}^{(1)} \cap (\mathscr{R}^{(2)}+h)\} = \frac{1}{Q^2}.$$

Hence, since each of $\mathscr{V}^{(1)}$, $\mathscr{V}^{(2)}$ may be represented as a union of congruence classes modulo Q, the value of the left-hand side of (6.7) is simply $(\mathbf{d}\mathscr{V}^{(1)})(\mathbf{d}\mathscr{V}^{(2)})$. Accordingly,

$$(\mathbf{d}\mathscr{V}^{(1)})(\mathbf{d}\mathscr{V}^{(2)}) \leqslant \mathbf{d}(\mathscr{V}^{(1)} \cap \mathscr{V}^{(2)}).$$

Rewriting this inequality in the form

$$(1-\mathbf{d}\mathscr{V}^{(1)})(1-\mathbf{d}\mathscr{V}^{(2)}) \leqslant 1-\mathbf{d}(\mathscr{V}^{(1)} \cup \mathscr{V}^{(2)}),$$

we have (6.4) as desired.

(11) F. Behrend, 'Generalization of an inequality of Heilbronn and Rohrbach', *Bull. Am. math. Soc.* **54** (1948) 681–4.

In the following lemmas, ϵ is a small positive number, and $\eta_0(\epsilon)$, $n_0(\epsilon)$ are large numbers depending at most on ϵ. Lemma 6 below is obtained from Theorem 1 of Chapter IV on writing $x = X$ and taking y to be the least integer exceeding η^ϵ.

LEMMA 6. *Let $X > \eta^{\frac{1}{2}}$, where $\eta > \eta_0(\epsilon)$. Then the number of natural numbers not exceeding X, which are composed entirely of prime factors $p > \eta^\epsilon$, is*

$$\ll X \prod_{p \leqslant \eta^\epsilon} \left(1 - \frac{1}{p}\right).$$

LEMMA 7. *Let $n > n_0(\epsilon)$ and let $m = m^{(1)}m^{(2)}$ be the unique representation† of m as the product of a number $m^{(1)}$ composed entirely of prime factors $p \leqslant n^{\epsilon^2}$ and a number $m^{(2)}$ composed entirely of prime factors $p > n^{\epsilon^2}$. Let $N(n, \epsilon)$ be the number of natural numbers m, not exceeding n, for which $m^{(1)} \geqslant n^{\frac{1}{2}\epsilon}$. Then*

$$N(n, \epsilon) \ll \epsilon n. \tag{6.8}$$

Proof. The exact power of p dividing $n!$ is

$$\sum_{\nu=1}^{\infty} \left[\frac{n}{p^\nu}\right] \ll \frac{n}{p}.$$

Hence, by Lemma 2 of the Appendix,

$$\sum_{m=1}^{n} \log m^{(1)} \ll \sum_{p < (n^\epsilon)^\epsilon} \frac{n}{p} \log p \ll n\epsilon \log n^\epsilon.$$

Since

$$\sum_{m=1}^{n} \log m^{(1)} \geqslant \tfrac{1}{2} N(n, \epsilon) \log n^\epsilon,$$

this proves the lemma.

Proof of Theorem 13. The proof of the necessity is quite easy. If $\mathbf{d}\mathscr{B}(\mathscr{A}) = \mathbf{b}$ (as must be the case, by Theorem 12, if the density exists), we have by (5.5), that

$$\lim_{n\to\infty} \frac{1}{n} B(n) = \mathbf{b} = \sum_{i=1}^{\infty} \mathbf{b}^{(i)}.$$

Hence, since

$$B(n) = \sum_{a_i \leqslant n} B^{(i)}(n), \tag{6.9}$$

$$L_m = \lim_{n\to\infty} \frac{1}{n} \sum_{m < a_i \leqslant n} B^{(i)}(n) = \lim_{n\to\infty} \frac{1}{n} B(n) - \sum_{a_i \leqslant m} \mathbf{b}^{(i)}$$
$$= \sum_{a_i > m} \mathbf{b}^{(i)},$$

† Empty prime products $m^{(1)}$ or $m^{(2)}$ are taken to be 1.

so that $L_m \to 0$ as $m \to \infty$. But

$$\limsup_{n\to\infty} \frac{1}{n} \sum_{n^{1-\epsilon}<a_i\leqslant n} B^{(i)}(n) \leqslant L_m$$

for every m, and therefore (6.1) is satisfied.

The proof of the sufficiency is much more difficult. We write

$$P(x) = \prod_{a_j\leqslant x} a_j,$$

and let $\phi = \phi(n)$ be a function which tends to infinity with n, but sufficiently slowly to ensure that

$$P_n = P\{\phi(n)\} = o(n). \tag{6.10}$$

Let

$$n = qP_n+r, \qquad 0 \leqslant r < P_n. \tag{6.11}$$

If $a_i \leqslant \phi(n)$, we have $B^{(i)}(qP_n) = qP_n \mathbf{b}^{(i)}$, since, for each $j \leqslant i$, the set of multiples of a_j can be represented as a union of congruence classes $\bmod P_n$. Thus, by (6.10) and (6.11),

$$\begin{aligned} \sum_{a_i\leqslant\phi(n)} B^{(i)}(n) &\leqslant \sum_{a_i\leqslant\phi(n)} B^{(i)}(qP_n) + \sum_{a_i\leqslant n} \{B^{(i)}(n)-B^{(i)}(qP_n)\} \\ &= qP_n \sum_{a_i\leqslant\phi(n)} \mathbf{b}^{(i)} + \{B(n)-B(qP_n)\} \\ &= \{n+o(n)\} \sum_{a_i\leqslant\phi(n)} \mathbf{b}^{(i)} + O(P_n) \\ &= n\mathbf{b}+o(n). \end{aligned} \tag{6.12}$$

It is required to prove that (6.1) suffices to ensure the existence of $\lim n^{-1}B(n)$. Writing

$$R(n,\epsilon) = \sum_{\phi<a_i\leqslant n^{1-\epsilon}} B^{(i)}(n), \tag{6.13}$$

equations (6.9) and (6.12) show that it will suffice to prove

$$\lim_{\epsilon\to+0} \limsup_{n\to\infty} n^{-1}R(n,\epsilon) = 0. \tag{6.14}$$

For a given set of natural numbers $d_1,\ldots, d_k$, we denote by

$$T(n) = T(n; d_1,\ldots, d_k)$$

the number of natural numbers not exceeding n, which are not divisible by any of the numbers $d_1,\ldots, d_k$. As before (see Lemma 5), we write

$$\mathbf{t} = \mathbf{t}(d_1,\ldots, d_k) = \lim_{n\to\infty} \frac{1}{n} T(n; d_1,\ldots, d_k),$$

and we observe that, on putting

$$d_j^{(i)} = a_j/(a_i, a_j) \quad (j = 1,2,\ldots, i-1),$$

we obtain

$$B^{(i)}(n) = T(n/a_i;\, d_1^{(i)},\dots, d_{i-1}^{(i)}) = T^{(i)}(n/a_i) \tag{6.15}$$

and therefore

$$a_i\mathbf{b}^{(i)} = \mathbf{t}(d_1^{(i)},\dots, d_{i-1}^{(i)}) = \mathbf{t}^{(i)} \tag{6.16}$$

say.

From now on we shall suppose that $\epsilon > 0$ is sufficiently small, that $n > n_0(\epsilon)$ where $n_0(\epsilon)$ is sufficiently large, and for convenience we put

$$\tau = n^{\epsilon}. \tag{6.17}$$

According to (6.13), $R(n, \epsilon)$ is equal to the number of b's not exceeding n which arise as multiples of those a_i satisfying $\phi(n) < a_i \leqslant n^{1-\epsilon}$. By Lemma 7 the contribution to $R(n, \epsilon)$ of those b's for which $b^{(1)} \geqslant \tau^{\frac{1}{2}}$ is $\ll \epsilon n$. Employing the notation introduced in (6.15), we have therefore that

$$\begin{aligned} R(n, \epsilon) &= \sum_{\phi < a_i \leqslant n\tau^{-1}} T^{(i)}\left(\frac{n}{a_i}\right) \\ &\ll \sum_{\phi < a_i \leqslant n\tau^{-1}} T^{*(i)}\left(\frac{n}{a_i}\right) + \epsilon n, \end{aligned} \tag{6.18}$$

where $T^{*(i)}(n/a_i)$ represents the number of natural numbers $m = m^{(1)}m^{(2)}$ satisfying

$$m \leqslant n/a_i, \qquad m^{(1)} < \tau^{\frac{1}{2}} \tag{6.19}$$

and

$$m \text{ is not divisible by any } d_j^{(i)}. \tag{6.20}$$

The condition

$$m \text{ is not divisible by any } d_j^{(i)} \text{ satisfying } d_j^{(i)} \leqslant \tau^{\epsilon} \tag{6.21}$$

is weaker than (6.20), and we may therefore estimate the number of $m = m^{(1)}m^{(2)}$ satisfying (6.19) and (6.21) to obtain an estimate for $T^{*(i)}(n/a_i)$.

For each value of $m^{(1)}$ satisfying $m^{(1)} < \tau^{\frac{1}{2}}$ we estimate the number $M(m^{(1)})$ of natural numbers $m = m^{(1)}m^{(2)}$ satisfying the first inequality in (6.19). We apply Lemma 6 with $\eta = \tau$, $X = n/(a_i m^{(1)})$. The condition $X > \eta^{\frac{1}{2}}$ is satisfied since $a_i \leqslant n\tau^{-1}$, and accordingly we obtain the estimate

$$M(m^{(1)}) \ll \frac{n}{a_i m^{(1)}} \prod_{p \leqslant \tau^{\epsilon}} \left(1 - \frac{1}{p}\right).$$

Hence

$$T^{*(i)}\left(\frac{n}{a_i}\right) \ll \frac{n}{a_i} \sum_{(6.21)} \frac{1}{m^{(1)}} \prod_{p \leqslant \tau^{\epsilon}} \left(1 - \frac{1}{p}\right). \tag{6.22}$$

where the summation on the right is over all $m^{(1)}$ satisfying (6.21). (The omission of the condition of summation $m^{(1)} < \tau^{\frac{1}{2}}$ merely increases the sum.)

Now the product on the right of (6.22) is the density of all integers of type $m^{(2)}$, and $\sum 1/m^{(1)}$ converges; hence the sum on the right-hand side of (6.22) represents the density of all integers $m = m^{(1)}m^{(2)}$ with $m^{(1)}$ satisfying (6.21). In other words, since the conditions $d \leqslant \tau^{\epsilon}$, $d \mid m$ imply $d \mid m^{(1)}$,

$$\sum_{(6.21)} \frac{1}{m^{(1)}} \prod_{p \leqslant \tau^{\epsilon}} \left(1-\frac{1}{p}\right) = \lim_{u\to\infty} \frac{1}{u} T'^{(i)}(u) = \mathbf{t}'^{(i)}, \tag{6.23}$$

where
$$T'^{(i)}(u) = T'(u;\, d_1^{(i)},\dots, d_{i-1}^{(i)})$$
is defined in the same way as $T^{(i)}(u)$ (cf. (6.15)), except that the dash signifies that all the $d_j^{(i)} > \tau^{\epsilon}$ are ignored. It follows from (6.18), (6.22), and (6.23) that

$$R(n,\epsilon) \leqslant n \sum_{\phi < a_i \leqslant n\tau^{-1}} \frac{1}{a_i} \mathbf{t}'^{(i)} + \epsilon n. \tag{6.24}$$

We write
$$T''^{(i)}(u) = T''(u;\, d_1^{(i)},\dots, d_{i-1}^{(i)}),$$
$$\mathbf{t}''^{(i)} = \lim_{u\to\infty} \frac{1}{u} T''^{(i)}(u),$$

where in this case the $d_j^{(i)} \leqslant \tau^{\epsilon}$ are ignored. Since the sets of $d_j^{(i)}$ significant in the definitions of $\mathbf{t}'^{(i)}$, $\mathbf{t}''^{(i)}$ respectively are complementary we have by Lemma 5 that

$$\mathbf{t}'^{(i)}\mathbf{t}''^{(i)} \leqslant \mathbf{t}^{(i)}. \tag{6.25}$$

Now $\mathbf{t}''^{(i)}$ is certainly not less than the density of the set of all integers $u = xy$ with $x \leqslant \tau^{\epsilon}$ and y composed entirely of prime factors $p > n$ (integers of this form are not divisible by any d in the interval $\tau^{\epsilon} < d \leqslant n$). Thus

$$\mathbf{t}''^{(i)} \geqslant \sum_{x \leqslant \tau^{\epsilon}} \frac{1}{x} \prod_{p \leqslant n} \left(1-\frac{1}{p}\right) \gg \epsilon^2$$

on applying Lemma 4 of the Appendix, and hence by (6.25) and (6.16)

$$\mathbf{t}'^{(i)} \ll \epsilon^{-2}\mathbf{t}^{(i)} = \epsilon^{-2} a_i \mathbf{b}^{(i)}.$$

Thus (6.24) implies

$$R(n,\epsilon) \ll n\epsilon^{-2} \sum_{\phi < a_i \leqslant n\tau^{-1}} \mathbf{b}^{(i)} + \epsilon n. \tag{6.26}$$

Since $\phi = \phi(n)$ tends to infinity with n,

$$\sum_{\phi < a_i} \mathbf{b}^{(i)} = o(1) \quad \text{as } n \to \infty,$$

and therefore, by (6.26),

$$\limsup_{n\to\infty} \frac{1}{n} R(n,\epsilon) \ll \epsilon.$$

This implies (6.14) and completes the proof of Theorem 13.

7. The set of multiples of a special sequence

To prove Theorem 15 we shall require the following extension of Theorem 10.

THEOREM 16 (Erdös[12]). *Let $\mathscr{J}(T,\epsilon)$ denote the set of integers t lying in the interval*

$$T^{1-\epsilon} < t \leqslant T. \tag{7.1}$$

Then $\mathbf{d}\mathscr{B}\{\mathscr{J}(T,\epsilon)\} \to 0$ *as* $\epsilon \to +0$ *and* $T \to \infty$ *independently.*

In other words, there exists a function $\Delta_1(\epsilon)$ which tends to 0 as $\epsilon \to +0$, and a function $\Delta_2(T)$ which tends to 0 as $T \to \infty$, such that

$$\mathbf{d}\mathscr{B}\{\mathscr{J}(T,\epsilon)\} < \Delta_1(\epsilon)+\Delta_2(T). \tag{7.2}$$

If we put $\epsilon = (\log 2)/\log T$ we obtain Theorem 10. The proof of Theorem 16 runs along somewhat similar lines to the argument which led to Theorem 10, but the analysis involved is much more delicate; the whole of this section is devoted to the proof.

We suppose that ϵ is positive and small, and write

$$\xi = \log(1/\epsilon). \tag{7.3}$$

We shall prove that there exists an absolute constant c such that

$$\mathbf{d}\mathscr{B}\{\mathscr{J}(T,\epsilon)\} < c\xi^{-1} = \Delta_1(\epsilon) \tag{7.4}$$

for all T satisfying

$$T^{\epsilon} \geqslant 2. \tag{7.5}$$

To see that this suffices to prove the theorem, we note that, for fixed T, $\mathbf{d}\mathscr{B}(\mathscr{J})$ increases with ϵ; thus, if $\epsilon < \epsilon_T = \log 2/\log T$ and T is large,

$$\mathbf{d}\mathscr{B}\{\mathscr{J}(T,\epsilon)\} \leqslant \mathbf{d}\mathscr{B}\{\mathscr{J}(T,\epsilon_T)\} < \Delta_1(\epsilon_T) = \Delta_2(T).$$

Accordingly, we assume throughout that (7.5) is satisfied (so that, in particular, T is large).

The primes p satisfying

$$T^{\epsilon} < p \leqslant T \tag{7.6}$$

will play a special role in the proof; we denote the set of all such primes by $\mathscr{P}$. We also denote by $\mathscr{Q}$ the set of all natural numbers q composed entirely of prime factors from $\mathscr{P}$, and by $\mathscr{U}$ the set of all natural numbers

(12) P. Erdös, 'Generalization of a theorem of Besicovitch', *J. Lond. math. Soc.* **11** (1936) 92–8.

u not divisible by any member of $\mathscr{P}$. We include the number 1 in both $\mathscr{Q}$ and $\mathscr{U}$, so that every n has a unique representation of the form $n = qu$.

We note that

$$\mathbf{d}\mathscr{U} = \prod_{p\in\mathscr{P}}\left(1-\frac{1}{p}\right) = \left\{\prod_{p\leqslant T}\left(1-\frac{1}{p}\right)\right\}\Big/\left\{\prod_{p\leqslant T^\epsilon}\left(1-\frac{1}{p}\right)\right\},$$

so that, by (7.5),

$$\mathbf{d}\mathscr{U} \ll (\log T^\epsilon)/(\log T) = \epsilon. \tag{7.7}$$

As in Theorem 10, we use $E_N(tm)$ to denote the number of natural numbers not exceeding N, representable in the form tm, where t satisfies (7.1) and m is unrestricted. Thus

$$\mathbf{d}\mathscr{B}(\mathscr{J}) = \lim_{N\to\infty} N^{-1}E_N(tm). \tag{7.8}$$

We shall prove first of all that, for $N > N_0(\epsilon, T)$,

$$E_N(tm) \ll E_N(q^*u)+N\xi^{-1}, \tag{7.9}$$

where $E_N(q^*u)$ denotes the number of natural numbers not exceeding N, representable in the form q^*u, as u runs through all elements of $\mathscr{U}$ and q^* runs through all those elements of $\mathscr{Q}$ satisfying both the conditions (a^*) and (b^*) below:

(a^*) *the number* $\nu(q^*)$ *of prime factors of* q^* *(counted according to their multiplicities) satisfies*

$$\nu(q^*) \leqslant \tfrac{4}{3}\xi; \tag{7.10}$$

(b^*) *the integer* q^* *is representable in the form* $q^* = q'q''$, *where*

$$T^{1-\xi\epsilon} < q' \leqslant T. \tag{7.11}$$

Let t satisfy (7.1) and m be unrestricted. We write

$$t = q'u' \tag{7.12}$$

and

$$m = q''u'',$$

so that

$$tm = qu, \quad \text{where} \quad q = q'q'' \quad \text{and} \quad u = u'u''. \tag{7.13}$$

If $q' \leqslant T^{1-\xi\epsilon}$, then, by (7.1) and (7.12), $T \geqslant u' > T^{\xi\epsilon-\epsilon} > T^{\frac{1}{2}\xi\epsilon}$. Thus, if q is not a q^* satisfying both conditions (a^*) and (b^*), at least one of the following conditions (α) and (β) must be satisfied:

$$(\alpha) \qquad \nu(q) > \tfrac{4}{3}\xi,$$

$$(\beta) \qquad T^{\frac{1}{2}\xi\epsilon} < u' \leqslant T.$$

Hence

$$E_N(tm) \leqslant E_N(q^*u)+E_N(q^{(\alpha)}u)+E_N(u^{(\beta)}m), \tag{7.14}$$

where $q^{(\alpha)}$ runs through all the elements q of $\mathcal{Q}$ satisfying (α) and $u^{(\beta)}$ runs through all the elements u' of $\mathcal{U}$ satisfying (β).

To obtain (7.9) we estimate the last two terms on the right-hand side of (7.14). To estimate $E_N(q^{(\alpha)}u)$ we apply the corollary to Lemma 7 of the Appendix, with the set $\mathcal{P}$ defined by (7.6) above. Supposing $N \geqslant T$, we obtain (see (7.3), and (7.15) below)

$$\sum_{qu \leqslant N} \left\{\nu(q) - \sum_{p \in \mathcal{P}} \frac{1}{p}\right\}^2 \ll N \sum_{p \in \mathcal{P}} \frac{1}{p}.$$

But by (7.5) and Lemma 3 of the Appendix,

$$\sum_{p \in \mathcal{P}} \frac{1}{p} = \log\log T - \log\log T^{\epsilon} + O(1) = \xi + O(1). \tag{7.15}$$

Thus, since $\nu(q^{(\alpha)}) \geqslant \frac{4}{3}\xi$ always,

$$\{\tfrac{1}{3}\xi + O(1)\}^2 E_N(q^{(\alpha)}u) \ll N\xi,$$

and it follows that

$$E_N(q^{(\alpha)}u) \ll N\xi^{-1}. \tag{7.16}$$

To estimate $E_N(u^{(\beta)}m)$ we use the fact that every u' satisfying (β) is composed entirely of prime factors p satisfying $p \leqslant T^{\epsilon}$. We denote by $V(n)$ the greatest divisor of n composed entirely of such primes. Then, since $V(u^{(\beta)}m) > T^{\frac{1}{2}\xi\epsilon}$ always, we have

$$\sum_{n=1}^{N} \log V(n) > \tfrac{1}{2}\xi(\log T^{\epsilon}) E_N(u^{(\beta)}m). \tag{7.17}$$

On the other hand, if $p^{\theta} \,\|\, N!$, then $\theta = \theta(N, p)$ is given by

$$\theta(N, p) = \sum_{\nu=1}^{\infty} [N/p^{\nu}] \ll N/p.$$

Hence, by (7.5),

$$\sum_{n=1}^{N} \log V(n) = \sum_{p \leqslant T^{\epsilon}} \theta(N, p) \log p \ll N \sum_{p \leqslant T^{\epsilon}} \frac{\log p}{p} \ll N \log T^{\epsilon}.$$

Combining this with (7.17) we obtain

$$E_N(u^{(\beta)}m) \ll N\xi^{-1}, \tag{7.18}$$

and (7.9) now follows from (7.14), (7.16), and (7.18).

Since $\sum q^{-1}$, summed over all q of $\mathcal{Q}$, converges, we have

$$\limsup_{N \to \infty} N^{-1} E_N(q^* u) \leqslant (\mathbf{d}\mathcal{U}) \sum \frac{1}{q^*} \ll \epsilon \sum \frac{1}{q^*}$$

by (7.7). Recalling (7.8) and (7.9), it is now clear that in order to establish (7.4) it will suffice to prove

$$\sum \frac{1}{q^*} \ll \epsilon^{-1}\xi^{-1}. \tag{7.19}$$

We denote by q_r an element of $\mathscr{Q}$ containing exactly r *distinct* prime factors (so that $r < \nu(q_r)$ unless q_r is square-free), with the convention that $q_0 = 1$. Since each q^* satisfies the conditions (a^*) and (b^*), we have

$$\sum \frac{1}{q^*} < \sum_{0 \leqslant r+s \leqslant \frac{8}{3}\xi} \sum \frac{1}{q'_r} \sum \frac{1}{q_s}, \tag{7.20}$$

where the middle sum on the right is taken over those q'_r satisfying

$$T^{1-\xi\epsilon} < q'_r \leqslant T. \tag{7.21}$$

To estimate the right-hand side of (7.20) we require the following lemmas, of which the first is an immediate consequence of the corollary to Lemma 3 of the Appendix. (The contribution corresponding to terms with $\nu \geqslant 2$ in Lemma 8 below is $O(YX^{-1}) = O(X^{-\frac{1}{2}})$.)

LEMMA 8. *If $X > Y^2 > 2$, where $Y > 0$, then*

$$\sum_{XY^{-1} < p^\nu \leqslant X} \frac{1}{p^\nu} \ll \frac{\log Y}{\log X},$$

the sum being taken over all primes p and all positive integers ν satisfying the condition of summation.

LEMMA 9. *For every $s \leqslant \frac{4}{3}\xi$,*

$$\sum \frac{1}{q_s} \ll \frac{\xi^s}{s!}.$$

Proof. If $s = 0$, the sum contains the single term 1, so that we may suppose that $s \geqslant 1$. Each term $1/q_s$ occurs at least $s!$ times in the formal expansion of

$$\Big\{\sum_{p \in \mathscr{P}} \sum_{\nu=1}^{\infty} \frac{1}{p^\nu}\Big\}^s.$$

But, by (7.15), there exists a positive absolute constant c such that

$$\sum_{p \in \mathscr{P}} \sum_{\nu} \frac{1}{p^\nu} \leqslant \sum_{p \in \mathscr{P}} \frac{1}{p} + O(1) \leqslant \xi + c,$$

and, since $s \leqslant \frac{4}{3}\xi$,

$$(\xi+c)^s = \xi^s(1+c\xi^{-1})^s \leqslant \xi^s(1+c\xi^{-1})^{\frac{4}{3}\xi} \leqslant \xi^s e^{\frac{4}{3}c} \ll \xi^s.$$

LEMMA 10. *For every r satisfying $r \leqslant \frac{4}{3}\xi$,*

$$\sum \frac{1}{q'_r} \ll \xi^2 \epsilon \frac{\xi^r}{r!};$$

the summation (as in (7.20)) being over those q'_r which satisfy (7.21).

Proof. Since $T^{1-\xi\epsilon} > 1$, the sum is empty when $r = 0$, and we may therefore suppose that $r \geqslant 1$. Let p^ν be the largest prime power factor in the prime decomposition of q_r', so that $p^\nu \geqslant (q_r')^{1/r}$ and

$$q_r' = p^\nu q_{r-1}. \tag{7.22}$$

Writing $X = X(q_{r-1}) = T/q_{r-1}$, we have, by (7.21) and (7.22),

$$XT^{-\epsilon\xi} < p^\nu \leqslant X$$

and

$$X \geqslant p^\nu \geqslant (q_r')^{1/r} > T^{(1-\xi\epsilon)/r} > T^{1/(2r)}.$$

Hence, by (7.22),

$$\sum \frac{1}{q_r'} \leqslant \sum_{X > T^{1/(2r)}} \frac{1}{q_{r-1}} \sum_{XT^{-\xi\epsilon} < p^\nu \leqslant X} \frac{1}{p^\nu}, \tag{7.23}$$

where the outer sum on the right is taken over those q_{r-1} satisfying $X(q_{r-1}) > T^{1/(2r)}$.

Since $r \leqslant \frac{4}{3}\xi$, we have $X > T^{1/(2r)} > (T^{\xi\epsilon})^2$, so that we can apply Lemma 8, with $Y = T^{\xi\epsilon}$, to the inner sum on the right-hand side of (7.23). We obtain

$$\sum \frac{1}{q_r'} \ll \frac{\xi\epsilon \log T}{\log T^{1/(2r)}} \sum \frac{1}{q_{r-1}} \ll \xi^2\epsilon \sum \frac{1}{q_{r-1}},$$

using (7.5) and the fact that $r \leqslant \frac{4}{3}\xi$. The proof of the lemma is now complete since, by Lemma 9,

$$\sum \frac{1}{q_{r-1}} \ll \frac{\xi^{r-1}}{(r-1)!} \ll \frac{\xi^r}{r!}.$$

We are now in a position to estimate the right-hand side of (7.20). Applying Lemmas 9 and 10, we obtain

$$\sum \frac{1}{q^*} \ll \xi^2\epsilon \sum_{0 \leqslant r+s \leqslant \frac{4}{3}\xi} \frac{\xi^{r+s}}{r!\,s!} = \xi^2\epsilon \sum_{l=0}^{[4\xi/3]} \frac{(2\xi)^l}{l!}.$$

But, for $l \leqslant 2\xi$, $(2\xi)^l/l!$ increases with l, so that

$$\sum \frac{1}{q^*} \ll \xi^2\epsilon (2\xi)^{\frac{4}{3}\xi} \{\Gamma(\tfrac{4}{3}\xi)\}^{-1}.$$

By Stirling's formula,

$$(\Gamma(\tfrac{4}{3}\xi))^{-1} \ll \xi^{\frac{1}{2}} (\tfrac{3}{4}e\xi^{-1})^{\frac{4}{3}\xi};$$

thus

$$\sum \frac{1}{q^*} \ll \xi^4\epsilon (\tfrac{3}{2}e)^{\frac{4}{3}\xi} = \xi^4\epsilon^{1-c}, \tag{7.24}$$

where c is defined by $(\frac{3}{2}e)^{\frac{4}{3}} = e^c$. To see that (7.24) implies (7.19), we need only verify that $c < 2$. This is clearly the case, for $(\frac{3}{2}e)^{\frac{4}{3}} < e^2$ follows from $e > \frac{9}{4}$.

As was remarked earlier, (7.19) implies (7.4), so that the proof of Theorem 16 is complete.

8. Proof of Theorem 15

We now proceed to prove Theorem 15, by constructing a sequence $\mathscr{A}$, satisfying (6.3), whose set of multiples does not possess asymptotic density. The construction will be somewhat similar in nature to the one used to prove Theorem 11; Theorem 16 replacing Theorem 10 as the principal tool.

We may restate Theorem 16 in the following way:

If $\mathscr{J}(T,\epsilon)$ denotes the set of integers t in the interval

(8.1) $$T^{1-\epsilon} < t \leqslant T,$$

then, for any given $\delta > 0$,

(8.2) $$\mathbf{d}\mathscr{B}\{\mathscr{J}(T,\epsilon)\} < \delta \quad \textit{provided} \quad T > T_0(\delta) \quad \textit{and} \quad \epsilon < \epsilon_0(\delta).$$

On the basis of this formulation we set about constructing a certain sequence of intervals $\mathscr{J}_r = \mathscr{J}(T_r, \epsilon_r)$ $(r = 1, 2, 3,\ldots)$ from which the elements of $\mathscr{A}$ will be selected. First of all, let ϵ_1, ϵ_2,... be a sequence of positive numbers, each less than $\frac{1}{4}$, which tend to 0 sufficiently rapidly to satisfy

(8.3) $$\epsilon_r < \epsilon_0(2^{-r-3}) \quad (r = 1, 2, 3,\ldots).$$

Next, let T_1, T_2,... be a sequence of natural numbers which tend to infinity sufficiently rapidly to satisfy the following conditions (for $r = 1, 2, 3,\ldots$):

(i) $T_{r+1} > (T_r!)^2$,

(ii) each T_r is sufficiently large as a function of ϵ_r; so that, in particular,

(8.4) $$T_r > T_0(2^{-r-3}),$$

(8.5) $$\psi(T_r^{1-\epsilon_r}) > \epsilon_r^{-2}$$

(where ψ is the function introduced in the statement of Theorem 15),

(8.6) $$T_r^{\epsilon_r} \to \infty \quad \text{as} \quad r \to \infty.$$

For each interval $\mathscr{J}_r$ $(r = 1, 2, 3,\ldots)$ we write $\mathscr{B}_r = \mathscr{B}(\mathscr{J}_r)$ $(r = 1, 2, 3,\ldots)$ and note that, having imposed conditions (8.3) and (8.4), we may apply (8.2) with $\delta = 2^{-r-3}$, yielding $\mathbf{d}\mathscr{B}_r < 2^{-r-3}$ and consequently

(8.7) $$\sum_{r=1}^{\infty} \mathbf{d}\mathscr{B}_r < \tfrac{1}{8}.$$

We denote by $\mathscr{J}'_r$ the subset of $\mathscr{J}_r$ consisting of all those integers m of $\mathscr{J}_r$ composed entirely of prime factors $p > T_r^{\epsilon_r^2}$, and we define $\mathscr{A}$ by the relation

$$\mathscr{A} = \bigcup_{r=1}^{\infty} \mathscr{J}'_r.$$

We shall prove that $\mathscr{A}$ possesses the properties required for Theorem 15, and we begin by showing that (6.3) is satisfied. Suppose that m is large and that $s = s(m)$ denotes the largest suffix r for which $T_r^{1-\epsilon_r} < m$. Since $A(m) \leqslant T_s$, we may suppose that $T_s > m^{\frac{1}{2}}$. We write

$$\tau_r = T_r^{\epsilon_r}. \tag{8.8}$$

We estimate the number $J_s'(m)$ of elements of $\mathscr{J}_s'$ which do not exceed m by applying Lemma 6 with $X = m$, $\epsilon = \epsilon_s$, $\eta = \tau_s$; the lemma is applicable in view of (ii) and $m > T_s^{1-\epsilon_s}$. We obtain

$$J_s'(m) \ll m \prod_{p \leqslant \tau_s^{\epsilon_s}} \left(1-\frac{1}{p}\right) \ll \epsilon_s^{-2} \frac{m}{\log m} \ll \psi(T_s^{1-\epsilon_s}) \frac{m}{\log m} \tag{8.9}$$
$$\ll \psi(m) \frac{m}{\log m},$$

by virtue of (8.5) and the monotonicity of ψ. Since $A(m) \leqslant T_{s-1} + J_s'(m)$ and $T_{s-1} \ll m/\log m$ by (i), (6.3) follows from (8.9).

It remains to show that $\mathbf{d}\mathscr{B}(\mathscr{A})$ does not exist, and we shall do so by proving that

$$\underline{\mathbf{d}}\mathscr{B}(\mathscr{A}) \leqslant \tfrac{1}{2}, \qquad \bar{\mathbf{d}}\mathscr{B}(\mathscr{A}) = 1. \tag{8.10}$$

We have, if $l > (T_r!)$ (cf. a similar statement early in the proof of Theorem 11), that

$$B_r(l) \leqslant 2l\mathbf{d}\mathscr{B}_r \quad (r = 1, 2, 3, \ldots);$$

hence, by (i) and (8.7),

$$B(T_s \tau_s^{-1}) < 2T_s \tau_s^{-1} \sum_{r=1}^{s-1} \mathbf{d}\mathscr{B}_r < \tfrac{1}{2} T_s \tau_s^{-1},$$

and the first assertion in (8.10) has been proved.

Any integer m in $\mathscr{J}_r$ is uniquely representable in the form $m = m_r^{(1)} m_r^{(2)}$, where $m_r^{(1)}$ is composed entirely of primes $p \leqslant \tau_r^{\epsilon_r}$ and $m_r^{(2)}$ is composed entirely of primes $p > \tau_r^{\epsilon_r}$; empty prime products $m_r^{(1)}$ or $m_r^{(2)}$ are taken to be 1. If $T_r \tau_r^{-\frac{1}{2}} < m \leqslant T_r$ and $m_r^{(1)} < \tau_r^{\frac{1}{2}}$, then by (8.1) m is divisible by the element $a = m_r^{(2)}$ of $\mathscr{A}$, since $m_r^{(2)} = m/m_r^{(1)} > T_r \tau_r^{-1}$. Hence, in the notation of Lemma 7, the number $B(T_r)$ of elements of $\mathscr{B}(\mathscr{A})$ not exceeding T_r satisfies

$$B(T_r) \geqslant T_r - T_r \tau_r^{-\frac{1}{2}} - N(T_r, \epsilon_r).$$

By (ii), Lemma 7 is applicable with $n = T_r$, $\epsilon = \epsilon_r$, and (6.8) yields

$$T_r^{-1} B(T_r) \geqslant 1 - \tau_r^{-\frac{1}{2}} + O(\epsilon_r). \tag{8.11}$$

Since $\epsilon_r \to 0$ and, by (8.6), $\tau_r^{-\frac{1}{2}} \to 0$ as $r \to \infty$, (8.11) implies the second assertion of (8.10). This completes the proof of Theorem 15.

APPENDIX

As was explained in the Preface to the book, this appendix is intended for those readers with no knowledge of number theory. Apart from two direct references† (in the text of the book) to outside sources, the appendix contains every number-theoretic result assumed in the book, no matter how elementary or well known. We shall give outside references for the proofs of some of these results (the others will be proved here), but such references are restricted to a single text: namely the classical work by G. H. Hardy and E. M. Wright, *An Introduction to Theory of Numbers*, 5th ed. (Clarendon Press, Oxford, 1979). *We shall refer to the Hardy–Wright text as* HW for short. Although we mention additional sources, these will relate only to refinements which are not required in our book.

1. The Fibonacci sequence $\{f_j\}$

This sequence, given by

(1.1) $$f_1 = 1, \qquad f_2 = 2$$

and

(1.2) $$f_j = f_{j-1}+f_{j-2} \quad (j = 3, 4, 5,...)$$

features in the proof of Theorem 12 (Cassels) of Chapter I (§ 5). There we need to know that

(1.3) $$(f_j, f_{j+1}) = (f_j, f_{j+2}) = 1 \quad \text{for } j = 1, 2, 3,...,$$

and that

(1.4) $$f_j = A\rho^{-j}+B(-\rho)^j \quad (j = 1, 2, 3,...)$$

where ρ satisfies

(1.5) $$\rho^2+\rho-1 = 0, \qquad 0 < \rho < 1,$$

and A, B are the (non-zero) constants given by

(1.6) $$A\rho^{-1}-B\rho = 1, \qquad A\rho^{-2}+B\rho^2 = 2.$$

We begin by proving (1.3). It is obvious from the form of the recurrence relation (1.2), with $j+2$ replacing j, that, for each $j = 1, 2, 3,...$,

$$(f_j, f_{j+1}) = (f_{j+1}, f_{j+2}) = (f_{j+2}, f_j);$$

it therefore suffices to prove that $(f_{j+1}, f_{j+2}) = 1$. We have just remarked, however, that $(f_{j+1}, f_{j+2}) = (f_j, f_{j+1})$ for $j = 1, 2, 3,...$, so that, by (1.1),

$$(f_{j+1}, f_{j+2}) = (f_1, f_2) = (1, 2) = 1.$$

This completes the proof of (1.3).

Next we turn to the formula (1.4), where ρ, A, and B are given by (1.5) and (1.6). (The only significance of the condition $0 < \rho < 1$ in (1.5) is that we have chosen to work with the positive root of $x^2+x-1 = 0$.) To prove (1.4) we could apply to (1.2) and (1.1) the standard technique for solving linear difference

† See the footnote relating to (3.17) in the proof of Theorem 5 of Chapter II (§ 3), and also the reference (7) in § 14 of Chapter III.

equations; instead, we shall argue by induction on j. For $j = 1$ and 2, the truth of (1.4) (subject to (1.6)) is ensured by (1.6). Let us assume then that (1.4) holds for all integers j satisfying $1 \leqslant j < i$, where $i \geqslant 3$, and prove that (1.4) holds also for $j = i$. We have

$$\begin{aligned} f_i = f_{i-1}+f_{i-2} &= A\{\rho^{-i+1}+\rho^{-i+2}\}+B\{(-\rho)^{i-1}+(-\rho)^{i-2}\} \\ &= A\rho^{-i}(\rho+\rho^2)+B(-\rho)^{i-2}(-\rho+1) \\ &= A\rho^{-i}+B(-\rho)^i \end{aligned}$$

using (1.5); and this completes the proof of (1.4).

Relations (1.3), (1.4) contain all the information (about the Fibonacci sequence) required in the proof of Theorem 12 of Chapter I. The Fibonacci sequence has many interesting properties, and for an account of these the reader is referred to, for example, N. N. Vorobyov, *The Fibonacci Numbers* (Heath, Boston; Topics in Mathematics series, 1963. Translated from the Russian).

2. The distribution of prime numbers

Let p denote a variable prime, and let $\pi(x)$ have its usual meaning as the number of primes not exceeding x. The following properties of the sequence of primes, stated below in Lemmas 1–5, are frequently required in the book.

LEMMA 1. $$\frac{x}{\log x} \ll \pi(x) \ll \frac{x}{\log x} \quad \textit{for } x \geqslant 2.$$

This result, first proved by Tchebycheff in 1850, is Theorem 7 of HW (see inequalities (22.4.1), (22.4.2)). Although superseded by the prime number theorem (Lemma 5 below), Tchebycheff's result is much easier to prove and suffices for many applications.

LEMMA 2. $$\sum_{p \leqslant x} \frac{\log p}{p} = \log x + O(1) \quad \textit{for } x \geqslant 2.$$

LEMMA 3. $$\sum_{p \leqslant x} \frac{1}{p} = \log\log x + a + O\left(\frac{1}{\log x}\right) \quad \textit{for } x \geqslant 2,$$

where a is a positive absolute constant.

Lemma 2 is Theorem 425 of HW, and Lemma 3 is a combination† of equations (22.7.3), (22.7.4) and Theorem 428 of HW.

COROLLARY (to Lemma 3). *If $X \geqslant Y^2 > 2$, where $Y > 0$, then*

$$\sum_{XY^{-1} < p \leqslant X} \frac{1}{p} = \frac{\log Y}{\log X} + O\left\{\frac{1}{\log X} + \left(\frac{\log Y}{\log X}\right)^2\right\}.$$

† It is not stated explicitly in HW that a is positive. However, Theorem 428 of HW gives

$$a = \gamma + \sum_p \left\{\log\left(1-\frac{1}{p}\right)+\frac{1}{p}\right\}$$

where γ is Euler's constant. Hence (see proof of Lemma 4 below)

$$a = \gamma - \sum_p \left(\frac{1}{2p^2}+\frac{1}{3p^3}+\dots\right) > \gamma - \tfrac{1}{2}\sum_p \frac{1}{p(p-1)} > \gamma - \tfrac{1}{2}\sum_{n=2}^{\infty} \frac{1}{n(n-1)} = \gamma - \tfrac{1}{2}$$

and, since $\gamma = 0{\cdot}5772\dots$, it follows that $a > 0$.

Deduction of the corollary. On applying the lemma, we obtain

$$\sum_{XY^{-1}<p\leqslant X}\frac{1}{p} = (\log\log X+a)-\{\log\log(XY^{-1})+a\}+O\left(\frac{1}{\log X^{\frac{1}{2}}}\right)$$

$$= \log\left(\frac{\log X}{\log X-\log Y}\right)+O\left(\frac{1}{\log X}\right)$$

$$= -\log\left(1-\frac{\log Y}{\log X}\right)+O\left(\frac{1}{\log X}\right)$$

$$= \frac{\log Y}{\log X}+O\left\{\frac{1}{\log X}+\left(\frac{\log Y}{\log X}\right)^2\right\}$$

as required.

LEMMA 4. $$\log x \ll \prod_{p\leqslant x}\left(1-\frac{1}{p}\right)^{-1} \ll \log x \quad \textit{for } x\geqslant 2.$$

Proof. The product is discussed in sections 22.7, 22.8 of HW, with a view to proving Mertens's theorem (Theorem 429, HW)

$$\prod_{p\leqslant x}\left(1-\frac{1}{p}\right) \sim \frac{e^{-\gamma}}{\log x} \quad \text{as} \quad x\to\infty,$$

where γ is Euler's constant. As the result of Lemma 4 is not stated explicitly in HW, we give an independent proof here.

For each prime p, let

$$s(p) = -\log\left(1-\frac{1}{p}\right)-\frac{1}{p} = \frac{1}{2p^2}+\frac{1}{3p^3}+\ldots,$$

so that $$\frac{1}{2p^2} < s(p) < \frac{1}{2}\left(\frac{1}{p^2}+\frac{1}{p^3}+\ldots\right) = \frac{1}{2(p-1)p}.$$

Then $$\sum_p s(p)$$

is a convergent series, and if we denote its sum by b, we obtain

$$\sum_{p\leqslant x}\left\{\log\left(1-\frac{1}{p}\right)^{-1}-\frac{1}{p}\right\} = b+o(1) \quad \text{as} \quad x\to\infty;$$

hence $$\prod_{p\leqslant x}\left(1-\frac{1}{p}\right)^{-1} = \exp\left\{\sum_{p\leqslant x}\frac{1}{p}+b+o(1)\right\}$$

$$= \{1+o(1)\}e^{a+b}\log x \quad \text{as } x\to\infty,$$

by Lemma 3. This is sharper than the required result, and it is clear that to prove Mertens's theorem it would suffice to show that $a+b=\gamma$ (see HW, Theorem 428).

LEMMA 5 (the prime number theorem).

$$\pi(x)\sim\frac{x}{\log x} \quad \textit{as} \quad x\to\infty.$$

This is Theorem 6 of HW (see §§ 22.14–22.16, where a technically elementary proof is given). The theorem was first proved in 1896 by Hadamard and de la Vallée Poussin (independently of each other, and by analytic methods). The first elementary proofs of the theorem were given in 1949 by P. Erdös[1]–A. Selberg and A. Selberg.[2]

The corollary below follows immediately in view of the relation $\pi(p_n) = n$.

COROLLARY (to Lemma 5). *If p_n denotes the n-th prime, then*

$$p_n \sim n \log n \quad as \quad n \to \infty.$$

In particular,

$$\lim_{n\to\infty} \frac{p_{n+1}}{p_n} = 1.$$

For further information about the distribution of primes, the reader is referred, for example, to *Cambridge Tracts in Mathematics*, Nos. 30 and 41, *The Distribution of Prime Numbers*, by A. E. Ingham and *Introduction to Modern Prime Number Theory* by T. Estermann. (See also the book by Titchmarsh cited below, and in particular Chapter III.)

3. Mean values of certain arithmetic functions

Let $d(m)$ denote the number of positive divisors of the natural number m.

LEMMA 6.
$$\sum_{m \leqslant x} d(m) \ll x \log x \quad for \quad x \geqslant 2.$$

Proof. We need only remark that $d(m)$ may be viewed as the number of representations of m as the product of an ordered pair of natural numbers. Thus the sum on the left is

$$\sum_{\substack{l,m\\ lm \leqslant x}} 1 = \sum_{l \leqslant x} \sum_{m \leqslant x/l} 1 = \sum_{l \leqslant x} \left[\frac{x}{l}\right] \leqslant x \sum_{l \leqslant x} \frac{1}{l} \ll x \log x.$$

By a slightly more complicated argument one can prove (see Theorem 320 of HW) that

$$\sum_{m \leqslant x} d(m) = x \log x + (2\gamma - 1)x + O(x^{\frac{1}{2}}).$$

For further information the reader is referred to Chapter XII of E. C. Titchmarsh's *The Theory of the Riemann Zeta Function* (Clarendon Press, Oxford, 1951).

The function $d(m)$ is an example of a *multiplicative* arithmetic function, in the sense that it satisfies $d(mn) = d(m)d(n)$ whenever $(m,n) = 1$. We have next to consider an arithmetic function of *additive* type; if f is such a function, then

(3.1) $$f(mn) = f(m) + f(n) \quad \text{whenever} \quad (m,n) = 1.$$

We shall indicate that p^α is the highest power of p dividing m by the notation $p^\alpha \| m$. It then follows from (3.1) that (in a self-explanatory notation)

(3.2) $$f(m) = \sum_{p^\alpha \| m} f(p^\alpha).$$

We shall consider in particular those real additive functions which satisfy also (for all primes p)

(3.3) $$f(p^\alpha) = \alpha f(p) \quad (\alpha = 2, 3, \ldots)$$

and

(3.4) $$0 \leqslant f(p) \leqslant 1.$$

(1) P. Erdös, 'On a new method in elementary number theory which leads to an elementary proof of the prime number theorem', *Proc. natn. Acad. Sci.* U.S.A. **35** (1949) 374–84.

(2) A. Selberg, 'An elementary proof of the prime number theorem', *Ann. Math.* **50** (1949) 305–13.

For example, when $f(p) = 1$ for all p, $f(m)$ is the number of prime factors of m, counted according to multiplicity; or if $f(p) = 1$ when p belongs to some subset $\mathscr{P}$ of the sequence of primes, and is 0 otherwise, $f(m)$ is the number of prime factors *from* $\mathscr{P}$ of m, again counted according to multiplicity. This latter case is the only one required for applications in Chapter V (see §§ 5 and 7), but the proof of Lemma 7 below gains in clarity, without any loss of simplicity, by dealing with the more general class of functions satisfying (3.4). The proof of Lemma 7 has much in common with the arguments in HW, §§ 22.10 and 22.11, but is sufficiently different to warrant an independent account.

LEMMA 7. *Let f be an additive function satisfying* (3.3) *and* (3.4), *and let*

$$A_n = \sum_{p \leqslant n} \frac{f(p)}{p}, \qquad B_n = \sum_{p \leqslant n} \frac{f^2(p)}{p}.$$

Then

$$\sum_{m=1}^{n} \{f(m)-A_n\}^2 \ll n(A_n+B_n+1) \quad \textit{for} \quad n \geqslant 2. \tag{3.5}$$

Proof. We introduce the associated arithmetic function

$$f'(m) = \sum_{p|m} f(p),$$

where the sum on the right extends over the *distinct* prime divisors of m. Then, by (3.2) and (3.3),

$$f(m)-f'(m) = \sum_{p} f(p) \sum_{\substack{k=2 \\ p^k|m}}^{\infty} 1,$$

and we begin by showing that

$$\sum_{m=1}^{n} \{f(m)-f'(m)\}^2 \ll n. \tag{3.6}$$

We see at once (using p and q to denote variable primes) that the sum on the left of (3.6) is equal to

$$\begin{aligned} S &= \sum_{p} \sum_{q} f(p)f(q) \sum_{k=2}^{\infty} \sum_{l=2}^{\infty} \Big(\sum_{\substack{m=1 \\ p^k|m,\, q^l|m}}^{n} 1 \Big) \\ &= \sum_{\substack{p \\ p \neq q}} \sum_{q} f(p)f(q) \sum_{k=2}^{\infty} \sum_{l=2}^{\infty} \left[\frac{n}{p^k q^l}\right] + \sum_{p} f^2(p) \sum_{k=2}^{\infty} \sum_{l=2}^{\infty} \left[\frac{n}{p^{\max(k,l)}}\right] \\ &\ll n\left\{\left(\sum_{p} \frac{f(p)}{p^2}\right)^2 + \sum_{p} f^2(p) \sum_{k=2}^{\infty} \frac{k}{p^k}\right\} \\ &\ll n \end{aligned}$$

by (3.4), thus proving (3.6).

Since $(a+b)^2 \leqslant 2a^2+2b^2$ for real a, b, it follows from (3.6) that

$$\begin{aligned} \sum_{m=1}^{n} \{f(m)-A_n\}^2 &\leqslant 2 \sum_{m=1}^{n} \{f(m)-f'(m)\}^2 + 2 \sum_{m=1}^{n} \{f'(m)-A_n\}^2 \\ &\ll n + \sum_{m=1}^{n} \{f'(m)-A_n\}^2. \end{aligned}$$

Thus, in order to prove (3.5), it suffices to show that

$$\sum_{m=1}^{n} \{f'(m)-A_n\}^2 \ll n(A_n+B_n+1). \tag{3.7}$$

We observe that, by (3.4),

$$(3.8)\quad \sum_{m=1}^{n} f'(m) = \sum_{m\leqslant n}\sum_{p|m} f(p) = \sum_{p} f(p) \sum_{\substack{m\leqslant n\\ p|m}} 1 = \sum_{p} f(p)\left[\frac{n}{p}\right] = nA_n + O(n)$$

and

$$\sum_{m=1}^{n}\{f'(m)\}^2 = \sum_{m\leqslant n}\sum_{p|m}\sum_{q|m} f(p)f(q)$$
$$= \sum_{p}\sum_{q} f(p)f(q) \sum_{\substack{m\leqslant n\\ p|m,q|m}} 1$$
$$= \sum_{p}\sum_{\substack{q\\ p\neq q}} f(p)f(q)\left[\frac{n}{pq}\right] + \sum_{p} f^2(p)\left[\frac{n}{p}\right]$$
$$= n\sum_{\substack{p\neq q\\ pq\leqslant n}} \frac{f(p)f(q)}{pq} + O\Big(\sum_{pq\leqslant n} 1\Big) + nB_n + O(n)$$
$$= n\sum_{pq\leqslant n} \frac{f(p)f(q)}{pq} + nB_n + O(n),$$

since every integer m has at most two representations $m = pq$. But

$$A_{n^{1/2}}^2 = \Big(\sum_{p\leqslant n^{1/2}} \frac{f(p)}{p}\Big)^2 \leqslant \sum_{pq\leqslant n} \frac{f(p)f(q)}{pq} \leqslant \Big(\sum_{p\leqslant n} \frac{f(p)}{p}\Big)^2 = A_n^2$$

and, by (3.4) and the corollary to Lemma 3,

$$A_n - A_{n^{1/2}} = \sum_{n^{1/2}<p\leqslant n} \frac{f(p)}{p} \ll \sum_{n^{1/2}<p\leqslant n} \frac{1}{p} \ll 1;$$

hence $\qquad A_{n^{1/2}}^2 = A_n^2 + O(A_n + A_{n^{1/2}}) = A_n^2 + O(A_n + 1)$

and we conclude that

$$(3.9)\quad \sum_{m=1}^{n}\{f'(m)\}^2 = nA_n^2 + O\{n(A_n + B_n + 1)\}.$$

Thus, by (3.8) and (3.9),

$$\sum_{m=1}^{n}\{f'(m) - A_n\}^2 = \sum_{m=1}^{n}\{f'(m)\}^2 - 2A_n\sum_{m=1}^{n} f'(m) + nA_n^2$$
$$= nA_n^2 - 2A_n\{nA_n + O(n)\} + nA_n^2 + O\{n(A_n + B_n + 1)\}$$
$$= O\{n(A_n + B_n + 1)\},$$

proving (3.7). As we have remarked already, (3.7) establishes the lemma.

COROLLARY (to Lemma 7). *Let $\mathscr{P}$ be a finite set of distinct primes p, such that*

$$\sum_{p\in\mathscr{P}} \frac{1}{p} \geqslant 1;$$

and let p^* be the greatest prime in $\mathscr{P}$. Suppose that the function f in Lemma 7 (which is additive and satisfies (3.3)) is defined by the relations

$$f(p) = 1 \quad for \quad p\in\mathscr{P}, \qquad f(p) = 0 \quad for \quad p\notin\mathscr{P};$$

(so that, on the set of all primes, $f(p)$ is the characteristic function of $\mathscr{P}$).

Then $$\sum_{m=1}^{n}\Big\{f(m) - \sum_{p\in\mathscr{P}} \frac{1}{p}\Big\}^2 \ll n\sum_{p\in\mathscr{P}} \frac{1}{p} \quad for\ n \geqslant p^*.$$

In particular, if k is positive and $N_n = N_n(k)$ is the number of natural numbers $m \leqslant n$ such that

$$\left|f(m) - \sum_{p \in \mathscr{P}} \frac{1}{p}\right| \geqslant k \sum_{p \in \mathscr{P}} \frac{1}{p},$$

then

$$N_n \ll k^{-2} n \left(\sum_{p \in \mathscr{P}} \frac{1}{p}\right)^{-1} \quad \textit{for} \quad n \geqslant p^*.$$

This is merely a special case of the lemma, since, for the function f in question,

$$A_n = B_n = \sum_{p \in \mathscr{P}} \frac{1}{p} \quad \text{for} \quad n \geqslant p^*.$$

The result of Lemma 7 can be generalized and sharpened in a number of ways. For an account of some of these extensions, the reader is referred to P. Turan, *J. Lond. math. Soc.* **11** (1936) 125–33, and H. Halberstam, ibid. **30** (1955) 43–53.

4. Miscellanea from elementary number theory

In addition to the results listed in the preceding three sections, to which specific reference has been made in the text, several basic theorems and notations from elementary number theory are assumed implicitly throughout the book, and it may help the reader who is not an expert if we conclude with a brief account of these.

Chief among the theorems is the so-called Fundamental Theorem of Arithmetic (HW, Theorem 2) which lies at the root of the multiplicative structure of the integers; it asserts that

every natural number greater than 1 *can be represented as a product of primes, and this representation is unique apart from the order of the factors.*

It follows at once that if

$$p_1 < p_2 < p_3 < \ldots$$

is the sequence of all primes, then every natural number n can be expressed uniquely in the *standard* form

$$n = p_1^\alpha p_2^\beta p_3^\gamma \ldots, \tag{4.1}$$

where the exponents $\alpha, \beta, \gamma, \ldots$ are non-negative integers and only finitely many are positive.

If all the positive exponents among $\alpha, \beta, \gamma, \ldots$ are equal to 1, so that n is the product of distinct primes, n is said to be *square-free*.

Let ν be a divisor of n, and suppose that n has prime decomposition (4.1). Then the prime decomposition of ν has standard form

$$p_1^{\alpha'} p_2^{\beta'} p_3^{\gamma'} \ldots$$

where

$$0 \leqslant \alpha' \leqslant \alpha, \qquad 0 \leqslant \beta' \leqslant \beta, \quad \ldots,$$

and the number $d(n)$ of positive integer divisors of n is clearly the total number of ways of selecting an ordered set of exponents $\alpha', \beta', \ldots$; that is, $d(n)$ is given by

$$d(n) = (\alpha+1)(\beta+1)(\gamma+1)\ldots \quad \text{when} \quad n = p_1^\alpha p_2^\beta p_3^\gamma \ldots.$$

It is at once apparent from this formula that

$$d(mn) = d(m)d(n) \quad \text{whenever} \quad (m, n) = 1,$$

so that (see § 3) d is a *multiplicative* arithmetic function. It also follows that if $\omega(n)$ denotes the number of distinct prime factors of n, then

$$2^{\omega(n)} \leqslant d(n),$$

with equality if and only if n is square-free.

If $n_1, n_2, \ldots, n_k$ are natural numbers and

$$n_i = p_1^{\alpha_i} p_2^{\beta_i} p_3^{\gamma_i} \ldots \quad (i = 1, \ldots, k)$$

is the prime decomposition of n_i in standard form (i.e. in the manner of (4.1)) then the *highest common factor* $(n_1, n_2, \ldots, n_k)$ and *lowest common multiple* $[n_1, n_2, \ldots, n_k]$ of $n_1, n_2, \ldots, n_k$ are given respectively by

$$(n_1, n_2, \ldots, n_k) = p_1^a p_2^b p_3^c \ldots$$

and

$$[n_1, n_2, \ldots, n_k] = p_1^A p_2^B p_3^C \ldots,$$

where

$$a = \min_i \alpha_i, \quad b = \min_i \beta_i, \quad \ldots$$

and

$$A = \max_i \alpha_i, \quad B = \max_i \beta_i, \quad \ldots.$$

When $k = 2$ (but not otherwise) we have the useful identity

$$n_1 n_2 = (n_1, n_2)[n_1, n_2].$$

It can be shown (HW, Theorem 25 generalized in an obvious way) that there exist integers $x_1, x_2, \ldots, x_k$ such that

$$n_1 x_1 + n_2 x_2 + \ldots + n_k x_k = (n_1, n_2, \ldots, n_k).$$

When we regard the set of all integers in the context of division by a fixed natural number $n > 1$, we have by Euclid's Algorithm (HW, § 12.3) that

every integer a can be expressed in the form

$$a = nq + r, \qquad 0 \leqslant r < n,$$

where the quotient q and the remainder r are unique.

In particular, n divides a (or, written symbolically, $n \mid a$) if and only if $r = 0$. It is clear from the above algorithm that, to each n, there corresponds a subdivision of the integers into n mutually exclusive sets, where two integers belong to the same set if and only if both leave the same remainder r on division by n. These sets are the so-called *residue classes* (or *congruence classes*) modulo n. If a, b belong to the same residue class modulo n, we say that '*a is congruent to b modulo n*' and express this symbolically by

$$a \equiv b \pmod{n}.$$

We observe that congruence modulo n between two integers is an equivalence relation in the sense that

$$a \equiv a,$$

$$a \equiv b \text{ implies } b \equiv a$$

and

$$a \equiv b,\ b \equiv c \text{ imply that } a \equiv c.$$

We note also that $a \equiv b \pmod{n}$ if and only if $n \mid (a-b)$ and that the congruence notation has the following easily confirmed properties (HW, §§ 5.3, 5.4).

If $\quad a \equiv b \pmod{n}, \quad a' \equiv b' \pmod{n},$

then

$$a + a' \equiv b + b' \pmod{n}$$

and

$$aa' \equiv bb' \pmod{n};$$

moreover, if $(k,n) = 1$ and $ka \equiv kb \pmod{n}$,

then $a \equiv b \pmod{n}$.

It is obvious but perhaps worth remarking that the terms of the arithmetic progression

$$a,\ a+n,\ a+2n,\ a+3n,\ldots$$

all belong to the same residue class modulo n.

Any set of n integers, one from each of the n residue classes mod n, is called a *complete set of residues modulo n*; for instance, the numbers

$$1,\ 2,\ldots,\ n$$

constitute such a set. It is easy to confirm (HW, Theorem 56) that if k, l are fixed integers with $(k,l) = 1$, and x runs through a complete set of residues mod n, then $kx+l$ also runs through a complete set of residues mod n.

If one element of a residue class mod n is coprime with n, so are all elements of this residue class, and we refer to the whole class as being coprime with n. We denote by $\phi(n)$ the number of those residue classes modulo n that are coprime with n; the function ϕ defined in this way is known as *Euler's function*. We refer to any set of $\phi(n)$ numbers, one from each of these residue classes, as a *reduced set of residues modulo n*; for example, the numbers among

$$1,\ 2,\ 3,\ldots,\ n$$

that are coprime with n constitute a reduced set of residues mod n. It can be shown (HW, Theorems 60, 61) that if m, n are fixed coprime natural numbers and x runs through a reduced set of residues mod m while y runs through a reduced set of residues mod n, then $nx+my$ runs through a reduced set of residues mod mn; one consequence of this is that

$$\phi(mn) = \phi(m)\phi(n) \quad \text{whenever} \quad (m,n) = 1, \tag{4.2}$$

so that ϕ is a multiplicative arithmetic function.

If n has prime decomposition (4.1), it follows from (4.2) that

$$\phi(n) = \phi(p_1^{\alpha})\phi(p_2^{\beta})\ldots.$$

To evaluate $\phi(n)$ for any natural number n it suffices, therefore, to calculate $\phi(p^r)$. But among the natural numbers $1, 2,\ldots, p^r$ there are exactly p^{r-1} multiples of p, and the rest are coprime with p^r; hence

$$\phi(p^r) = p^r - p^{r-1} = p^r\left(1-\frac{1}{p}\right),$$

and we arrive at the formula

$$\phi(n) = n\prod_{p|n}\left(1-\frac{1}{p}\right). \tag{4.3}$$

Another result which follows readily from the multiplicativity of ϕ is (HW, Theorem 63)

$$\sum_{\nu|n}\phi(\nu) = n; \tag{4.4}$$

although (4.4) can be proved without appeal to (4.2) by a simple classification argument. We need only divide the numbers $1, 2,\ldots, n$ into classes, one class to each positive integer divisor ν of n, in such a way that each element of the class corresponding to ν has H.C.F. with n equal to ν. The class corresponding to ν

then contains $\phi(n/\nu)$ members, and (4.4) follows at once after an obvious transformation of the variable of summation from ν to n/ν.

Actually, formulae (4.3) and (4.4) are connected by an important reciprocity relation which we shall now describe.

If we imagine the product on the right of (4.3) multiplied out, we find that the resulting sum may be regarded as a sum over the divisors of n, by inserting zero terms corresponding to those divisors of n that are not square-free. This sum is conveniently expressed in terms of a certain arithmetic function μ, named after Moebius (who introduced it in 1832 although it had already occurred implicitly in the work of Euler as early as 1748) and given by

$$\mu(\nu) = \begin{cases} 1 & \text{if } \nu = 1, \\ (-1)^r & \text{if } \nu \text{ is square-free and } \omega(\nu) = r, \\ 0 & \text{if } \nu > 1 \text{ and } \nu \text{ is not square-free;} \end{cases}$$

it is clear that μ is a multiplicative function. Using this function, (4.3) may be rewritten as

$$\phi(n) = n \sum_{\nu|n} \frac{\mu(\nu)}{\nu}$$

or as

$$\phi(n) = \sum_{\nu|n} \mu(\nu)\frac{n}{\nu} = \sum_{\nu|n} \mu\left(\frac{n}{\nu}\right)\nu; \tag{4.5}$$

and in this form it appears to be the outcome of inverting the relation (4.4) with μ acting as 'kernel' of the inversion. In fact, not only can this remark be justified, but one can prove that the relation between (4.4) and (4.5) is but a special case of the following general inversion principle, discovered by Moebius (HW, Theorem 266).

If f, g are two arithmetic functions connected by the relations ($n = 1, 2,\ldots$)

$$f(n) = \sum_{\nu|n} g(\nu), \tag{4.6}$$

then

$$g(n) = \sum_{\nu|n} \mu\left(\frac{n}{\nu}\right) f(\nu) = \sum_{\nu|n} \mu(\nu) f\left(\frac{n}{\nu}\right). \tag{4.7}$$

In the special case $f(n) = n$, $g(n) = \phi(n)$ we therefore find that (4.5) follows from (4.4) by the Moebius inversion formula.

There is a companion inversion formula of Moebius associated with a different type of summation (HW, Theorem 268).

If f, g are two functions of a real variable, connected by the relation

$$f(x) = \sum_{\nu \leqslant x} g\left(\frac{x}{\nu}\right) \quad \textit{for} \quad x \geqslant 1, \tag{4.8}$$

then

$$g(x) = \sum_{\nu \leqslant x} \mu(\nu) f\left(\frac{x}{\nu}\right) \quad \textit{for} \quad x \geqslant 1. \tag{4.9}$$

Both the inversion formulae of Moebius are consequences of the fact that (HW, Theorem 263)

$$\sum_{\nu|n} \mu(\nu) = \begin{cases} 1 & \text{if } n = 1, \\ 0 & \text{otherwise;} \end{cases}$$

and this characteristic property of the μ-function can be proved by a simple application of the binomial theorem.

It is important to note that the above inversion formulae may be 'reversed' in the sense that the relations (4.7) imply (4.6) (HW, Theorem 267) and the relation (4.9) implies (4.8) (HW, Theorem 269).

There are several applications of the Moebius inversion principle in Chapter IV; for example, the statement in § 7 of that chapter, that (7.5) implies (7.7), is simply the 'reversed' form of the first of the above inversion formulae. In fact, two further inversion formulae are established in § 2 of Chapter IV, and are applied in later sections of that chapter.

REFERENCES

CHAPTER I

(1) L. G. Schnirelmann, 'Über additive Eigenschaften von Zahlen', *Math. Annln* **107** (1933) 649–90.

(2) H. B. Mann, 'A proof of the fundamental theorem on the density of sums of sets of positive integers', *Ann. Math.* (2) **43** (1942) 523–7.

(3) A. Khintchin, 'Zur additiven Zahlentheorie', *Mat. Sb.* n.s. **39** (1932) 27–34.

(4) A. S. Besicovitch, 'On the density of the sum of two sequences of integers', *J. Lond. math. Soc.* **10** (1935) 246–8.

(5) Ju. V. Linnik, 'Elementare Lösung des Waringschen Problems mit der Methode von Schnirelmann', *Mat. Sb.* n.s. **12** (**54**) (1943) 225–30.

(6) A. Stöhr, 'Gelöste und ungelöste Fragen über Basen der natürlichen Zahlenreihe, I', *J. reine angew. Math.* **194** (1955) 40–65.

(7) A. Khintchin, 'Über ein metrisches Problem der additiven Zahlentheorie', *Mat. Sb.* n.s. **40** (1933) 180–9.

(8) P. Erdös, 'On the arithmetical density of the sum of two sequences one of which forms a basis for the integers', *Acta arith.* **1** (1936) 197–200.

(9) E. Landau, *Vorlesungen über Zahlentheorie*, Bd. 1 (Hirzel, Leipzig, 1927), 114–25.

(10) E. Landau, 'Über einige neuere Fortschritte der additiven Zahlentheorie', *Cambridge Tracts in Mathematics*, No. 35 (Cambridge, 1937) 60–2.

(11) P. Erdös, 'Über einige Probleme der additiven Zahlentheorie', *J. reine angew. Math.* **206** (1961) 61–6.

(12) A. Brauer, 'Über die Dichte der Summe zweier Mengen, deren eine von positiver Dichte ist', *Math. Z.* **44** (1939) 213–32.

(13) F. Kasch, 'Abschätzung der Dichte von Summenmengen', I and III, *Math. Z.* **62** (1955) 368–87; **66** (1956/57) 164–72.

(14) S. Selberg, 'Note on a metrical problem in the additive theory of numbers', *Arch. Math. Naturv.* **47** (1944) 111–8.

(15) H. Plünnecke, 'Über ein metrisches Problem der additiven Zahlentheorie', *J. reine angew. Math.* **197** (1957) 97–103.

(16) P. Erdös, 'Some results on additive number theory', *Proc. Am. math. Soc.* **5** (1954) 847–53.

(17) G. Lorentz, 'On a problem of additive number theory', *Proc. Am. math. Soc.* **5** (1954) 838–41.

(18) E. Artin and P. Scherk, 'On the sum of two sets of integers', *Ann. Math.* (2) **44** (1943) 138–42.

(19) F. Dyson, 'A theorem on the densities of sets of integers', *J. Lond. math. Soc.* **20** (1945) 8–14.

(20) H. B. Mann, *Addition Theorems*, Interscience Publishers, New York, 1965, Chapter 3.

(21) J. van der Corput, 'On sets of integers', I, II, III, *Proc. K. ned. Akad. Wet.* **50** (1947) 252–61, 340–50, 429–35.

(22) N. Romanoff, 'Über einige Sätze der additiven Zahlentheorie', *Math. Annln* **109** (1934) 668–78.

(23) A. Stöhr, 'Gelöste und ungelöste Fragen über Basen der natürlichen Zahlenreihe', II, *J. reine angew. Math.* **194** (1955) 111–40.
(24) P. Erdös, 'Einige Bemerkungen zur Arbeit von A. Stöhr', *J. reine angew. Math.* **197** (1957) 216–9.
(25) Ju. V. Linnik, 'On Erdös's theorem on the addition of numerical sequences', *Mat. Sb.* N.S. **10** (**52**) (1942) 67–78.
(26) A. Stöhr and E. Wirsing, 'Beispiele von wesentlichen Komponenten die keine Basen sind', *J. reine angew. Math.* **196** (1956) 96–8.
(27) A. Stöhr, 'Eine Basis h-ter Ordnung für die Menge aller natürlichen Zahlen', *Math. Z.* **42** (1937) 739–43; 'Anzahlabschätzung einer bekannten Basis h-ter Ordnung', ibid. **47** (1942) 778–87 and **48** (1942/3) 792.
(28) D. Raikov, 'Über die Basen der natürlichen Zahlenreihe', *Mat. Sb.* N.S. **2** (**44**) (1937) 595–7.
(29) J. W. S. Cassels, 'Über Basen der natürlichen Zahlenreihe', *Abh. math. Semin. Univ. Hamburg* **21** (1957) 247–57.
(30) H. Ostmann, *Additive Zahlentheorie*, I, II (Ergebnisse Series; Springer, 1956).
(31) H. Rohrbach, 'Einige neuere Untersuchungen über die Dichte in der additiven Zahlentheorie', *Jber. dt. Mat.-Verein.* **48** (1938) 199–236.
(32) A. L. Cauchy, 'Recherches sur les nombres', *J. Éc. polytech.* **9** (1813) 99–116.
(33) H. Davenport, 'On the addition of residue classes', *J. Lond. math. Soc.* **10** (1935) 30–2; 'A historical note', ibid. **22** (1947) 100–1.
(34) I. Chowla, 'A theorem on the addition of residue classes', *Proc. Indian Acad. Sci.* **2** (1935) 242–3.
(35) J. H. B. Kemperman and P. Scherk, 'Complexes in abelian groups', *Can. J. Math.* **6** (1954) 230–7.
(36) A. G. Vosper, 'The critical pairs of sub-sets of a group of prime order', *J. Lond. math. Soc.* **31** (1956) 200–5; Addendum 280–2.
(37) J. H. B. Kemperman, 'On small sumsets in an abelian group', *Acta math.* **103** (1960) 63–88.
(38) M. Kneser, 'Abschätzungen der asymptotischen Dichte von Summenmengen', *Math. Z.* **58** (1953) 459–84.
(39) L. Danzer, 'Über eine Frage von G. Hanani aus der additiven Zahlentheorie', *J. reine angew. Math.* **214/215** (1964) 392–4.

CHAPTER II

(1) S. Sidon, 'Ein Satz über trigonometrische Polynome und seine Anwendung in der Theorie der Fourier-Reihen', *Math. Annln* **106** (1932) 536–9.
(2) P. Erdös and W. H. J. Fuchs, 'On a problem of additive number theory', *J. Lond. math. Soc.* **31** (1956) 67–73.
(3) J. Singer, 'A theorem in finite projective geometry and some applications to number theory', *Trans. Am. math. Soc.* **43** (1938) 377–85.
(4) P. Erdös and P. Turan, 'On a problem of Sidon in additive number theory, and some related problems', *J. Lond. math. Soc.* **16** (1941) 212–5; Addendum (by P. Erdös) ibid. **19** (1944) 208.
(5) S. Chowla, 'Solution of a problem of Erdös and Turan in additive number theory', *Proc. natn. Acad. Sci. India* **14** (1944) 1–2.

(6) R. C. Bose, 'An affine analogue of Singer's theorem', *J. Indian math. Soc.* (new series) **6** (1942) 1–15.

(7) R. C. Bose and S. Chowla, 'Theorems in the additive theory of numbers', *Comment. math. helvet.* **37** (1962–63) 141–7.

(8) A. Stöhr,† 'Gelöste und ungelöste Fragen über Basen der natürlichen Zahlenreihe', II, *J. reine angew. Math.* **194** (1955) 111–40.

(9) F. Krückeberg, 'B_2-Folgen und verwandte Zahlenfolgen', *J. reine angew. Math.* **206** (1961) 53–60.

(10) P. Erdös, 'Problems and results in additive number theory', *Colloque sur la Théorie des Nombres*, Bruxelles, 1956, 127–37.

(11) A. Mian and S. Chowla, 'On the B_2-sequences of Sidon', *Proc. natn. Acad. Sci. India*, Sect. A **14** (1944) 3–4.

(12) P. T. Bateman, E. F. Kohlbecker, and J. P. Tull, 'On a theorem of Erdös and Fuchs in additive number theory', *Proc. Am. math. Soc.* **14** (1963) 278–84.

(13) G. A. Dirac, 'Note on a problem in additive number theory', *J. Lond. math. Soc.* **26** (1951) 312–3.

(14) H.-E. Richert, 'Zur multiplikativen Zahlentheorie', *J. reine angew. Math.* **206** (1961) 31–8.

CHAPTER III

(1) P. Erdös, 'Problems and results in additive number theory', *Colloque sur la Théorie des Nombres (CBRM)*, Bruxelles, 1956, 127–37.

(2) P. Erdös and A. Rényi, 'Additive properties of random sequences of positive integers', *Acta arith.* **6** (1960) 83–110.

(3) P. Erdös, 'Some results on additive number theory', *Proc. Am. math. Soc.* **5** (1954) 847–53.

(4) A. O. L. Atkin, Cambridge doctoral thesis, 1952: 'On pseudo-squares', *Proc. London Math. Soc.* **14A** (1965) 22–7.

(5) A. N. Kolmogorov, *Grundbegriffe der Wahrscheinlichkeitsrechnung* (Springer, Berlin, 1933).

(6) P. Erdös and A. Rényi, 'On Cantor's series with convergent $\sum 1/q_n$', *Annls Univ. Scient. Bpest. Rolando Eötvös*, sect. math. **2** (1959) 93–109.

(7) G. Hoheisel, 'Primzahlprobleme in der Analysis', *S.B. preuss. Akad. Wiss.* (Berlin, 1930) 580–8.

CHAPTER IV

(1) F. Mertens, 'Ein Beitrag zur analytischen Zahlentheorie', *J. reine angew. Math.* **78** (1874) 46–62.

(2) V. Brun, 'Le crible d'Eratosthène et le théorème de Goldbach', *Skr. Vidensk. Selsk. Christ.* **1**, No. **3** (1920).

(3) A. Selberg, 'On an elementary method in the theory of primes', *K. norske Vidensk. Selsk. Forh. Trondhjem* **19** (1947) 64–7.

(4) H. Rademacher, 'Beiträge zur Viggo Brunschen Methode in der Zahlentheorie', *Hamburger Abh.* **3** (1924) 12–30.

(5) T. Estermann, 'Eine neue Darstellung und neue Anwendungen der Viggo Brunschen Methode', *J. reine angew. Math.* **168** (1932) 106–16.

† Two results due to Erdös were communicated by him to Stöhr for inclusion in Stöhr's survey. Accordingly, a number of references in the text of Chapter II are to Erdös.[8]

(6) A. A. Buchstab, 'Neue Verbesserungen in der Methode des Eratosthenischen Siebes', *Mat. Sb.* n.s. **4** (1938) 375–87. 'On the representation of integers as sums of two numbers each the product of a finite number of primes', *Dokl. Akad. Nauk SSSR*, **29** (1940) 544–8. 'On an additive representation of integers', *Mat. Sb.* n.s. **10** (**52**): 1–2 (1942) 87–91.

(7) A. Rényi, 'On the representation of an even number as the sum of a prime and an almost prime', *Izv. Akad. Nauk SSSR*, ser. mat. **12** (1948), 57–78; see also *Am. math. Soc. Transl.* ser. 2, **19** (1962) 299–321.

(8) M. B. Barban, 'The density of zeros of Dirichlet *L*-series and the problem of sums of prime and "almost prime" numbers', *Mat. Sb.* n.s. **61** (**103**) (1963) 418–25.

(9) B. V. Levin, 'Distribution of "almost prime" numbers in polynomial sequences', *Mat. Sb.* n.s. **61** (**103**) (1963) 389–407.

(10) Wang Yuan, 'On the representation of a large integer as a sum of a prime and an almost prime', *Acta math. sin.* **10** (1960) 168–81; see also *Sci. sin.* **11** (1962) 1033–54.

(11) A. Selberg, 'On elementary methods in prime number theory and their limitations', *Den 11-te Skand. Mat.-Kongress Trondhjem* (1949), 13–22.

(12) A. I. Vinogradov, 'The application of $\zeta(s)$ to the sieve of Eratosthenes', *Mat. Sb.* n.s. **41** (**83**) (1957) 49–80; ibid. 415–6; see also *Am. math. Soc. Transl.* ser. 2, **13** (1960), 29–60.

(13) Wang Yuan, 'On the representation of a large even number as a sum of two almost primes', *Sci. Rec.* n.s. **1** (1957) 291–5.

(14) Ju. V. Linnik, 'The large sieve', *Dokl. Akad. Nauk SSSR*, n.s. **30** (1941) 292–4.

(15) A. Rényi, 'Un nouveau théorème concernant les fonctions indépendantes et ses applications à la théorie des nombres', *J. Math. pures appl.* **28** (1949) 137–49.

(16) A. Rényi, 'Sur un théorème général de probabilité', *Annls Inst. Fourier Univ. Grenoble* **1** (1950) 43–52. 'On the large sieve of Ju. V. Linnik', *Compositio math.* **8** (1950) 68–75. 'On a general theorem in probability theory and its application in the theory of numbers', *Zprávy o spolecnem 3 sjezdu matematiku československych a 7 sjezdu matematiku Polskych, Praha* (1950) 167–74. 'On the probabilistic generalization of the large sieve of Linnik', *Magy. tudom. Akad. mat. Kut. Intez. Közl.* **3** (1958) 199–206. 'New version of the probabilistic generalization of the large sieve', *Acta math. hung.* **10** (1959) 217–26.

(17) A. Rényi, 'Probabilistic methods in number theory' (in Chinese), *Prog. Math. Sci.* **4** (1958) 465–510.

(18) I. M. Vinogradov, *The Method of Trigonometrical Sums in the Theory of Numbers* (1947), English translation by K. F. Roth and A. Davenport (Interscience, London, 1954).

(19) A. A. Buchstab, 'Asymptotische Abschätzungen einer allgemeinen zahlentheoretischen Function', *Mat. Sb.* n.s. **44** (1937) 1239–46.

(20) N. G. De Bruijn, 'On the number of uncancelled elements in the sieve of Eratosthenes', *Indag. math.* **12** (1950) 247–56.

(21) L. K. Hua, 'Die Abschätzung von Exponentialsummen und ihre Anwendung in der Zahlentheorie', *Enzyklopädie der Math. Wissenschaften* (Teubner, Leipzig, 1959), vol. I2, book 13, part 1, Art. 29, § 24.

(22) A. Selberg, 'The general sieve method and its place in prime number theory', *Proc. Intern. Congr. Math., Cambridge, Mass.* 1950, **1**, 286–92 (Am. math. Soc., 1952).

(23) P. Kuhn, 'Zur Viggo Brun'schen Siebmethode', 1, *K. norske Vidensk. Selsk. Forh.* **39** (1942) 145–7; also *Tolfte Skand. Mat. Kongress. Lund* (1953), 160–8.

CHAPTER V

(1) P. Erdös, 'On primitive abundant numbers', *J. Lond. math. Soc.* **10** (1935) 49–58.

(2) P. Erdös, 'Note on sequences of integers no one of which is divisible by any other', *J. Lond. math. Soc.* **10** (1935) 126–8.

(3) A. S. Besicovitch, 'On the density of certain sequences of integers', *Math. Annln* **110** (1934) 336–41.

(4) H. Davenport and P. Erdös, 'On sequences of positive integers', *Acta arith.* **2** (1937) 147–51; also *J. Indian math. Soc.* **15** (1951) 19–24.

(5) F. Behrend, 'On sequences of integers not divisible one by another', *J. Lond. math. Soc.* **10** (1935) 42–4.

(6) E. Sperner, 'Ein Satz über Untermengen einer endlichen Menge', *Math. Z.* **27** (1928) 544–8.

(7) S. Pillai, 'On numbers which are not multiples of any other in the set', *Proc. Indian Acad. Sci.* A **10** (1939) 392–4.

(8) N. G. De Bruijn, C. van E. Tengbergen, and D. Kruyswijk, 'On the set of divisors of a number', *Nieuw Arch. f. Wisk.* Ser. II, **23** (1949–51) 191–3.

(9) P. Erdös, 'Integers with exactly *k* prime factors', *Ann. Math.* II **49** (1948) 53–66.

(10) P. Erdös, 'On the density of some sequences of integers', *Bull. Am. math. Soc.* **54** (1948) 685–92.

(11) F. Behrend, 'Generalization of an inequality of Heilbronn and Rohrbach', *Bull. Am. math. Soc.* **54** (1948) 681–4.

(12) P. Erdös, 'Generalization of a theorem of Besicovitch', *J. Lond. math. Soc.* **11** (1936) 92–8.

POSTSCRIPT

THE subject matter of this book has been enriched by many new contributions since its appearance in 1966. To give anything like a complete account of more recent developments would be a formidable task and we shall not attempt it. We restrict ourselves to mentioning only a few references that seem to us to describe significant progress with the main themes of our book.

Chapter I. G. A. Freĭman's book *Foundations of a structural theory of set addition* (Amer. Math. Soc. translations of Math. monographs, v. 37, 1973) represents a serious and partially successful attempt to go beyond the 'naïve' Schnirelmann theory on the addition of general integer sequences by introducing more varied and sophisticated machinery.

A propos of the construction of thin essential components, a virtually best possible result of this kind (cf. references to Linnik and Erdös–Roth on p. 35) has been obtained by E. Wirsing, 'Thin essential components', *Topics in Number Theory* (Colloq., Math. Soc. János Bolyai, Debrecen, North Holland 1976) 429–42.

Chapter II. R. C. Vaughan, 'On the addition of sequences of integers', *J. Number Theory* 4 (1972), 1–16, has generalized the Erdös–Fuchs theorem (Theorem 11) to more than two summands.

Chapter IV. '*Sieve Methods*' (Academic Press 1974) by H. Halberstam and H.-E. Richert contains a systematic and up-to-date account of developments in the subject matter of sections 2–9. For progress with 'large' sieves (section 10), see H. Davenport, '*Multiplicative Number Theory*' (Second Edition, revised by H. L. Montgomery, Springer-Verlag 1980), Chapter 27.

Chapter V. There is a good account of developments, with references, and still open problems in the survey article by P. Erdös, A. Sarközi, and E. Szemeredi, 'On divisibility properties of sequences of integers', *Number Theory* (Colloq., Math. Soc. János Bolyai, Debrecen, North Holland 1970), 35–49.

The collection of problems referred to at the end of the Introduction has been updated in Chapter 10 of P. Erdös and R. L. Graham, 'Old and new problems and results in combinatorial number theory', *Monographies de l'Enseignement Mathématique*, No. 28, Genève. See also Chapter E of R. K. Guy, 'Unsolved problems in number theory', *Unsolved problems in intuitive mathematics*, Vol. 1, Springer-Verlag, New York Heidelberg Berlin, 1980.

AUTHOR INDEX